T0212621

Birkhäuser

ISNM
International Series of Numerical Mathematics

Volume 159

Managing Editors:
K.-H. Hoffmann, München
G. Leugering, Erlangen-Nürnberg

Associate Editors:
Z. Chen, Beijing
R. H. W. Hoppe, Augsburg/Houston
N. Kenmochi, Chiba
V. Starovoitov, Novosibirsk

Honorary Editor:
J. Todd, Pasadena †

Exact and Truncated Difference Schemes for Boundary Value ODEs

Ivan P. Gavrilyuk
Martin Hermann
Volodymyr L. Makarov
Myroslav V. Kutniv

Birkhäuser

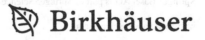

Ivan P. Gavrilyuk
Staatliche Studienakademie Thüringen
Berufsakademie Eisenach
(University of Cooperative Education)
Am Wartenberg 2
99817 Eisenach
Germany
ipg@ba-eisenach.de

Martin Hermann
Friedrich Schiller University
Institute of Applied Mathematics
Ernst-Abbe-Platz 2
07743 Jena
Germany
martin.hermann@uni-jena.de

Volodymyr L. Makarov
National Academy of Sciences of Ukraine
Institute of Mathematics
Tereshchenkivska 3
01601 Kiev-4
Ukraine
makarov@imath.kiev.ua

Myroslav V. Kutniv
Lviv Polytechnical National University
Institute of Applied Mathematics
S. Bandery 12
79013 Lviv
Ukraine
kutniv@yahoo.com

ISBN 978-3-0348-0341-0 ISBN 978-3-0348-0107-2 (eBook)
DOI 10.1007/978-3-0348-0107-2

2010 Mathematics Subject Classification: 65L10, 65L12, 65L20, 65L50, 65L70

© Springer Basel AG 2011
Softcover reprint of the hardcover 1st edition 2011
This work is subject to copyright. All rights are reserved, whether the whole or part of the material is concerned, specifically the rights of translation, reprinting, re-use of illustrations, broadcasting, reproduction on microfilms or in other ways, and storage in data banks. For any kind of use whatsoever, permission from the copyright owner must be obtained.

Cover design: deblik, Berlin

Printed on acid-free paper

Springer Basel AG is part of Springer Science+Business Media

www.birkhauser-science.com

This book is dedicated to
the memory of Aleksandr Andreevich Samarskii

Contents

Preface

Work is much more fun than fun.

Noel Coward (1899–1973)

In this book we present a first unified theory of finite difference methods for the solution of linear and nonlinear boundary value problems (BVPs) of ordinary differential equations (ODEs). The aim of the authors is to describe the state-of-the-art in this important field of numerical analysis. New numerical algorithms for BVPs are developed and evaluated which have the same efficiency and accuracy as the well-known and contemporary methods for the solution of initial value problems (IVPs). Moreover, the tools from different areas of mathematics (e.g. theory of ODEs, functional analysis, theory of difference schemes, systems of nonlinear algebraic equations) are formalized and unified to construct and justify numerical methods of a high order of accuracy for BVPs.

The basis of the present text is the theory of difference schemes established by the famous Russian mathematician Aleksandr Andreevich Samarskii. Some of the authors were Samarskii's students. The new class of exact and truncated difference schemes treated in this book can be considered as further developments of the results presented in his highly respected books [71, 72, 75]. It is shown that the new Samarskii-like techniques open horizon for the numerical treatment of more complicated problems.

The main focus of this text is the theory of numerical methods for BVPs that are based on the exact difference schemes (EDS) as well as on the well-studied and sophisticated solvers for IVPs. The EDS themselves are only of theoretical interest. However, they are used as a starting point for the construction of so-called truncated difference schemes (TDS) suitable for implementation on a computer. It is shown that the combination of the EDS with the modern IVP-solvers results in numerical TDS-algorithms which are highly efficient. The theory of the EDS and TDS permits the construction of a posteriori error estimators which in turn are the basis for adaptive algorithms.

The EDS/TDS and the well-known multiple shooting methods (see e.g. [35, 39, 40, 79, 84]) are based on similar concepts. However, compared with the shooting methods, the advantage of the new EDS and TDS algorithms is their detailed theoretical analysis and the possibility of estimating the error of the approximated solution by suitable a priori and a posteriori estimators. The reason for that is the good conformity of the structure of the difference equations with the original differential equations.

The book is addressed to graduate students of mathematics and physics, as well as to working scientists and engineers as a self-study tool and reference. Researchers dealing with BVPs will find appropriate and effective numerical algorithms for their needs.

The book can be used as a textbook for a one- or two-semester course on numerical methods for ODEs. Advanced calculus, basic numerical analysis, some ordinary differential equations, and a smattering of nonlinear functional analysis are assumed. Mathematically experienced engineers and scientists who are only interested in the practical use of the algorithms on a computer can skip without loss of understanding, the proofs of the theorems and the more theoretical parts of the book.

We now outline the contents of the book. The introductionary chapter is designed to acquaint the reader with some types of difference schemes that occur in the literature. This historical overview places the book in a proper perspective. Chapter 2 deals with BVPs for systems of first-order nonlinear ODEs on a finite interval. The existence of two-point EDS on an arbitrary non-equidistant grid is shown and its structure is described. These EDS are the basis for the construction of two-point TDS with an order of accuracy which can be given by the user. For the computation of the coefficients of these schemes (which represent systems of nonlinear equations) the well-known IVP-solvers (see e.g. [31, 8]) are used. Its accuracy defines the accuracy of the TDS. Two *a posteriori* error estimators (which are based on the Runge principle and on embedded IVP-solvers) are proposed along with appropriate grid generators. Some numerical examples confirming the theoretical results are given. Chapter 3 is devoted to BVPs for second-order nonlinear ODEs with a monotone operator. The appropriate EDS and TDS are constructed, theoretically analyzed and practically tested. In Chapter 4 the results of Chapter 3 are generalized to BVPs for systems of second-order nonlinear ODEs with monotone operators. The developed algorithms have been tested on a variety of numerical examples. The corresponding results indicate a good conformity with the theoretical predictions. Chapter 5 gives some results on EDS and TDS for BVPs on the half-axis. These techniques are a real alternative to those methods which are based on transformation of the original problems into a singular BVP on a finite interval, and the subsequent solution of this singular problem by various special approaches. The results of some numerical examples are presented. The knowledge gained with this book can be checked in Chapter 6 where 32 exercises and the corresponding solutions are given.

Some FORTRAN codes of the numerical methods presented in this book can be obtained from the authors.

Acknowledgment. The authors thank Dr. Dieter Kaiser for interesting comments and suggestions. Many thanks also to Ivo Hedtke for his assistence concerning the preparation in LaTeX. We gratefully acknowledge the financial support provided by the German Research Foundation (DFG) for our joint research projects. Our thanks also go to all persons and institutions of the Friedrich Schiller University at Jena that have supported the stays of our colleagues from the Ukraine. We especially wish to thank our wives for tolerating the (many) years of effort that have gone into writing of this book. It has been a pleasure working with the Birkhäuser Basel publication staff, in particular Dr. Barbara Hellriegel.

May 2011, IPG, MH, MVK and MLM

Chapter 1

Introduction and a short historical overview

He who never starts, never finishes.

William Shakespeare (1564–1616)

One of the important fields of application for modern computers is the numerical solution of diverse problems arising in science, engineering, industry, etc. Here, mathematical models have to be solved which describe e.g. natural phenomena, industrial processes, nonlinear vibrations, nonlinear mechanical structures or phenomena in hydrodynamics and biophysics. A lot of such mathematical models can be formulated as initial value problems (IVPs) or boundary value problems (BVPs) for systems of nonlinear ordinary differential equations (ODEs). However, it is not possible in general to determine the solution of nonlinear problems in a closed form. Therefore the exact solution must be approximated by numerical techniques. Considerable progress has been made in developing the theory and numerical analysis of IVPs and there exist many effective IVP-solvers (see e.g. [8, 31, 32, 35]). On the other hand, there are no references in current literature providing a general approach that allows the construction of difference schemes and the associated algorithms of a prescribed order of accuracy for BVPs. The development and analysis of numerical methods for nonlinear BVPs that can be used to solve new classes of problems or that are better than the existing ones remains an actual problem of numerical analysis and scientific computing.

In the last decade so-called compact difference schemes of a high order of accuracy were frequently used [41]. These schemes use a stencil consisting of $k+1$ grid nodes for ODEs of the order k. An important property of compact difference schemes is their low complexity, i.e. the computational costs for their solution are low. On the other hand, if the order of accuracy is high enough these difference schemes produce very accurate approximations on relatively rough grids. Finally, it was shown in [41] that the compactness of a difference scheme implies its stability.

The aim of this book is to provide an overview of the theoretical and numerical state of research on compact difference schemes of a high order of accuracy (prescribed by the user) for nonlinear BVPs on finite intervals and on the half axis. Since the given problems and the corresponding difference schemes are nonlinear, our analysis uses the linearization technique, the principle of contracting maps and the theory of monotone operators. All of our new algorithms are based on compact exact difference schemes (EDS). In order to determine the coefficients and the right-hand side of an EDS at an arbitrary node of the underlying grid, some auxiliary IVPs on a small interval around this node must be solved. The existence and uniqueness of the solutions of the EDS is proven. Effective implementations of the EDS which are based on so-called compact truncated difference schemes (TDS) and on the well-studied robust IVP-solvers are developed. The convergence of the associated iteration methods (fixed point iteration, Newton's iteration) is shown. The effectiveness of the proposed approaches is illustrated by a series of numerical examples.

Note that the theory of numerical methods for IVPs is already very advanced today. A variety of reliable and effective implementations of these methods is available. They can be used to solve non-stiff as well as stiff IVPs. In this monograph the authors will show how numerical algorithms for BVPs can be constructed which have the same accuracy and performance level as the modern IVP-solver. The idea is to use an IVP-solver of the order p to obtain a difference scheme for BVPs of the same order p.

1.1 BVPs, Grids, Differences, Difference Schemes

Without loss of generality, in the following we consider $[0,1]$ as the basic interval. To develop a difference scheme some *grid points* or *nodes* $x_0, \ldots, x_N$ have to be chosen in $[0,1]$ which define a *grid* ω_h.

Definition 1.1. In this text the following grids are used:

- the equidistant (uniform) grid on $[0,1]$:

$$\bar{\omega}_h \stackrel{\text{def}}{=} \{x_j \in [0,1], \quad x_j = j\,h, \quad j = 0, 1, \ldots, N\}, \tag{1.1}$$

  where $h = 1/N$ is the constant step-size. Moreover, we introduce the following sub-grids of $\bar{\omega}_h$:

$$\begin{aligned} \omega_h &\stackrel{\text{def}}{=} \{x_j = jh, \quad j = 1, 2, \ldots, N-1\}, \\ \omega_h^+ &\stackrel{\text{def}}{=} \{x_j = jh, \quad j = 1, 2, \ldots, N\}, \\ \omega_h^- &\stackrel{\text{def}}{=} \{x_j = jh, \quad j = 0, 1, \ldots, N-1\}, \\ \gamma_h &\stackrel{\text{def}}{=} \{x_0 = 0, \quad x_N = 1\}; \end{aligned} \tag{1.2}$$

- the non-equidistant (irregular) grid on $[0,1]$:

$$\hat{\bar{\omega}}_h \stackrel{\text{def}}{=} \{x_j \in [0,1], \quad x_0 = 0,\ x_N = 1,\ x_j = x_{j-1} + h_j, \quad j = 1,\dots,N\}, \tag{1.3}$$

where $h_j > 0$ is the local step-size and $h_1 + h_2 + \cdots + h_N = 1$. In the case of a non-equidistant grid we set $\boldsymbol{h} \stackrel{\text{def}}{=} (h_1, h_2, \dots, h_N)^T$ and define

$$h \stackrel{\text{def}}{=} \|\boldsymbol{h}\|_\infty = \max_{1\le j\le N} h_j \quad \text{or} \quad h \stackrel{\text{def}}{=} \|\boldsymbol{h}\|_2 = \left(\sum_{j=1}^{N} h_j^2\right)^{1/2}.$$

The corresponding sub-grids are

$$\begin{aligned} \hat{\omega}_h &\stackrel{\text{def}}{=} \{x_j \in \hat{\bar{\omega}}_h, \quad j = 1,2,\dots,N-1\}, \\ \hat{\omega}_h^+ &\stackrel{\text{def}}{=} \{x_j \in \hat{\bar{\omega}}_h, \quad j = 1,2,\dots,N\}, \\ \hat{\omega}_h^- &\stackrel{\text{def}}{=} \{x_j \in \hat{\bar{\omega}}_h, \quad j = 0,1,\dots,N-1\}; \end{aligned} \tag{1.4}$$

- the quasi-uniform grid

$$\hat{\bar{\omega}} \stackrel{\text{def}}{=} \{x_j^{(N)} = \xi(t_j^{(N)}) \in (0,1), \quad j = 0,1,\dots,N\}, \tag{1.5}$$

where $x = \xi(t) : t \in [0,1] \to x \in [0,1]$ is a function such that $\xi(t) \in \mathbb{C}^2[0,1]$, $\xi'(t) \ge \varepsilon > 0$ and the grid consisting of the nodes $t_j^{(N)} = i/N$ is equidistant with the step-size $\tau = i/N$. Under these assumptions we have for the step-size of the grid (1.5) $h_i = x_i^{(N)} - x_{i-1}^{(N)} \approx \xi'(t_i^{(N)})/N$ and it holds that $h_i - h_{i-1} \approx \xi''(t_i^{(N)})/N^2$, i.e. the difference between two neighboring step-sizes is much smaller than the step-size itself so that they are almost equal. On the other hand, the quotient of two remote steps $h_i/h_j \approx \xi'(t_i^{(N)})/\xi'(t_j^{(N)})$ might be rather large. In order to densify a quasi-uniform grid one should choose a denser t-grid or, which is the same, N should be increased.
For example, if for a given function less emphasis is placed near the center of the underlying interval $(0,1)$ and more emphasis when x is near the boundary points 0 and 1, one can choose

$$x = \left(e^{\alpha t} - 1\right)\left(e^{\alpha} - 1\right).$$

Obviously, for $\alpha > 0$ the x-grid is denser on the left end and for $\alpha < 0$ it is denser on the right end. The step-sizes build a geometric progression with the quotient $q \stackrel{\text{def}}{=} h_i/h_{i-1} = e^{\alpha/N}$. The quotient $\tilde{q} \stackrel{\text{def}}{=} h_1/h_N \approx e^{\alpha}$ can be rather large for large values of α. □

Since it is only possible to work with discrete values of a function $u(x)$ on a computer we need the concept of the discretization of $u(x)$.

Definition 1.2. The *discretization* of a function $u(x)$, $0 \le x \le 1$, is the projection of $u(x)$ onto the underlying grid $\hat{\omega}_h$. The result is a sequence $\{u(x_j)\}_{j=0}^N$, with $x_j \in \hat{\omega}_h$. A function which is only defined on a grid is called a *grid function*. □

To simplify the representation we use the abbreviation $u_j \stackrel{\text{def}}{=} u(x_j)$. If $u(x)$ denotes the exact solution of a BVP, the aim of each numerical method is to determine an approximated grid function $y_h \stackrel{\text{def}}{=} y \stackrel{\text{def}}{=} \{y_j\}_{j=0}^N$, with $y_j \approx u_j$.

The set of grid functions forms a linear space H_h. If a grid contains only a finite number of nodes, then the corresponding space is finite-dimensional. Its dimension is equal to the number of grid points.

The development of difference schemes for BVPs is based on the approximation of the derivatives by divided differences which are defined recursively.

Definition 1.3. The (*backward*) *divided difference* of first order is

$$u_{\bar{x},j} \stackrel{\text{def}}{=} u_{\bar{x}}(x_j) \stackrel{\text{def}}{=} \frac{u_j - u_{j-1}}{h_j} \tag{1.6}$$

and the (*forward*) *divided difference* of first order (using indices) is

$$u_{x,j} \stackrel{\text{def}}{=} u_x(x_j) \stackrel{\text{def}}{=} \frac{u_{j+1} - u_j}{h_{j+1}}. \tag{1.7}$$

A further divided difference of first order is

$$u_{\hat{x},j} \stackrel{\text{def}}{=} \frac{u_{j+1} - u_j}{\hbar_j}, \quad \hbar_j \stackrel{\text{def}}{=} \frac{h_{j+1} + h_j}{2}. \tag{1.8}$$

Without the use of indices these differences can also be written in the form

$$\begin{aligned}
u_{\bar{x}} &\stackrel{\text{def}}{=} u_{\bar{x}}(x) \stackrel{\text{def}}{=} \frac{u(x) - u(x - h_-)}{h_-}, \quad h_- > 0,\\
u_x &\stackrel{\text{def}}{=} u_x(x) \stackrel{\text{def}}{=} \frac{u(x + h_+) - u(x)}{h_+}, \quad h_+ > 0,\\
u_{\hat{x}} &\stackrel{\text{def}}{=} u_{\bar{x}}(x) \stackrel{\text{def}}{=} \frac{u(x + h_+) - u(x)}{\hbar}, \quad \hbar = \frac{h_- + h_+}{2}.
\end{aligned} \tag{1.9}$$

The differences (1.6) – (1.8) are based on the 2-point stencils (x_{j-1}, x_j) and (x_j, x_{j+1}). We have the relations

$$u_{x,j} = u_{\bar{x},j+1}, \quad u_{x,j} = \frac{\hbar_j}{h_{j+1}} u_{\hat{x},j}$$

which result from (1.6) – (1.8). Using the divided differences of first order we now define divided differences of second order on a 3-point stencil (x_{j-1}, x_j, x_{j+1}):

$$u_{\bar{x}\hat{x},j} \stackrel{\text{def}}{=} \frac{1}{\hbar_j}(u_{x,j} - u_{\bar{x},j}) = \frac{1}{\hbar_j}\left[\frac{u_{j+1} - u_j}{h_{j+1}} - \frac{u_j - u_{j-1}}{h_j}\right]. \tag{1.10}$$

Note, that the equidistant grid (1.1) is a special case of the non-equidistant grid (1.3) with $h_j \equiv h$. Thus, on the equidistant grid (1.1) the divided differences take the form

$$u_{\bar{x},j} = \frac{u_j - u_{j-1}}{h}, \quad u_{x,j} = \frac{u_{j+1} - u_j}{h},$$
$$u_{\bar{x}x,j} = \frac{1}{h}\left(u_{x,j} - u_{\bar{x},j}\right) = \frac{u_{j+1} - 2u_j + u_{j-1}}{h^2}. \tag{1.11}$$

The other divided differences of second and higher order can be generated in the same recursive way. □

There are discrete analogues of the well-known formula

$$(uv)' = u'v + uv'$$

for the differentiation of a product of two functions (prove it!):

$$(uv)_{\bar{x},i} = u_{\bar{x},i}v_i + u_{i-1}v_{\bar{x},i} = u_{\bar{x},i}v_{i-1} + u_i v_{\bar{x},i},$$
$$(uv)_{x,i} = u_{x,i}v_i + u_{i+1}v_{x,i} = u_{x,i}v_{i+1} + u_i v_{x,i}, \tag{1.12}$$
$$(uv)_{\hat{x},i} = u_{\hat{x},i}v_i + u_{i+1}v_{\hat{x},i} = u_{\hat{x},i}v_{i+1} + u_i v_{\hat{x},i}.$$

Using the formulas (1.12) one obtains easily the partial summation formulas

$$\sum_{i=1}^{N-1} u_i v_{\hat{x},i}\hbar_i = u_N v_N - u_0 v_1 - \sum_{i=1}^{N} u_{\bar{x},i} v_i h_i,$$
$$\sum_{i=1}^{N-1} u_i v_{\bar{x},i} h_i = u_N v_{N-1} - u_0 v_0 - \sum_{i=0}^{N-1} u_{\hat{x},i} v_i \hbar_i, \tag{1.13}$$

which are the discrete analogues of the integration by parts,

$$\int_0^1 uv'\,dx = uv|_0^1 - \int_0^1 u'v\,dx.$$

Introducing the following scalar products on the space of grid function H_h which are defined on $\hat{\bar{\omega}}_h$,

$$(u,v)_{\hat{\omega}_h} \overset{\text{def}}{=} \sum_{i=1}^{N-1} u_i v_i \hbar_i, \quad (u,v)_{\hat{\omega}_h^+} \overset{\text{def}}{=} \sum_{i=1}^{N} u_i v_i h_i, \tag{1.14}$$

the formulas (1.13) can be written as

$$(u, v_{\hat{x}})_{\hat{\omega}_h} = u_N v_N - u_0 v_1 - (u_{\bar{x}}, v)_{\hat{\omega}_h^+}. \tag{1.15}$$

Let us consider the linear BVP

$$\begin{aligned} L\,u(x) &= f(x), \quad x \in (0,1), \\ l\,u(x) &= \mu(x), \quad x \in \gamma = \{0,1\}, \end{aligned} \tag{1.16}$$

where L denotes a linear differential operator and l is a linear boundary operator. Assume that (1.16) is approximated by a difference scheme

$$L_h\, y_h(x) = \varphi_h(x), \quad x \in \hat{\omega}_h, \tag{1.17}$$

$$l_h\, y_h(x) = \chi_h(x), \quad x \in \gamma_h = \{0,1\}. \tag{1.18}$$

Here, L_h is an appropriate difference operator and l_h is a boundary difference operator which are defined on the grid $\hat{\bar{\omega}}_h \stackrel{\text{def}}{=} \hat{\omega}_h \cup \gamma_h$ covering the interval $[0,1]$, and y_h, φ_h, χ_h are grid functions. The exact solution $u(x)$ of the BVP (1.16) does not satisfy the difference scheme (1.17), (1.18). Let u_h be the projection of the solution $u(x)$ of (1.16) onto the grid $\hat{\bar{\omega}}_h$. Obviously, the error $z_h \stackrel{\text{def}}{=} y_h - u_h$ satisfies the problem

$$\begin{aligned} L_h\, z_h(x) &= \psi_h(x), \quad x \in \hat{\omega}_h, \\ l_h\, z_h(x) &= \nu_h(x), \quad x \in \gamma_h, \end{aligned} \tag{1.19}$$

where ψ_h is the so-called *truncation error* of the discretized ODE (1.17) and ν_h is the *truncation error* of the discretized boundary condition (1.18).

In the next definition the concept of consistency is introduced. Consistency is the minimum requirement that a difference scheme should fulfill.

Definition 1.4. The difference scheme (1.17), (1.18) is said to be *consistent of the order* p if p denotes the largest positive integer such that the truncation errors satisfy

$$\|\psi_h\|_{\hat{\omega}_h} = \mathcal{O}(h^p), \quad \|\nu_h\|_{\gamma_h} = \mathcal{O}(h^p), \quad h \to 0, \tag{1.20}$$

where $\|\cdot\|_{\hat{\omega}_h}$ and $\|\cdot\|_{\gamma_h}$ denote appropriate grid norms. *Consistency* normally means that the order $p \geq 1$. □

Another important property of a difference scheme is its stability.

Definition 1.5. The difference scheme (1.17), (1.18) is *stable* if there exists a $h_0 > 0$ such that for all $h \leq h_0$ and for arbitrary grid functions φ_h and χ_h problem (1.17), (1.18) possesses a unique solution and

$$\|y_h\|_{\hat{\bar{\omega}}_h} \leq c_1 \|\varphi_h\|_{\hat{\omega}_h} + c_2 \|\chi_h\|_{\gamma_h}, \tag{1.21}$$

where the constants c_1, c_2 do not depend on h, nor on φ_h or χ_h. □

A straightforward design of difference approximations for derivatives naturally leads to consistent approximations of the underlying ODEs. However, our real objective is *convergence* but not consistency.

Definition 1.6. We say that the exact solution y_h of the difference scheme (1.17), (1.18) converges to the exact solution $u(x)$ of the BVP (1.16) if $\|z_h\|_{\hat{\hat{\omega}}_h} \to 0$ as $h \to 0$, where $\|\cdot\|_{\hat{\hat{\omega}}_h}$ denotes some grid norm on $\hat{\hat{\omega}}_h = \hat{\omega}_h \cup \gamma_h$. The difference scheme (1.17), (1.18) has the *order of accuracy* p if there exists an $h_0 > 0$ such that for all $h \le h_0$ it holds that $\|z_h\|_{\hat{\hat{\omega}}_h} = \mathcal{O}(h^p)$ or $\|z_h\|_{\hat{\hat{\omega}}_h} \le c\,h^p$, where c is a constant independent of h. □

The order of accuracy is also called the *degree of accuracy.* In the following we use both notations.

The main theorem of the linear theory of difference schemes says that in the case of a stable difference scheme the order of consistency coincides with the *order of accuracy,* i.e. we have $\|y_h - u_h\|_{\hat{\hat{\omega}}_h} = \mathcal{O}(h^p)$.

For nonlinear problems there does not exist such a general statement. However, it is often possible to show that a nonlinear scheme is consistent and convergent.

As an example for the construction of difference schemes let us consider the following nonlinear second-order BVP

$$u''(x) = f(x, u(x)), \quad x \in (0,1), \quad u(0) = \mu_1, \quad u(1) = \mu_2. \tag{1.22}$$

To obtain an approximation $\{y_j\}_{j=0}^N$ of the grid function $\{u_j\}_{j=0}^N$ of the exact solution $u(x)$ we consider the ODE at $x = x_j$, where x_j is a node of the equidistant grid (1.1). Then we replace the second derivative $u''(x_j)$ on the left-hand side by the divided difference (1.11). It results in the well-known 3-point difference scheme

$$\begin{aligned} &\frac{y_{j+1} - 2y_j + y_{j-1}}{h^2} = f(x_j, y_j), \quad j = 1, 2, \ldots, N-1, \\ &y_0 = \mu_1, \quad y_N = \mu_2. \end{aligned} \tag{1.23}$$

If the function f is sufficiently smooth, the scheme (1.23) has an order of accuracy 2, i.e. it holds that $\|y - u\|_{\omega_h} = \max_{1 \le j \le N-1} |y_j - u_j| = \mathcal{O}(h^2)$.

If the accuracy of the scheme (1.23) is not sufficient for a certain application the question arises, how one can obtain a difference scheme whose order is higher than 2. There are two main approaches to answer this question. The first one is to use a difference approximation on a stencil with more than 3 points. For example, the 5-point difference scheme for the BVP (1.22),

$$\begin{aligned} &\frac{-y_{j+2} + 16y_{j+1} - 30y_j + 16y_{j-1} - y_{j-2}}{(12h)^2} = f(x_j, y_j), \\ &y_0 = \mu_1, \quad y_N = \mu_2, \quad j = 1, 2, \ldots, N-1, \end{aligned} \tag{1.24}$$

possesses an order of accuracy 4. This approach can be used for the construction of difference approximations of an arbitrary order. However, to enhance the order of accuracy of a scheme, the number of points used in the corresponding stencils must be increased significantly.

The second approach is based on the following idea. Using the Taylor expansion of the second finite difference (1.11),

$$u_{\bar{x}x,j} = u''(x_j) + \frac{h^2}{12}u^{(4)}(x_j) + \frac{h^4}{360}u^{(6)}(x_j) + O(h^6), \tag{1.25}$$

and the relation $u^{(4)} = f''(x, u)$ (see formula (1.22)), we get the difference approximation

$$y_{\bar{x}x,j} = f(x_j, y_j) + \frac{h^2}{12}f''(x_j, y_j), \quad j = 1, 2, \ldots, N-1,$$

which possesses the order of accuracy 4. Replacing the second derivative of the function f on the right-hand side by the divided difference of second order

$$f''(x_j, y_j) \approx \frac{1}{h^2}\left(f(x_{j+1}, y_{j+1}) - 2f(x_j, y_j) + f(x_{j-1}, y_{j-1})\right),$$

we obtain the well-known difference scheme of Numerov

$$\begin{aligned} y_{\bar{x}x,j} &= \frac{1}{12}\left[f(x_{j+1}, y_{j+1}) + 10f(x_j, y_j) + f(x_{j-1}, y_{j-1})\right], \\ y_0 &= \mu_1, \quad y_N = \mu_2, \quad j = 1, 2, \ldots, N-1. \end{aligned} \tag{1.26}$$

Unfortunately, the idea of the second approach cannot be extended to the construction of difference approximations of an order of accuracy higher than 4. It is still possible to replace the derivative $u^{(6)}$ by $f^{(4)}$ in formula (1.25), but there does not exist a 3-point approximation for $f^{(4)}(x_j, u_j)$ which uses only values of the function f.

Since a difference scheme for nonlinear BVPs represents a nonlinear system of algebraic equations, an iteration method is typically used to approximate its solution. The method of choice is often Newton's method which requires at each iteration step the solution of a system of linear algebraic equations. In case of the difference scheme (1.24) the coefficient matrix of the linear system is a 5-diagonal matrix and in the case of the difference scheme (1.26) it is a tridiagonal matrix. The solution of a system of linear equations with a tridiagonal coefficient matrix by an elimination method requires $8N + 1$ flops, whereas the solution of a system with a 5-diagonal coefficient matrix requires $19N - 10$ flops. Thus, for the 3-point difference scheme (1.26) the amount of work is significantly lower than for the 5-point scheme (1.24). This motivates us to introduce the following definition.

Definition 1.7. A difference scheme for ODEs of the order k is called *compact* if it uses only $k + 1$ values of the grid function. □

For example the schemes (1.23) and (1.26) are compact whereas the scheme (1.24) is not compact.

Since the growth behavior of the solution of a BVP can be very different on various subintervals of $[0, 1]$ the step-size should be controlled automatically. For this, instead of the grid (1.1), the non-equidistant grid (1.3) or a quasi-uniform grid has to be used and the corresponding difference approximations must be constructed on this grid.

It can be proved that the 3-point difference scheme on the grid (1.3)

$$y_{\bar{x}\hat{x},j} = f(x_j, y_j), \quad j = 1, 2, \ldots, N-1, \quad y_0 = \mu_1, \quad y_N = \mu_2,$$

where $y_{\bar{x}\hat{x},j}$ is defined in (1.10), has only the order of accuracy 1. This simple example shows that the construction of difference approximations of a high order of accuracy on non-equidistant grids is more complicated than for equidistant ones. However, compact difference schemes on non-uniform grids are of great importance for many applications.

In the past there were some attempts to develop difference schemes of a high order of accuracy for special classes of BVPs. For example, in [38] a 3-point difference scheme of the order of accuracy 4 has been developed for BVPs of the form

$$a(x)\frac{d^2u}{dx^2} + b(x)\frac{du}{dx} = f(x,u), \quad a(x) > 0, \quad x \in (0,1),$$

$$u(0) = \mu_1, \quad u(1) = \mu_2.$$

However, the theoretical foundation of this difference scheme is only given for the case $a(x) = \text{const} > 0$ and $b(x) = \text{const}$. For the BVP (1.22) a 3-point difference scheme of the order of accuracy 6 has been presented in [53]. But the approaches used in [38] and [53] can not be extended to general BVPs. Moreover, a disadvantage of these schemes is the use of an equidistant grid. Real progress in the direction of difference schemes for rather general BVPs whose order of accuracy can be prescribed arbitrarily has been achieved with the papers [26, 27, 42, 43, 55, 58, 59]. Here, the concept of an exact difference scheme plays an important role.

Definition 1.8. A difference scheme is called *exact* if its solution $\{y_j\}_{j=0}^N$ coincides with the grid function of the exact solution $u(x)$ of the given BVP, i.e., $y_j = u_j = u(x_j)$. □

Let us construct the exact difference scheme for the BVP

$$\begin{aligned} &u''(x) = f(x), \quad x \in (0,1), \\ &u(0) = \mu_1, \quad u(1) = \mu_2. \end{aligned} \tag{1.27}$$

For the discretization of (1.27) we use a non-equidistant grid $\hat{\omega}_h$. As can be easily seen, the solution of the BVP (1.27) coincides with the solution of the sequence of BVPs

$$\begin{aligned} &u''(x) = f(x), \quad x \in (x_{j-1}, x_{j+1}), \\ &u(x_{j-1}) = u_{j-1}, \quad u(x_{j+1}) = u_{j+1}, \quad j = 1, 2, \ldots, N-1. \end{aligned} \tag{1.28}$$

The solution of (1.28) can be represented in the form

$$u(x) = -\int_{x_{j-1}}^{x_{j+1}} G^j(x,\xi) f(\xi) d\xi + \frac{x - x_{j-1}}{x_{j+1} - x_{j-1}} u_{j+1} + \frac{x_{j+1} - x}{x_{j+1} - x_{j-1}} u_{j-1}, \quad x \in [x_{j-1}, x_{j+1}], \tag{1.29}$$

where

$$G^j(x,\xi) = \begin{cases} \dfrac{(x - x_{j-1})(x_{j+1} - \xi)}{x_{j+1} - x_{j-1}}, & x_{j-1} \le x \le \xi, \\ \dfrac{(x_{j+1} - x)(\xi - x_{j-1})}{x_{j+1} - x_{j-1}}, & \xi \le x \le x_{j+1}, \end{cases} \tag{1.30}$$

is Green's function for the BVP (1.28). Substituting $x = x_j$ into (1.29) we get the difference equation

$$u_j = \frac{h_j}{2\hbar_j} \int_{x_j}^{x_{j+1}} (x_{j+1} - \xi) f(\xi) d\xi - \frac{h_{j+1}}{2\hbar_j} \int_{x_{j-1}}^{x_j} (\xi - x_{j-1}) f(\xi) d\xi + \frac{h_j}{2\hbar_j} u_{j+1} + \frac{h_{j+1}}{2\hbar_j} u_{j-1}, \quad j = 1, 2, \ldots, N-1. \tag{1.31}$$

If this equation is multiplied by the factor $-2/(h_{j+1}h_j)$, the following difference scheme on the 3-point stencil (x_{j-1}, x_j, x_{j+1}) results:

$$u_{\bar{x}\hat{x},j} = -\frac{1}{\hbar_j h_{j+1}} \int_{x_j}^{x_{j+1}} (x_{j+1} - \xi) f(\xi) d\xi - \frac{1}{\hbar_j h_j} \int_{x_{j-1}}^{x_j} (\xi - x_{j-1}) f(\xi) d\xi, \quad j = 1, 2, \ldots, N-1. \tag{1.32}$$

Thus, the difference scheme

$$y_{\bar{x}\hat{x},j} = -\frac{1}{\hbar_j h_{j+1}} \int_{x_j}^{x_{j+1}} (x_{j+1} - \xi) f(\xi) d\xi - \frac{1}{\hbar_j h_j} \int_{x_{j-1}}^{x_j} (\xi - x_{j-1}) f(\xi) d\xi, \quad y_0 = \mu_1, \quad y_N = \mu_2, \quad j = 1, 2, \ldots, N-1, \tag{1.33}$$

is the exact difference scheme for the linear BVP (1.27).

In the nonlinear case an exact difference scheme represents a system of nonlinear algebraic equations containing nonlinear expressions (functionals) of the problem data. In general, these expressions cannot be evaluated directly, but as we will see later, they can be defined by the solutions of some associated IVPs. Therefore, the exact difference schemes provide the basis for the development of so-called truncated difference schemes of an arbitrary (given by the user) order of accuracy.

Definition 1.9. If in an exact difference scheme the parameters determined by nonlinear expressions (e.g., IVPs or nonlinear equations) are numerically approximated, a so-called *truncated* difference scheme results. The accuracy of a truncated difference scheme depends on how accurately these parameters are approximated. □

To illustrate the idea of EDS and TDS let us consider the simple BVP

$$\begin{aligned} &\frac{d}{dx}\left(k(x)\frac{du}{dx}\right) - q(x)u(x) = -f(x,u), \quad x \in (0,1), \\ &u(0) = \mu_1, \quad u(1) = \mu_2. \end{aligned} \tag{1.34}$$

On the interval $[0,1]$ we introduce a non-equidistant grid $\hat{\bar{\omega}}_h$ (see formula (1.3). We look for a difference scheme on this grid which has a similar form as the BVP (1.34), namely

$$\begin{aligned} &\left(a(x)y_{\bar{x}}\right)_{\hat{x}} - b(x)y(x) = -\varphi(x,y), \quad x \in \hat{\omega}_h, \\ &y_0 = \mu_1, \quad y_N = \mu_2, \end{aligned} \tag{1.35}$$

where for each $x \in \hat{\omega}_h$ the coefficients $a(x)$, $b(x)$ and $\varphi(x,y)$ are some functionals of the input data of problem (1.34). The special form of (1.35) allows us to study the difference scheme with similar well-developed analytical techniques as for the BVP (1.34). Thus, the scheme (1.35) can be carefully analyzed.

Our approach consists of three basic steps:

- For the given BVP (1.34) determine the coefficients $a(x)$, $b(x)$ and $\varphi(x,y)$, $x \in \hat{\omega}_h$, (in general, in closed form) and define the EDS;
- Approximate a, b and φ numerically and generate the corresponding TDS;
- Solve the resulting TDS, i.e. solve the nonlinear system of algebraic equations (1.35).

We show that for each $x \in \hat{\omega}_h$ the coefficients $a(x)$, $b(x)$ and $\varphi(x,y)$ are determined by the solutions of some IVPs for the homogeneous and the inhomogeneous ODE (1.34) on small subintervals (x_{i-1}, x_{i+1}). Therefore the coefficients of the TDS can be efficiently computed by an IVP-solver. Moreover, we show that the accuracy of the solution of (1.35) is determined by the accuracy of the corresponding IVP-solvers for the IVPs.

Definition 1.10. To improve clarity of presentation, throughout the text we will use the following abbreviations:

$$\begin{aligned} &\bar{e}_\alpha^j \stackrel{\text{def}}{=} [x_{j-2+\alpha}, x_{j-1+\alpha}], \quad \bar{e}^j \stackrel{\text{def}}{=} [x_{j-1}, x_j], \quad \bar{e}_2^N \stackrel{\text{def}}{=} [x_N, \infty), \\ &\gamma \stackrel{\text{def}}{=} j - 1 + \alpha, \quad \beta \stackrel{\text{def}}{=} j + (-1)^\alpha. \end{aligned}$$

The corresponding open intervals are denoted by e_α^j, e^j and e_2^N, respectively. To become familiar with these abbreviations we will alert the reader at the corresponding passages of the text. □

To construct and to analyze the compact difference schemes we use two main approaches: The theory of contractive mappings (Fixed Point Theorem) and the theory of monotone operators. In contrast to the first approach, the theory of monotone operators allows us to develop a strategy for the study of nonlinear problems with large Lipschitz constants. Let us give here a short outline of the theoretical results which will be used frequently in this book.

Theorem 1.1

(BANACH'S FIXED-POINT THEOREM)
Let X be a Banach space, M a closed nonempty set in X and $F : M \subset X \to X$. Suppose that:

1) M is mapped into itself by F, i.e., $F : M \subset X \to M$, and

2) the operator F is q-contractive, i.e.,

$$\|F(x) - F(y)\| \leq q\,\|x - y\| \tag{1.36}$$

for all $x, y \in M$ and for a fixed q, $0 \leq q < 1$.

Then we can conclude the following:

- Existence and uniqueness: Equation $x = F(x)$, $x \in M$, has exactly one solution x^*, i.e., F has exactly one fixed point on M;
- Convergence of the iteration: For an arbitrary choice of the initial point $x^{(0)} \in M$, the sequence $\{x^{(k)}\}$ constructed by

$$x^{(k+1)} = F(x^{(k)}), \quad k = 0, 1, \ldots, \tag{1.37}$$

converges to the unique solution x^* of the equation $x = F(x)$, $x \in M$;
- Error estimates: For all $k = 0, 1, \ldots$ we have the so-called a priori error estimate

$$\|x^{(k)} - x^*\| \leq \frac{q^n}{1-q}\,\|x^{(1)} - x^{(0)}\|,$$

and the so-called a posteriori error estimate

$$\|x^{(k+1)} - x^*\| \leq \frac{q}{1-q}\,\|x^{(k+1)} - x^{(k)}\|;$$

- Rate of convergence: For all $k = 0, 1, \ldots$ we have

$$\|x^{(k+1)} - x^*\| \leq q\,\|x^{(k)} - x^*\|.$$

Proof. See e.g. [86, p.17]. ■

Banach's Fixed Point Theorem can be modified as follows. Assume that the nonlinear operator F maps a closed set $M \subset X$ into itself. In that case, for each natural number n the n-th power of the operator F can be defined as follows. Let us set $F^2(x) \stackrel{\text{def}}{=} F(F(x))$ for all $x \in M$. Then, higher powers $F^n(x)$ are defined recursively by

$$F^n(x) = F(F^{n-1}(x)).$$

We now have the following result.

Theorem 1.2

Let X be a Banach space, M a closed nonempty set in X and $F : M \subset X \to X$. Suppose that:

1) M is mapped into itself by F, i.e., $F : M \subset X \to M$, and

2) for some natural number n the operator F^n is q-contractive on M.

Then, the sequence (1.37) converges to x^* for each given $x^{(0)} \in M$.

Proof. See e.g. [82, p.392–393]. ■

It is well known that the theory of contractive operators provides a simple framework to prove the existence and uniqueness of solutions for IVPs and BVPs of ODEs as well as for systems of nonlinear difference equations.

Another approach to prove the existence and uniqueness of (weak) solutions for BVPs of nonlinear ODEs is the theory of monotone operators. The central statement is formulated in the theorem of Browder and Minty.

Theorem 1.3

(Browder-Minty Theorem)
Let X be a reflexive Banach space and F an operator defined on X with values in the dual space X^*. Suppose that the following conditions are satisfied:

1) F is a bounded operator, i.e., the image of any bounded subset of X is a bounded subset of X^*;

2) the operator F is demicontinuous, i.e., for arbitrary $x^* \in X$ and any sequence $\{x_k\}_{k=1}^{\infty}$ of elements of the space X such that

$$x_k \to x^* \quad \text{in } X$$

we have

$$F(x_k) \rightharpoonup F(x^*) \quad \text{in } X^*;$$

3) the operator is coercive, i.e.,

$$\lim_{\|x\|_X \to \infty} \frac{\langle F(x), x \rangle}{\|x\|_X} = \infty;$$

4) the operator F is monotone on X, i.e., for all $x, y \in X$ we have

$$\langle F(x) - F(y), x - y \rangle \geq 0. \tag{1.38}$$

Then the equation

$$F(x) = f \tag{1.39}$$

has at least one solution $x \in X$ for every $f \in X^*$. If, moreover, inequality (1.38) is strict for all $x, y \in X$, $x \neq y$, then equation (1.39) has precisely one solution $x \in X$ for every $f \in X^*$.

Proof. See e.g. [18]. ■

Many statements about the existence of solutions are based on the assumption that the corresponding operator F is strongly monotone.

Definition 1.11. An operator $F : X \to X^*$ where X is a Banach space is said to be strongly monotone if there exists a constant $c > 0$ such that

$$\langle F(x) - F(y), x - y \rangle \geq c\|x - y\|^2.$$ □

If the operator $F : X \to X^*$ is strongly monotone, then F is coercive.

In many applications the space $\mathbb{R}^n$ plays an important role. This particularly applies to the discretization of ODEs.

Theorem 1.4

Let $F : \mathbb{R}^n \to \mathbb{R}^n$ be continuous everywhere in $\mathbb{R}^n$. Suppose that

$$(F(x) - F(y), x - y) \geq c\|x - y\|^2 \quad \textit{for all } x, y \in \mathbb{R}^n.$$

Then the equation $F(x) = 0$ has a unique solution $x^* \in \mathbb{R}^n$.

Proof. See e.g. [82, p.461–462]. ■

The existence and uniqueness of solutions of IVPs for ODEs is often shown with the following three theorems.

Theorem 1.5

(PICARD-LINDELÖF'S THEOREM)
Suppose that

1) $f(x, u)$, $u(x) \in \mathbb{R}^n$;

2) $f(x, u)$ is continuous on the parallelepiped

$$S \stackrel{\text{def}}{=} \{(x, u) \in \mathbb{R} \times \mathbb{R}^n : \ x : 0 \leq x \leq x_0 + a, \|u - u_0\| \leq b\} \tag{1.40}$$

and uniformly Lipschitz-continuous w.r.t. u;

3) $\|f(x,u)\| \le M$ on S.

Then, the IVP

$$u'(x) = f(x, u(x)), \quad u(x_0) = u_0 \tag{1.41}$$

has a unique solution on the interval $[x_0, x_0 + \alpha]$, where $\alpha \stackrel{\text{def}}{=} \min(a, b/M)$.

Proof. See e.g. [10]. ■

Theorem 1.6

(Peano's Theorem)
Suppose that

1) $f(x,u),\ u(x) \in \mathbb{R}^n$;
2) $f(x,u)$ is continuous on S;
3) $\|f(x,u)\| \le M$ on S.

Then, the IVP (1.41) has a solution on the interval $[x_0, x_0 + \alpha]$.

Proof. See e.g. [10]. ■

Definition 1.12. (Carathéodory conditions; see e.g. [9])
Let $G \subset \mathbb{R}^n$ be an open set and $J \stackrel{\text{def}}{=} [a, b] \subset \mathbb{R}$, where $a < b$. One says that $f : J \times G \to \mathbb{R}^m$ satisfies the Carathéodory conditions on $J \times G$, written as $f \in \mathrm{Car}(J \times G)$, if

1) $f(\cdot, x) : J \to \mathbb{R}^m$ is measurable for every $x \in G$,
2) $f(t, \cdot) : G \to \mathbb{R}^m$ is continuous for almost every $t \in J$, and
3) for each compact set $K \subset G$ the function

$$h_K(t) \stackrel{\text{def}}{=} \sup\{\|f(t,x)\| : x \in K\}$$

is Lebesgue integrable on J, where $\|\cdot\|$ is the norm on $\mathbb{R}^m$. □

Definition 1.13. (Solution in the extended sense; see e.g. [10])
A function $u(x)$ is called a solution in the extended sense of the IVP (1.41) if u is absolutely continuous, u satisfies the ODE almost everywhere and u satisfies the initial condition. The absolute continuity of u implies that its derivative exists almost everywhere. □

Theorem 1.7

(CARATHÉODORY'S THEOREM)
If the function $f(x, u)$ satisfies the Carathéodory conditions (see Definition 1.12), then the IVP (1.41) has a solution in the extended sense in a neighbourhood of the initial condition.

Proof. See e.g. [10]. ■

We want to conclude this section with two important inequalities.

Theorem 1.8

(CAUCHY-BUNYAKOVSKY-SCHWARZ INEQUALITY)
Let E be an inner product space and $x, y \in E$. Then

$$|(x, y)|^2 \leq (x, x) \cdot (y, y) \tag{1.42}$$

where $(\cdot, \cdot)$ is the inner product. Equivalently, by taking the square root on both sides, and referring to the norms of the vectors, the inequality is written as

$$|(x, y)| \leq \|x\| \cdot \|y\|. \tag{1.43}$$

If $x_1, \ldots, x_n \in \mathbb{C}$ and $y_1, \ldots, y_n \in \mathbb{C}$ are any complex numbers, the inequality may be restated as

$$\left|\sum_{i=1}^{n} x_i y_i\right|^2 \leq \sum_{j=1}^{n} |x_j|^2 \sum_{k=1}^{n} |y_k|^2.$$

Proof. See e.g. [76, p.10–11]. ■

Theorem 1.9

(MINKOWSKI'S INEQUALITY)
For all $1 < p < \infty$ and all functions f, g on an interval $[a, b]$ for which the integrals $\int_a^b |f(x)|^p dx$ and $\int_a^b |g(x)|^p dx$ exist, then the integral $\int_a^b |f(x)+g(x)|^p dx$ exists too, and the Minkowski's integral inequality states that

$$\left(\int_a^b |f(x)+g(x)|^p dx\right)^{1/p} \leq \left(\int_a^b |f(x)|^p dx\right)^{1/p} + \left(\int_a^b |g(x)|^p dx\right)^{1/p}. \tag{1.44}$$

Similarly, if $p > 1$ and $a_k, b_k > 0$, then Minkowski's sum inequality states that

$$\left(\sum_{k=1}^{n} |a_k + b_k|^p\right)^{1/p} \leq \left(\sum_{k=1}^{n} |a_k|^p\right)^{1/p} + \left(\sum_{k=1}^{n} |b_k|^p\right)^{1/p}. \tag{1.45}$$

Equality holds iff the sequences $a_1, a_2, \ldots$ and $b_1, b_2, \ldots$ are proportional.

Proof. See e.g. [15, p.11]. ■

1.2 Short history

To gain a better understanding of the approach used in this book, we present in this section a short overview of the history of EDS and TDS.

A first general approach to exact difference schemes for *linear* BVPs has been developed by A. N. Tikhonov and A. A. Samarskii in the early 1960s. In the papers [80, 81] dealing with linear second-order ODEs with piece-wise continuous coefficients they have developed a theory of the exact 3-point (compact) difference schemes.

In [80] an exact 3-point difference scheme (EDS) is proposed for the BVP

$$L^{(k,q)}u \stackrel{\text{def}}{=} \frac{d}{dx}\left(k(x)\frac{du}{dx}\right) - q(x)u = -f(x), \quad 0 < x < 1, \qquad u(0) = \mu_1, \quad u(1) = \mu_2, \tag{1.46}$$

where

$$0 < c_1 \le k(x), \quad k(x), q(x), f(x) \in Q^{(0)}[0,1], \quad q(x) \ge 0, \tag{1.47}$$

and $Q^{(0)}[0,1]$ is the class of piece-wise continuous functions with a finite number of discontinuity points of first kind. Moreover, an algorithm is given that uses truncated 3-point difference schemes (TDS) of rank m for the implementation of the exact difference scheme. For an arbitrarily given m the resulting method has the same order of accuracy. These results have been generalized in [81] for a non-equidistant grid and boundary conditions of the third kind. The idea of this approach is the following: we use the non-equidistant grid (1.4,a) and assume that the set ρ of discontinuity points of $k(x)$, $q(x)$ and $f(x)$ is a subset of this grid: $\rho \subseteq \hat{\omega}_h$. The exact solution of the BVP (1.46) satisfies the following continuity conditions at these discontinuity points:

$$u(x_i - 0) = u(x_i + 0), \quad k(x)\frac{du}{dx}\bigg|_{x=x_i-0} = k(x)\frac{du}{dx}\bigg|_{x=x_i+0}, \quad x_i \in \rho. \tag{1.48}$$

We introduce the stencil functions $v_\alpha^j(x)$, $\alpha = 1, 2$, as those solutions of the IVPs [81]

$$L^{(k,q)}v_\alpha^j(x) = 0, \quad x_{j-1} < x < x_{j+1},$$
$$v_\alpha^j(x_\beta) = 0, \quad k(x)\frac{dv_\alpha^j}{dx}\bigg|_{x=x_\beta} = (-1)^{\alpha+1}, \quad \alpha = 1, 2, \quad j = 1, 2, \ldots, N-1, \tag{1.49}$$

which satisfy the continuity conditions (1.48), too. For the definition of the index β see Definition 1.10.

These stencil functions possess the following properties:

1) $v_1^j(x) > 0$ is monotonically increasing on $(x_{j-1}, x_{j+1}]$, and the function $v_2^j(x) > 0$ is monotonically decreasing on $[x_{j-1}, x_{j+1})$;

2) it holds that

$$v_1^j(x_{j+1}) = v_2^j(x_{j-1}), \quad v_2^j(x_j) = v_1^{j+1}(x_{j+1});$$

3) the relation

$$\begin{aligned} v_1^j(x_{j+1}) &= v_1^j(x_j) + v_2^j(x_j) \\ &+ v_2^j(x_j)\int_{x_{j-1}}^{x_j} v_1^j(x)q(x)dx + v_1^j(x_j)\int_{x_j}^{x_{j+1}} v_2^j(x)q(x)dx \end{aligned} \tag{1.50}$$

is satisfied.

The solution of problem (1.46) can be represented in the form

$$u(x) = A_j v_1^j(x) + B_j v_2^j(x) + v_3^j(x), \quad x_{j-1} \le x \le x_{j+1}, \tag{1.51}$$

where A_j, B_j are constants and $v_3^j(x)$ is a particular solution of (1.46) which satisfies

$$\begin{aligned} L^{(k,q)} v_3^j(x) &= -f(x), \quad x_{j-1} < x < x_{j+1}, \\ v_3^j(x_{j-1}) &= v_3^j(x_{j+1}) = 0. \end{aligned} \tag{1.52}$$

Setting in (1.51) $x = x_{j+1}$ as well as $x = x_{j-1}$, we find

$$A_j = \frac{u(x_{j+1})}{v_1^j(x_{j+1})} \quad \text{and} \quad B_j = \frac{u(x_{j-1})}{v_2^j(x_{j-1})}, \tag{1.53}$$

respectively. The function $v_3^j(x)$ can be written as

$$v_3^j(x) = \int_{x_{j-1}}^{x_{j+1}} G(x,\xi)f(\xi)d\xi, \quad x_{j-1} \le x \le x_{j+1}, \tag{1.54}$$

where $G(x,\xi)$ is Green's function of problem (1.52) defined by

$$G(x,\xi) = \frac{1}{v_1^j(x_{j+1})} \begin{cases} v_1^j(x)v_2^j(\xi), & x_{j-1} \le x \le \xi, \\ v_1^j(\xi)v_2^j(x), & \xi \le x \le x_{j+1}. \end{cases} \tag{1.55}$$

Substituting (1.55) into (1.54) and setting $x = x_j$ we obtain

$$v_3^j(x_j) = \frac{1}{v_1^j(x_{j+1})}\left[v_2^j(x_j)\int_{x_{j-1}}^{x_j} v_1^j(\xi)f(\xi)d\xi + v_1^j(x_j)\int_{x_j}^{x_{j+1}} v_2^j(\xi)f(\xi)d\xi\right]. \tag{1.56}$$

Using (1.53), (1.50) and (1.56) we obtain from (1.51) the EDS

$$\begin{aligned} &(a\,u_{\bar{x}})_{\hat{x}} - b\,u = -\varphi(x), \quad x \in \hat{\omega}_h, \\ &u(0) = \mu_1, \quad u(1) = \mu_2, \end{aligned} \tag{1.57}$$

where

$$a(x_j) \stackrel{\text{def}}{=} \left[\frac{1}{\hbar_j} v_1^j(x_j)\right]^{-1}, \quad b(x_j) \stackrel{\text{def}}{=} \hat{T}^{x_j}(q), \quad \varphi(x_j) \stackrel{\text{def}}{=} \hat{T}^{x_j}(f),$$

$$\hat{T}^{x_j}(w) \stackrel{\text{def}}{=} \left[\hbar_j v_1^j(x_j)\right]^{-1} \int_{x_{j-1}}^{x_j} v_1^j(\xi) w(\xi) d\xi + \left[\hbar_j v_2^j(x_j)\right]^{-1} \int_{x_j}^{x_{j+1}} v_2^j(\xi) w(\xi) d\xi. \tag{1.58}$$

This EDS cannot be used directly because we still need an algorithm to compute its coefficients. For this aim we introduce at the point $x = x_j$ a local coordinate system by choosing

$$x = x_j + \hbar_j(s - \Delta_j), \quad \Delta_j \stackrel{\text{def}}{=} \frac{h_j - h_{j+1}}{h_{j+1} + h_j} = \frac{h_j - h_{j+1}}{2\hbar_j}, \quad -1 \le s \le 1.$$

Thus $x = \bar{x}_j + \hbar_j s$, with $\bar{x}_j = x_j - \hbar_j \Delta_j$. The interval $[x_{j-1}, x_{j+1}]$ is transformed into the reference interval $-1 \le s \le 1$, where the point $x = x_j$ corresponds to the value $s = \Delta_j$. We set

$$\begin{aligned} v_1^j(x) &= v_1^j\left(x_j + \hbar_j(s - \Delta_j)\right) \stackrel{\text{def}}{=} \hbar_j \alpha^j(s, \hbar_j), \\ v_2^j(x) &= v_2^j\left(x_j + \hbar_j(s - \Delta_j)\right) \stackrel{\text{def}}{=} \hbar_j \beta^j(s, \hbar_j), \quad -1 \le s \le 1. \end{aligned} \tag{1.59}$$

The stencil functions $\alpha^j(s, \hbar_j)$ and $\beta^j(s, \hbar_j)$ satisfy the equations

$$\frac{d}{ds}\left(\bar{k}(s)\frac{d\alpha^j}{ds}\right) - \hbar_j^2 \bar{q}(s)\alpha^j = 0, \quad -1 < s < 1,$$

$$\alpha^j(-1, \hbar_j) = 0, \quad \bar{k}(s)\frac{d\alpha^j}{ds}\bigg|_{s=-1} = 1,$$

$$\frac{d}{ds}\left(\bar{k}(s)\frac{d\beta^j}{ds}\right) - \hbar_j^2 \bar{q}(s)\beta^j = 0, \quad -1 < s < 1,$$

$$\beta^j(1, \hbar_j) = 0, \quad \bar{k}(s)\frac{d\beta^j}{ds}\bigg|_{s=1} = -1,$$

where

$$\bar{k}(s) \stackrel{\text{def}}{=} k(x_j + \hbar_j(s - \Delta_j)), \quad \bar{q}(s) \stackrel{\text{def}}{=} q(x_j + \hbar_j(s - \Delta_j)).$$

The coefficients a, b and φ used in the exact difference scheme (1.57) are now given by

$$
\begin{aligned}
a(x_j) &= \left[\frac{\hbar_j}{h_j}\alpha^j(0,\hbar_j)\right]^{-1}, \\
b(x_j) &= \frac{1}{\alpha^j(0,\hbar_j)}\int_{-1}^{\Delta_j}\alpha^j(s,\hbar_j)\bar{q}(s)ds + \frac{1}{\beta^j(0,\hbar_j)}\int_{\Delta_j}^{1}\beta^j(s,\hbar_j)\bar{q}(s)ds, \qquad (1.60)\\
\varphi(x_j) &= \frac{1}{\alpha^j(0,\hbar_j)}\int_{-1}^{\Delta_j}\alpha^j(s,\hbar_j)\bar{f}(s)ds + \frac{1}{\beta^j(0,\hbar_j)}\int_{\Delta_j}^{1}\beta^j(s,\hbar_j)\bar{f}(s)ds.
\end{aligned}
$$

If $q(x) \neq 0$ the stencil functions (1.59) cannot be determined explicitly by integration. But due to the analyticity of $\alpha^j(s,\hbar_j)$ and $\beta^j(s,\hbar_j)$ with respect to $\hbar_j^2$, we can represent these functions by the power series

$$
\alpha^j(s,\hbar_j) = \sum_{i=0}^{\infty}\alpha_i^j(s)\hbar_j^{2i}, \quad \beta^j(s,\hbar_j) = \sum_{i=0}^{\infty}\beta_i^j(s)\hbar_j^{2i}, \qquad (1.61)
$$

with $\alpha_i^j(s)$ and $\beta_i^j(s)$ defined by the recurrence formulas

$$
\begin{aligned}
\alpha_i^j(s) &= \int_{-1}^{s}\frac{1}{\bar{k}(t)}\left[\int_{-1}^{t}\bar{q}(\lambda)\alpha_{i-1}^j(\lambda)d\lambda\right]dt, \quad i = 1,2,\dots,\\
\beta_i^j(s) &= \int_{s}^{1}\frac{1}{\bar{k}(t)}\left[\int_{t}^{1}\bar{q}(\lambda)\beta_{i-1}^j(\lambda)d\lambda\right]dt, \quad i = 1,2,\dots,\\
\alpha_0^j(s) &= \int_{-1}^{s}\frac{1}{\bar{k}(t)}dt, \quad \beta_0^j(s) = \int_{s}^{1}\frac{1}{\bar{k}(t)}dt.
\end{aligned}
$$

Let

$$
\alpha^{(m)j}(s,\hbar_j) \stackrel{\text{def}}{=} \sum_{i=0}^{m}\alpha_i^j(s)\hbar_j^{2i}, \quad \beta^{(m)j}(s,\hbar_j) \stackrel{\text{def}}{=} \sum_{i=0}^{m}\beta_i^j(s)\hbar_j^{2i}
$$

be the truncated sums of the series (1.61). Replacing in formula (1.60) $\alpha^j(s,\hbar_j)$ and $\beta^j(s,\hbar_j)$ by $\alpha^{(m)j}(s,\hbar_j)$ and $\beta^{(m)j}(s,\hbar_j)$, respectively, we obtain coefficients $a^{(m)}$, $b^{(m)}$ and $\varphi^{(m)}$ which are used in the following truncated difference scheme of rank m (abbreviated as m-TDS)

$$
\begin{aligned}
&\left(a^{(m)}y_{\bar{x}}^{(m)}\right)_{\hat{x}} - b^{(m)}y^{(m)} = -\varphi^{(m)}(x), \quad x \in \hat{\omega}_h,\\
&y^{(m)}(0) = \mu_1, \quad y^{(m)}(1) = \mu_2.
\end{aligned} \qquad (1.62)
$$

The next statement was proved in [81] and represents a convergence result for the above m-TDS.

Theorem 1.10

Let $k(x)$, $q(x)$ and $f(x)$ belong to the class $Q^{(0)}[0,1]$ and assume that

$$0 < C_1 \leq \frac{h_{j+1}}{h_j} \leq C_2.$$

Then, the m-TDS (1.62) has the order of accuracy $2m+2$, i.e. it holds that

$$\left\|y^{(m)} - u\right\|_{0,\infty,\hat{\omega}_h} \stackrel{\text{def}}{=} \max_{1\leq j\leq N-1} \left|y_j^{(m)} - u_j\right| \leq M\, h^{2m+2}, \quad h \leq h_0, \tag{1.63}$$

where C_1, C_2, M and h_0 are positive coefficients which do not depend on m and the grid.

If we want to solve a BVP with boundary conditions of second or third kind we need the exact and truncated difference equations which correspond to these boundary conditions. Let us consider the following BVP with boundary conditions of third kind

$$\begin{aligned} &L^{(k,q)}u = -f(x), \quad 0 < x < 1, \\ &k(x)\left.\frac{du}{dx}\right|_{x=0} - \sigma_1 u(0) = -\mu_1, \quad u(1) = \mu_2. \end{aligned} \tag{1.64}$$

For the discretization of (1.64) we use the non-equidistant grid $\hat{\omega}_h$ (see formula (1.3)).

At first, let us find an algebraic boundary condition (in the following referred to as *difference boundary condition*) at $x = 0$ that any solution of problem (1.64) satisfies. According to the superposition principle the exact solution of the ODE (1.64,a) can be represented on an interval $[0, x_1]$ in the form

$$u(x) = \frac{v_2^0(x)}{v_2^0(0)} u_0 + \frac{v_1^0(x)}{v_1^0(x_1)} u_1 + v_3^*(x), \tag{1.65}$$

where $v_1^0(x)$, $v_2^0(x)$ and $v_3^*(x)$ are the solutions of the problems

$$L^{(k,q)}v_1^0(x) = 0, \quad 0 < x < x_1, \quad v_1^0(0) = 0, \quad k(x)\left.\frac{dv_1^0}{dx}\right|_{x=0} = 1, \tag{1.66}$$

$$L^{(k,q)}v_2^0(x) = 0, \quad 0 < x < x_1, \quad v_2^0(x_1) = 0, \quad k(x)\left.\frac{dv_2^0}{dx}\right|_{x=x_1} = -1, \tag{1.67}$$

$$L^{(k,q)}v_3^*(x) = -f(x), \quad 0 < x < x_1, \quad v_3^*(0) = v_3^*(x_1) = 0. \tag{1.68}$$

Let us demand that the function $u(x)$ in (1.65) satisfies also the left boundary condition in (1.64,b), i.e.,

$$k(x)\left.\frac{du}{dx}\right|_{x=0} - \sigma_1 u(0) = -\mu_1. \tag{1.69}$$

Substituting (1.65) into (1.69) and taking into account that $v_2^0(0) = v_1^0(x_1)$, we obtain the exact difference boundary condition

$$a_1 u_{\bar{x},1} - \bar{\sigma}_1 u_0 = -\bar{\mu}_1, \tag{1.70}$$

with

$$\begin{gathered} a_1 \stackrel{\text{def}}{=} a(x_1) = \left[\frac{1}{h_1} v_2^0(0)\right]^{-1}, \\ \bar{\sigma}_1 = h_1 a_1 \int_0^{x_1} q(x) v_2^0(x)dx + \sigma_1, \quad \bar{\mu}_1 = \mu_1 + k(x)\frac{dv_3^*}{dx}\bigg|_{x=0}. \end{gathered} \tag{1.71}$$

Therefore, the EDS for problem (1.64) takes the form

$$\begin{gathered} (a\, u_{\bar{x}})_{\hat{x}} - b\, u = -\varphi(x), \quad x \in \hat{\omega}_h, \\ a_1 u_{\bar{x},1} - \bar{\sigma}_1 u_0 = -\bar{\mu}_1, \quad u_N = \mu_2, \end{gathered} \tag{1.72}$$

where $a(x_j)$, $b(x_j)$, $\varphi(x_j)$ are defined by formulas (1.58), and $\bar{\sigma}_1$, $\bar{\mu}_1$ by formula (1.71).

To develop a TDS we introduce the local coordinate $s = x/h_1$ at $x = 0$, i.e. we have $x = sh_1$, $0 \le s \le 1$. We change the dependent variable by

$$v_2^0(x) = v_2^0(sh_1) = h_1 \beta^0(s, h_1), \quad v_3^*(x) = v_3^*(sh_1) = h_1 \gamma^0(s, h_1).$$

The stencil functions $\beta^0(s, h_1)$ and $\gamma^0(s, h_1)$ satisfy the equations

$$\begin{aligned} &\frac{d}{ds}\left(\bar{k}(s)\frac{d\beta^0}{ds}\right) - h_1^2 \bar{q}(s)\beta^0 = 0, \quad 0 < s < 1, \\ &\beta^0(1, h_1) = 0, \quad \bar{k}(s)\frac{d\beta^0}{ds}\bigg|_{s=1} = -1, \\ &\frac{d}{ds}\left(\bar{k}(s)\frac{d\gamma^0}{ds}\right) - h_1^2 \bar{q}(s)\beta^0 = h_1^2 f(s), \quad 0 < s < 1, \\ &\beta^0(1, h_1) = 0, \quad \bar{k}(s)\frac{d\beta^0(s, h_1)}{ds}\bigg|_{s=-1} = -1, \end{aligned}$$

where

$$\bar{k}(s) \stackrel{\text{def}}{=} k(sh_1), \quad \bar{q}(s) \stackrel{\text{def}}{=} q(sh_1), \quad \bar{f}(s) \stackrel{\text{def}}{=} f(sh_1).$$

The coefficients of the exact difference boundary condition of third kind take now the form

$$\bar{\sigma}_1 = a_1 \int_0^1 \bar{q}(s)\beta^0(s, h_1)ds, \quad \bar{\mu}_1 = \mu_1 + h_1\, \bar{k}(s)\frac{d\gamma^0}{ds}\bigg|_{s=0}, \quad a_1 = \left[\frac{1}{h_1} v_2^0(0)\right]^{-1}.$$

Replacing the functions $\beta^0(s,h_1)$ and $\gamma^0(s,h_1)$ by the polynomials

$$\beta^{(m)0}(s,h_1) = \sum_{i=0}^{m} \beta_i^0(s)h_1^{2i}, \quad \gamma^{(m)0}(s,h_1) = \sum_{i=0}^{m} \gamma_i^0(s)h_1^{2i},$$

we obtain the truncated boundary condition of rank m,

$$a_1^{(m)} y_{\bar{x},1}^{(m)} - \bar{\sigma}_1^{(m)} y_0^{(m)} = -\bar{\mu}_1^{(m)}.$$

In [81] it is shown that the solution of the m-TDS

$$\begin{aligned} &\left(a^{(m)} y_{\bar{x}}^{(m)}\right)_{\hat{x}} - b^{(m)} y^{(m)} = -\varphi^{(m)}(x), \quad x \in \hat{\omega}_h, \\ &a_1^{(m)} y_{\bar{x},1}^{(m)} - \bar{\sigma}_1^{(m)} y_0^{(m)} = -\bar{\mu}_1^{(m)}, \quad y_N^{(m)} = \mu_2 \end{aligned} \tag{1.73}$$

possesses the order of accuracy $2m+2$. An analogous approach for the construction of EDS and TDS for the generalized BVP of third kind with minimal requirements on the smoothness of the coefficient was used in [23]. The papers [21, 22] deal with EDS and TDS for a class of variational inequalities as well as with the development of efficient algorithms for their implementation. The TDS enable the computation of approximations of the exact solution with an order of accuracy which can be prescribed by the user for arbitrary piece-wise continuous coefficients $k(x)$, $q(x)$ and $f(x)$.

The practical implementation of TDS of a higher order of accuracy obtained within this approach requires in the case of non-polynomial coefficients in the equation (1.46) the computation of multidimensional integrals. This can be done, for example, with Monte-Carlo methods but represents, in fact, a serious computational problem.

In order to overcome this drawback another efficient algorithmic approach for the construction of TDS of an arbitrarily given order of accuracy for problem (1.46), (1.47) with piece-wise smooth coefficients $k(x), q(x), f(x)$ was proposed in [60, 73]. The EDS from [80, 81] was the basis of this approach too. It was shown that at each node x_j of a non-equidistant grid $\hat{\omega}_h$ the coefficients $a(x)$, $b(x)$ and the right-hand side $\varphi(x)$ can be represented by the solutions of four auxiliary IVPs on the (small!) subinterval (x_{j-1}, x_{j+1}). From (1.58) we obtain immediately the following representations for $a(x)$ and $b(x)$:

$$a(x_j) = \left[\frac{1}{h_j} v_1^j(x_j)\right]^{-1}, \quad b(x_j) = \hbar_j^{-1} \sum_{\alpha=1}^{2} [v_\alpha^j(x_j)]^{-1}[(-1)^{\alpha+1} m_\alpha^j(x_j) - 1], \tag{1.74}$$

where

$$m_\alpha^j(x) \stackrel{\text{def}}{=} k(x)\frac{dv_\alpha^j(x)}{dx}, \quad \alpha = 1, 2.$$

In order to obtain the representation for $\varphi(x)$, two auxiliary functions $w_\alpha^j(x)$, $\alpha = 1, 2$, were introduced as the solutions of the IVPs

$$
\begin{aligned}
&L^{(k,q)} w_\alpha^j(x) = -f(x), \quad x_{j-2+\alpha} < x < x_{j-1+\alpha}, \\
&w_\alpha^j(x_{j+(-1)^\alpha}) = \left. \frac{dw_\alpha^j}{dx} \right|_{x=x_{j+(-1)^\alpha}} = 0, \quad \alpha = 1, 2.
\end{aligned} \tag{1.75}
$$

It was shown that the right-hand side $\varphi(x)$ of the EDS can be represented in the following way:

$$
\varphi(x_j) = \hbar_j^{-1} \sum_{\alpha=1}^{2} (-1)^\alpha \left[l_\alpha^j(x_j) - m_\alpha^j(x_j) \frac{w_\alpha^j(x_j)}{v_\alpha^j(x_j)} \right], \tag{1.76}
$$

where

$$
l_\alpha^j(x) = k(x) \frac{dw_\alpha^j(x)}{dx}, \quad \alpha = 1, 2.
$$

One can see from (1.74), (1.76) that, for the computation of the coefficients $a(x_j)$, $b(x_j)$ and the right-hand side $\varphi(x_j)$ of the EDS, the four IVPs have to be solved for each $x_j \in \hat{\omega}_h$: (1.117), (1.75) with $\alpha = 1$ on the subinterval $[x_{j-1}, x_j]$ and (1.49), (1.75) with $\alpha = 2$ on the subinterval $[x_j, x_{j+1}]$. These IVPs can be solved by executing only one step with an arbitrary one-step method, e.g. with the Taylor series method or a Runge-Kutta method of the order of accuracy $\bar{m} = 2[(m+1)/2]$, where $[\cdot]$ denotes the entire part of the argument in these brackets. We label the thus calculated solutions with an additional index m: $v_\alpha^{(m)j}(x_j)$, $m_\alpha^{(\bar{m})j}(x_j)$, $w_\alpha^{(\bar{m})j}(x_j)$ and $l_\alpha^{(m)j}(x_j)$. Instead of the three-point EDS (1.57), (1.74), (1.76) we now have the three-point TDS

$$
\begin{aligned}
&(a^{(m)} y_{\bar{x}}^{(m)})_{\hat{x}} - b^{(m)} y^{(m)} = -\varphi^{(m)}(x), \quad x \in \hat{\omega}_h, \\
&y^{(m)}(0) = \mu_1, \quad y^{(m)}(1) = \mu_2,
\end{aligned} \tag{1.77}
$$

where the coefficients are given by

$$
\begin{aligned}
a^{(m)}(x_j) &= \left[\frac{1}{h_j} v_1^{(m)j}(x_j) \right]^{-1}, \\
b^{(m)}(x_j) &= \hbar_j^{-1} \sum_{\alpha=1}^{2} (-1)^{\alpha+1} [v_\alpha^{(m)j}(x_j)]^{-1} [m_\alpha^{(\bar{m})j}(x_j) + (-1)^\alpha], \\
\varphi^{(m)}(x_j) &= \hbar_j^{-1} \sum_{\alpha=1}^{2} (-1)^\alpha \left[l_\alpha^{(m)j}(x_j) - m_\alpha^{(\bar{m})j}(x_j) \frac{w_\alpha^{(\bar{m})j}(x_j)}{v_\alpha^{(m)j}(x_j)} \right].
\end{aligned} \tag{1.78}
$$

The next statement characterizes the order of accuracy of this scheme (see e.g. [73]).

Theorem 1.11

Let $k(x) \in Q^{(m+1)}[0, 1]$, $q(x), f(x) \in Q^{(m)}[0, 1]$, and suppose that the homogeneous BVP (1.46), (1.47) and (1.117) possesses only the trivial solution. Then *the error of TDS (1.77) satisfies*

$$\left\|y^{(m)}-u\right\|^{*}_{1,\infty,\hat{\omega}_h}=\max\left\{\left\|y^{(m)}-u\right\|_{0,\infty,\hat{\omega}_h},\left\|k\frac{dy^{(m)}}{dx}-k\frac{du}{dx}\right\|_{0,\infty,\hat{\omega}_h}\right\}\le M\,h^m,$$

where h is small enough, the constant M is independent of h and

$$k(x_j)=\frac{dy^{(m)}(x_j)}{dx}$$
$$=\left[\sum_{\alpha=1}^{2}(-1)^{\alpha}v_{\alpha}^{(m)j}(x_j)m_{3-\alpha}^{(\bar m)j}(x_j)\right]^{-1}\sum_{\alpha=1}^{2}\left\{m_{\alpha}^{(\bar m)j}(x_j)y^{(m)}(x_{j+(-1)^{\alpha}})\right.$$
$$\left.+(-1)^{\alpha+1}\left[m_1^{(\bar m)j}(x_j)m_2^{(\bar m)j}(x_j)w_{\alpha}^{(\bar m)j}(x_j)+m_{\alpha}^{(\bar m)j}(x_j)v_{3-\alpha}^{(m)j}(x_j)l_{3-\alpha}^{(m)j}(x_j)\right]\right\}.$$

The advantage of the scheme (1.77) compared with the schemes from [19] for piece-wise smooth $k(x)$, $q(x)$ and $f(x)$ is that besides an approximation for the exact solution one obtains also a three-point approximation for $k(x)du/dx$ of the same accuracy (measured in the Chebyshev norm) without additional computational costs (in [19] an estimate for an approximation of $u_{\bar{x}}$ is given).

A new approach for the construction of 3-point EDS and the corresponding TDS of an arbitrarily given order of accuracy m for *nonlinear problems* of the form

$$\frac{d}{dx}\left[k(x)\frac{du}{dx}\right]=-f(x,u),\quad x\in(0,1),\quad u(0)=\mu_1,\quad u(1)=\mu_2,\qquad(1.79)$$

was published in 1990 (see [59]). These results were generalized and further developed in [45, 49]. In [42] monotone ODEs were considered. Under the assumption that the following conditions are fulfilled,

$$0<c_1\le k(x)\quad\text{for all } x\in[0,1],\quad k(x)\in Q^1[0,1],\qquad(1.80)$$

$$f_u(x)\stackrel{\text{def}}{=}f(x,u)\in Q^0[0,1],\quad |f(x,u)|\le K\quad\text{for all } x\in[0,1],\ u\in\Omega([0,1],r),\qquad(1.81)$$

$$|f(x,u)-f(x,v)|\le L\,|u-v|\quad\text{for all } x\in[0,1],\ u,v\in\Omega([0,1],r),\qquad(1.82)$$

$$q\stackrel{\text{def}}{=}L/c_1<1,\qquad(1.83)$$

a new implementation of the three-point EDS on an arbitrary non-equidistant grid $\hat{\omega}_h$ using a TDS of a desired order of accuracy was introduced in [46]. Here $Q^p[0,1]$ denotes the class of functions which are piece-wise differentiable up to the order p and possess a finite number of discontinuity points of the first kind. $\Omega([0,1],r)$ is the set

$$\Omega([0,1],r)$$

$$\stackrel{\text{def}}{=} \left\{u(x) : u(x) \in W^1_\infty(0,1),\ u(x),\ k(x)\frac{du}{dx} \in \mathbb{C}[0,1], \left\|u - u^{(0)}\right\|_{1,\infty,(0,1)} \le r\right\}, \tag{1.84}$$

where

$$\|u\|_{0,\infty,(0,1)}$$

$$\stackrel{\text{def}}{=} \operatorname*{vrai\,max}_{x\in(0,1)} |u(x)|, \quad \|u\|_{1,\infty,(0,1)} \stackrel{\text{def}}{=} \max\left\{\|u\|_{0,\infty,(0,1)}, \left\|\frac{du}{dx}\right\|_{0,\infty,(0,1)}\right\},$$

$$r \stackrel{\text{def}}{=} \frac{K}{c_1},\ V_1(x) \stackrel{\text{def}}{=} \int_0^x \frac{dt}{k(t)},\ V_2(x) \stackrel{\text{def}}{=} \int_x^1 \frac{dt}{k(t)},\ u^{(0)}(x) \stackrel{\text{def}}{=} \frac{V_2(x)}{V_1(1)}\mu_1 + \frac{V_1(x)}{V_1(1)}\mu_2.$$

The three-point EDS for (1.79) is

$$(au_{\bar{x}})_{\hat{x}} = -\varphi(x_j,u) \quad x \in \hat{\omega}_h, \quad u(0) = \mu_1, \quad u(1) = \mu_2, \tag{1.85}$$

where

$$a(x_j) = \left[\frac{1}{h_j}V_1^j(x_j)\right]^{-1},$$

$$\varphi(x_j,u) = \hbar_j^{-1}\sum_{\alpha=1}^{2}(-1)^\alpha \left[l_\alpha^j(x_j,u) + (-1)^\alpha \frac{w_\alpha^j(x_j,u)}{V_\alpha^j(x_j)}\right].$$

The functions $w_\alpha^j(x_j,u)$, $l_\alpha^j(x_j,u)$, $\alpha = 1,2$, are the solutions of the following IVPs on small intervals (the interval length can be controlled):

$$\frac{d}{dx}w_\alpha^j(x,u) = \frac{l_\alpha^j(x,u)}{k(x)}, \quad \frac{d}{dx}l_\alpha^j(x,u) = -f\left(x, Y_\alpha^j(x,u)\right), \quad x \in e_\alpha^j, \tag{1.86}$$

$$w_\alpha^j(x_\beta,u) = l_\alpha^j(x_\beta,u) = 0, \quad \alpha = 1,2,$$

where e_α^j and the index β are defined in Definition 1.10, and

$$Y_\alpha^j(x,u) = \hat{u}(x) + w_\alpha^j(x,u) - \frac{V_\alpha^j(x)}{V_\alpha^j(x_j)}w_\alpha^j(x_j,u), \quad x \in \bar{e}_\alpha^j, \quad \alpha = 1,2,$$

$$\hat{u}(x) = \left[u(x_j)V_1^j(x) + u(x_{j-1})V_2^{j-1}(x)\right]\cdot\left[V_1^j(x_j)\right]^{-1}, \quad x \in \bar{e}^j,$$

$$V_1^j(x) = \int_{x_{j-1}}^x \frac{dt}{k(t)}, \quad V_2^j(x) = \int_x^{x_{j+1}} \frac{dt}{k(t)}.$$

The practical realization of the EDS (1.85) for all $x_j \in \hat{\omega}_h$ (i.e. the computation of the corresponding coefficients as input data) can be achieved by the integration of the two IVPs (1.86). The first one has to be integrated forward on the interval $[x_{j-1}, x_j]$ with $\alpha = 1$, whereas the second one must be integrated backward on the interval $[x_j, x_{j+1}]$ with $\alpha = 2$. For this integration one can use an appropriate one-step method, e.g. a Runge-Kutta method of the order of accuracy $\bar{m}$.

Note that the solutions $w_\alpha^j(x,u)$ and $l_\alpha^j(x,u)$, $\alpha = 1, 2$, of the IVPs (1.86) depend on the parameters $b_\alpha^j(u) \stackrel{\text{def}}{=} w_\alpha^j(x_j, u)$, i.e., $w_\alpha^j(x,u) = w_\alpha^j(x,u,b_\alpha^j(u))$ and $l_\alpha^j(x,u) = l_\alpha^j(x,u,b_\alpha^j(u))$, $\alpha = 1, 2$.

Substituting the numerical solutions of the order $\bar{m}$ into the EDS (1.85) yields the TDS of the rank $\bar{m}$,

$$\begin{aligned}
&(a^{(\bar{m})} y_{\bar{x}}^{(\bar{m})})_{\hat{x}} = -\varphi^{(\bar{m})}(x, y^{(\bar{m})}), \quad x \in \hat{\omega}_h, \quad y^{(\bar{m})}(0) = \mu_1, \quad y^{(\bar{m})}(1) = \mu_2, \\
&a^{(\bar{m})}(x_j) = \left[\frac{1}{h_j} V_1^{(\bar{m})j}(x_j)\right]^{-1}, \\
&\varphi^{(\bar{m})}(x_j, u) = \hbar_j^{-1} \sum_{\alpha=1}^{2} (-1)^\alpha \left[l_\alpha^{(m)j}(x_j, u) + (-1)^\alpha \frac{w_\alpha^{(\bar{m})j}(x_j, u)}{V_\alpha^{(\bar{m})j}(x_j)} \right].
\end{aligned} \tag{1.87}$$

Then, the parameters in (1.87) can be written in the form

$$\begin{aligned}
b_\alpha^{(s-1)} &= w_\alpha^{(s-1)j}(x_j, u) \\
&= -\frac{h_\gamma^2}{2} \frac{f_\beta}{k_\beta} + \sum_{p=3}^{s-1} \frac{[(-1)^{\alpha+1} h_\gamma]^p}{p!} \frac{d^p w_\alpha^j(x_\beta, u, b_\alpha^{(s-2)})}{dx^p}, \qquad s = 3, 4, \ldots, \bar{m}, \\
w_\alpha^{(\bar{m})j}(x_j, u) &= -\frac{h_\gamma^2}{2} \frac{f_\beta}{k_\beta} + \sum_{p=3}^{\bar{m}} \frac{[(-1)^{\alpha+1} h_\gamma]^p}{p!} \frac{d^p w_\alpha^j(x_\beta, u, b_\alpha^{(\bar{m}-1)})}{dx^p}, \\
l_\alpha^{(m)j}(x_j, u) &= (-1)^\alpha h_\gamma f_\beta + \sum_{p=2}^{m} \frac{[(-1)^{\alpha+1} h_\gamma]^p}{p!} \frac{d^p l_\alpha^j(x_\beta, u, b_\alpha^{(\bar{m}-1)})}{dx^p}.
\end{aligned}$$

In the next theorem (see also [46]) the accuracy of the above TDS is characterized.

Theorem 1.12

Let the assumptions (1.80) – (1.83) be satisfied. Suppose that

$$k(x) \in Q^{m+2}[0,1], \quad f(x,u) \in \bigcup_{j=1}^{N} \mathbb{C}^{m+1}\left(\bar{e}^j \times \Omega([0,1], r + \Delta)\right).$$

Then there exists $h_0 > 0$ such that for $h \le h_0$ the TDS (1.87) has a unique solution. Moreover, the following estimate of the accuracy holds:

$$\left\| y^{(\bar{m})} - u \right\|_{1,\infty,\hat{\omega}_h}^* = \max \left\{ \left\| y^{(\bar{m})} - u \right\|_{0,\infty,\hat{\omega}_h}, \left\| k \frac{dy^{(\bar{m})}}{dx} - k \frac{du}{dx} \right\|_{0,\infty,\hat{\omega}_h} \right\} \le M\, h^{\bar{m}},$$

where

$$k(x_j)\frac{dy^{(\bar{m})}(x_j)}{dx} = \frac{h_j y^{(\bar{m})}_{\bar{x},j} - w_1^{(\bar{m})j}(x_j,y^{(\bar{m})})}{V_1^{(\bar{m})j}(x_j)} + l_1^{(m)j}(x_j,y^{(\bar{m})}),$$

and the constant M does not depend on h.

In the paper [47] an EDS of the type (1.87) is introduced and justified under the conditions

$$0 < c_1 \leq k(x) \leq c_2 \quad \text{for all } x \in [0,1], \quad k(x) \in Q^1[0,1],$$

$$f_u(x) \stackrel{\text{def}}{=} f(x,u) \in Q^0[0,1] \quad \text{for all } u \in \mathbb{R},$$

$$f_x(u) \stackrel{\text{def}}{=} f(x,u) \in C(\mathbb{R}) \quad \text{for all } x \in [0,1],$$

$$|f(x,u)| \leq g(x) + c|u| \quad \text{for all } x \in [0,1],\ u \in \mathbb{R}, \quad c \geq 0,$$

$$[f(x,u) - f(x,v)](u-v) \leq c_3 |u-v|^2 \quad \text{for all } x \in [0,1],\ u,v \in \mathbb{R},$$

$$0 \leq c_3 < \pi^2 c_1,$$

where $g(x) \in L_2(0,1)$, and c, c_1, c_2, c_3 are constants.

A drawback of the TDS of a high order of accuracy is that derivatives of the functions $w_\alpha^j(x,u)$ and $l_\alpha^j(x,u)$, $\alpha = 1,2$, have to be computed. This can be done by an analytical differentiation of the equations in (1.86), e.g. by well-known computer algebra tools like `Mathematica` and `Maple`.

EDS and TDS of an arbitrarily given order of accuracy for *systems of linear second order equations*

$$L^{(K,Q)}\boldsymbol{u} \stackrel{\text{def}}{=} \frac{d}{dx}\left(K(x)\frac{d\boldsymbol{u}}{dx}\right) - Q(x)\boldsymbol{u}(x) = -\boldsymbol{f}(x) \tag{1.88}$$

under the boundary conditions

$$\boldsymbol{u}(0) = \boldsymbol{u}(1) = 0 \tag{1.89}$$

were proposed in [57, 58]. The authors suppose that the matrices $K(x) = [k_{ij}(x)]_{i,j=1}^m$, $Q(x) = [q_{ij}(x)]_{i,j=1}^m$ and the vector $\boldsymbol{f}(x) = \{f_i(x)\}_{i=1}^m$ satisfy the conditions

$$C_1\|\boldsymbol{v}\|^2 \leq (K(x)\boldsymbol{v},\boldsymbol{v}), \quad C_1 > 0, \quad \text{for all } \boldsymbol{v} \in \mathbb{R}^m, \quad \text{for all } x \in [0,1], \tag{1.90}$$

$$k_{ij}(x) \in Q^{n+1}[0,1], \quad q_{ij}(x) \in Q^n[0,1], \quad f_i(x) \in Q^n[0,1] \tag{1.91}$$

and that at the discontinuity points of the coefficients x_i the solution of problem (1.88)–(1.91) satisfies the conditions

$$\boldsymbol{u}(x_i - 0) = \boldsymbol{u}(x_i + 0), \quad K(x)\frac{d\boldsymbol{u}}{dx}\bigg|_{x=x_i-0} = K(x)\frac{d\boldsymbol{u}}{dx}\bigg|_{x=x_i+0}.$$

In that case the EDS can be derived from the integral corollary of differential equations in the following way. Let us introduce the matrices and vectors $V_1^j(x)$, $V_2^j(x)$, $\boldsymbol{W}_1^j(x)$ and $\boldsymbol{W}_2^j(x)$ as the solutions of the following problems (compare this with the scalar case):

$$\begin{aligned} &L^{(K,Q)}V_\alpha^j(x) = 0, \quad x \in (x_{j-1}, x_{j+1}), \\ &V_\alpha^j(x_\beta) = 0, \quad K(x)\frac{dV_\alpha^j}{dx}\bigg|_{x=x_\beta} = (-1)^{\alpha+1}E, \end{aligned} \tag{1.92}$$

$$\begin{aligned} &L^{(K,Q)}\boldsymbol{W}_\alpha^j(x) = -\boldsymbol{f}(x), \quad x \in (x_{j-1}, x_{j+1}), \\ &\boldsymbol{W}_\alpha^j(x_\beta) = \boldsymbol{0}, \quad K(x)\frac{d\boldsymbol{W}_\alpha^j}{dx}\bigg|_{x=x_\beta} = \boldsymbol{0}, \quad \alpha = 1, 2 \end{aligned} \tag{1.93}$$

as well as the vector-function $\boldsymbol{W}^j(x)$ as the solution of the inhomogeneous BVP (on a small interval)

$$\begin{aligned} &L^{(K,Q)}\boldsymbol{W}^j(x) = \boldsymbol{f}(x), \quad x \in (x_{j-1}, x_{j+1}), \\ &\boldsymbol{W}^j(x_{j-1}) = \boldsymbol{W}^j(x_{j+1}) = \boldsymbol{0}. \end{aligned}$$

Let

$$M_\alpha^j(x) = K(x)\frac{dV_\alpha^j(x)}{dx}, \quad \boldsymbol{L}_\alpha^j(x) = K(x)\frac{d\boldsymbol{W}_\alpha^j}{dx}, \quad \alpha = 1, 2,$$

be the corresponding streams. We represent the Green's function of problem (1.88), (1.89) in the form

$$G^j(x,\xi) = \begin{cases} V_1^j(x)\tilde{V}_1^j(\xi), & x \le \xi, \\ V_2^j(x)\tilde{V}_2^j(\xi), & x \ge \xi, \end{cases} \qquad \xi \in [x_{j-1}, x_{j+1}],$$

where $\tilde{V}_\alpha^j(x)$, $\alpha = 1, 2$, are the solutions of the IVPs

$$\begin{aligned} &\tilde{L}^{(K,Q)}\tilde{V}_1^j(x) \stackrel{\text{def}}{=} \frac{d}{dx}\left[\frac{d\tilde{V}_1^j(x)}{dx}K(x)\right] - \tilde{V}_1^j(x)Q(x) = 0, \quad x \in (x_{j-1}, x_{j+1}), \\ &\tilde{V}_1^j(x_{j+1}) = 0, \quad \frac{d\tilde{V}_1^j}{dx}K(x)\bigg|_{x=x_{j+1}} = -\left[V_1^j(x_{j+1})\right]^{-1}, \end{aligned} \tag{1.94}$$

$$\begin{aligned} &\tilde{L}^{(K,Q)}\tilde{V}_2^j(x) = 0, \quad x \in (x_{j-1}, x_{j+1}), \\ &\tilde{V}_2^j(x_{j-1}) = 0, \quad \frac{d\tilde{V}_2^j}{dx}K(x)\bigg|_{x=x_{j-1}} = \left[V_2^j(x_{j-1})\right]^{-1}. \end{aligned} \tag{1.95}$$

The integral corollary of the equation (1.88) implies

$$\int_{x_{j-1}}^{x_{j+1}} G^j(x,\xi) L_\xi^{(K,Q)} \boldsymbol{u}(\xi) d\xi = -\int_{x_{j-1}}^{x_{j+1}} G^j(x,\xi) \boldsymbol{f}(\xi) d\xi = -\boldsymbol{W}^j(x).$$

By integration by parts of the left-hand side we obtain

$$\begin{aligned}
& h_{j+1} V_1^j(x_j) \left[V_1^j(x_{j+1})\right]^{-1} \boldsymbol{u}_{x,j} - h_j V_2^j(x_j) \left[V_2^j(x_{j-1})\right]^{-1} \boldsymbol{u}_{\bar{x},j} \\
& - \left(E - V_1^j(x_j) \left[V_1^j(x_{j+1})\right]^{-1} - V_2^j(x_j) \left[V_2^j(x_{j-1})\right]^{-1}\right) \boldsymbol{u}_j \\
& = -\boldsymbol{W}^j(x_j).
\end{aligned}$$

Note that the following holds:

$$\begin{aligned}
\boldsymbol{W}^j(x_j) &= \int_{x_{j-1}}^{x_{j+1}} G^j(x_j,\xi) \boldsymbol{f}(\xi) d\xi \\
&= V_1^j(x_j) \tilde{V}_1^j(x_j) \left\{ \left[\tilde{V}_2^j(x_j)\right]^{-1} \int_{x_{j-1}}^{x_j} \tilde{V}_2^j(\xi) \boldsymbol{f}(\xi) d\xi \right. \\
&\quad \left. + \left[\tilde{V}_1^j(x_j)\right]^{-1} \int_{x_j}^{x_{j+1}} \tilde{V}_1^j(\xi) \boldsymbol{f}(\xi) d\xi \right\},
\end{aligned}$$

and

$$\begin{aligned}
& E - V_1^j(x_j) \left[V_1^j(x_{j+1})\right]^{-1} - V_2^j(x_j) \left[V_2^j(x_{j-1})\right]^{-1} \\
&= V_1^j(x_j) \tilde{V}_1^j(x_j) \left\{ \left[\tilde{V}_2^j(x_j)\right]^{-1} \int_{x_{j-1}}^{x_j} \tilde{V}_2^j(\xi) Q(\xi) d\xi \right. \\
&\quad \left. + \left[\tilde{V}_1^j(x_j)\right]^{-1} \int_{x_j}^{x_{j+1}} \tilde{V}_1^j(\xi) Q(\xi) d\xi \right\}.
\end{aligned}$$

Thus, we arrive at the 3-point EDS (see [57])

$$\begin{aligned}
& \hbar_j^{-1}(A^j(x_j)\boldsymbol{u}_{x,j} - B^j(x_j)\boldsymbol{u}_{\bar{x},j}) - D_j \boldsymbol{u}(x_j) = -\boldsymbol{\Phi}_j, \quad x_j \in \hat{\omega}_h, \\
& \boldsymbol{u}_0 = \boldsymbol{u}_N = \boldsymbol{0},
\end{aligned} \tag{1.96}$$

where

$$\begin{aligned}
& A^j(x_j) = h_{j+1} \left[\tilde{V}_1^j(x_j)\right]^{-1} \left[V_1^j(x_{j+1})\right]^{-1}, \\
& B^j(x_j) = h_{j+1} \left[\tilde{V}_2^j(x_j)\right]^{-1} \left[V_2^j(x_{j-1})\right]^{-1}, \quad D_j = \hat{T}^j(Q), \quad \boldsymbol{\Phi}_j = \hat{T}^j(f), \\
& \hat{T}^j(\boldsymbol{w}) = \frac{1}{\hbar_j} \left[\tilde{V}_2^j(x_j)\right]^{-1} \int_{x_{j-1}}^{x_j} \tilde{V}_2^j(\xi) \boldsymbol{w}(\xi) d\xi + \frac{1}{\hbar_j} \left[\tilde{V}_1^j(x_j)\right]^{-1} \int_{x_j}^{x_{j+1}} \tilde{V}_1^j(\xi) \boldsymbol{w}(\xi) d\xi.
\end{aligned}$$

The existence of EDS for problem (1.88)–(1.91) has been proven in [58], but the possibility of its transformation into the form (1.96) was only shown for self-adjoint $K(x)$, $Q(x)$ and under the assumption $Q(x) \geq 0$. Besides, in [58] (see Lemma 4 and Theorem 2) it is shown that only the conditions

$$K(x) = K^*(x), \quad Q(x) = Q^*(x),$$
$$K(x)K(x_1) = K(x_1)K(x) \quad \text{for all } x, x_1 \in [x_{j-1}, x_{j+1}]$$

provide the scheme (1.96) in the divergence form

$$(A\boldsymbol{u}_{\bar{x}})_{\hat{x},j} - D_j \boldsymbol{u}(x_j) = -\boldsymbol{\Phi}_j, \quad x_j \in \hat{\omega}_h, \quad \boldsymbol{u}_0 = \boldsymbol{u}_N = \boldsymbol{0},$$

with $A_j \stackrel{\text{def}}{=} h_{j+1}[\tilde{V}_1^{j*}(\xi)]^{-1}$.

Let us introduce the matrix-valued functions

$$\bar{V}_2^j(x) \stackrel{\text{def}}{=} V_1^j(x_{j+1})\tilde{V}_1^j(x), \quad \bar{V}_1^j(x) \stackrel{\text{def}}{=} V_2^{j-1}(x_{j-1})\tilde{V}_2^j(x)$$

and the derivatives

$$\bar{M}_2^j(x) \stackrel{\text{def}}{=} \frac{d\bar{V}_2^j}{dx}K(x), \quad \bar{M}_1^j(x) \stackrel{\text{def}}{=} \frac{d\bar{V}_1^j}{dx}K(x).$$

Then (1.94) and (1.95) imply that the functions $\bar{V}_2^j(x)$, $\bar{V}_1^j(x)$ satisfy the following IVPs on the interval $[x_{j-1}, x_{j+1}]$:

$$\begin{aligned}
&\tilde{L}^{(K,Q)}\bar{V}_2^j(x) = 0, \quad \bar{V}_2^j(x_{j+1}) = 0, \quad \left.\frac{d\bar{V}_2^j(x)}{dx}K(x)\right|_{x=x_{j+1}} = -E, \\
&\tilde{L}^{(K,Q)}\bar{V}_1^j(x) = 0, \quad \bar{V}_1^j(x_{j-1}) = 0, \quad \left.\frac{d\bar{V}_1^j(x)}{dx}K(x)\right|_{x=x_{j-1}} = E.
\end{aligned} \tag{1.97}$$

The coefficients $A^j(x_j)$, $B^j(x_j)$, D_j and $\boldsymbol{\Phi}_j$ can be represented by the solutions of the IVPs (1.92), (1.93) and (1.97) in the form

$$A^j(x_j) = \left[h_{j+1}^{-1}\bar{V}_2^j(x_j)\right]^{-1}, \quad B^j(x_j) = \left[h_j^{-1}\bar{V}_1^j(x_j)\right]^{-1},$$

$$D_j = \frac{1}{\hbar_j}\left[\bar{V}_2^j(x_j)\right]^{-1}\left[-E - \left.\frac{d\bar{V}_2^j}{dx}\right|_{x=x_j} K(x_j+0)\right]$$
$$+ \frac{1}{\hbar_j}\left[\bar{V}_1^j(x_j)\right]^{-1}\left[\left.\frac{d\bar{V}_1^j}{dx}\right|_{x=x_j} K(x_j-0) - E\right],$$

$$\boldsymbol{\Phi}_j = \sum_{\alpha=1}^{2}(-1)^\alpha \left\{\boldsymbol{L}_\alpha^j(x_j) - \left[\bar{V}_\alpha^j(x_j)\right]^{-1}\bar{M}_\alpha^j(x_j)\boldsymbol{W}_\alpha^j(x_j)\right\}.$$

Starting from the EDS we can now introduce in analogy to the scalar case the n-TDS

$$\hbar_j^{-1}(A^{(n)j}(x_j)\boldsymbol{y}_{x,j}^{(n)} - B^{(n)j}(x_j)\boldsymbol{y}_{\bar{x},j}^{(n)}) - D_j^{(n)}\boldsymbol{y}_j = -\boldsymbol{\Phi}_j^{(n)}, \quad x_j \in \hat{\omega}_h,$$

$$\boldsymbol{y}_0^{(n)} = \boldsymbol{y}_N^{(n)} = 0,$$

where

$$A^{(n)j}(x_j) \stackrel{\text{def}}{=} \left[h_{j+1}^{-1}\bar{V}_2^{(n)j}(x_j)\right]^{-1}, \quad B^{(n)j}(x_j) \stackrel{\text{def}}{=} \left[h_j^{-1}\bar{V}_1^{(n)j}(x_j)\right]^{-1},$$

$$D_j^{(n)} = \frac{1}{\hbar_j}\sum_{\alpha=1}^{2}\left[\bar{V}_\alpha^{(n)j}(x_j)\right]^{-1}\left[-E-(-1)^\alpha \bar{M}_\alpha^{(n)j}(x_j)\right],$$

$$\boldsymbol{\Phi}_j^{(n)} \stackrel{\text{def}}{=} \sum_{\alpha=1}^{2}(-1)^\alpha\left\{\boldsymbol{L}_\alpha^{(n)j}(x_j) - \left[\bar{V}_\alpha^{(n)j}(x_j)\right]^{-1}\bar{M}_\alpha^{(\bar{n})j}(x_j)\boldsymbol{W}_\alpha^{(\bar{n})j}(x_j)\right\},$$

and $\bar{V}_\alpha^{(n)j}(x)$, $\bar{M}_\alpha^{(\bar{n})j}(x)$, $\boldsymbol{W}_\alpha^{(\bar{n})j}(x)$, $\boldsymbol{L}_\alpha^{(n)j}(x)$, $\alpha = 1, 2$, are approximations of the solutions of the IVPs (1.92), (1.93) and (1.97) obtained e.g. by the Taylor series method of the order of accuracy $\bar{n}$.

An a priori estimate for the n-TDS gives the following theorem (see [58]).

Theorem 1.13

Under the assumptions (1.90) *and* (1.91) *the following error estimate holds:*

$$\left\|\boldsymbol{u}-\boldsymbol{y}^{(n)}\right\|_{1,\infty,\hat{\omega}_h}^{*} = \max\left\{\left\|\boldsymbol{u}-\boldsymbol{y}^{(n)}\right\|_{0,\infty,\hat{\omega}_h}, \left\|K\frac{d\boldsymbol{u}}{dx} - K\frac{d\boldsymbol{y}^{(n)}}{dx}\right\|_{0,\infty,\hat{\omega}_h}\right\} \leq M\,h^n,$$

where

$$K(x_j)\frac{d\boldsymbol{y}_j^{(n)}}{dx} = \left\{\sum_{\alpha=1}^{2}(-1)^{\alpha+1}M_\alpha^{(\bar{n})j}(x_j)V_\alpha^{(n)j}(x_j)\right\}^{-1}$$

$$\times\sum_{\alpha=1}^{2}(-1)^\alpha\left\{\boldsymbol{L}_\alpha^{(n)j}(x_j) - M_\alpha^{(\bar{n})j}(x_j)\left[V_\alpha^{(n)j}(x_j)\right]^{-1}\boldsymbol{W}_\alpha^{(\bar{n})j}(x_j)\right\} + \boldsymbol{L}_2^{(n)j}(x_j), \tag{1.98}$$

and h is small enough.

EDS as well as their implementation in the form of TDS of an arbitrarily given order of accuracy for *nonlinear second-order ODEs* and Dirichlet boundary conditions

$$\frac{d}{dx}\left[K(x)\frac{d\boldsymbol{u}}{dx}\right] = -\boldsymbol{f}\left(x, \boldsymbol{u}, \frac{d\boldsymbol{u}}{dx}\right), \quad x \in (0,1), \quad \boldsymbol{u}(0) = \boldsymbol{\mu}_1, \quad \boldsymbol{u}(1) = \boldsymbol{\mu}_2, \tag{1.99}$$

with given $K(x) \in \mathbb{R}^{n\times n}$, $\boldsymbol{f}(x, \boldsymbol{u}, \boldsymbol{\xi})$, $\boldsymbol{\mu}_1$, $\boldsymbol{\mu}_2 \in \mathbb{R}^n$, and unknown $\boldsymbol{u}(x) \in \mathbb{R}^n$, were proposed and studied in [28, 29, 43, 44].

Two-point EDS and two-point TDS for *systems of first-order ODEs with non-separated boundary conditions*

$$\boldsymbol{u}'(x) + A(x)\boldsymbol{u} = \boldsymbol{f}(x, \boldsymbol{u}), \quad x \in (0,1), \quad B_0\boldsymbol{u}(0) + B_1\boldsymbol{u}(1) = \boldsymbol{d},$$

$$A(x), B_0, B_1, \in \mathbb{R}^{d\times d}, \quad \operatorname{rank}[B_0, B_1] = d, \quad \boldsymbol{f}(x, \boldsymbol{u}), \boldsymbol{d}, \boldsymbol{u}(x) \in \mathbb{R}^d$$

were introduced and analyzed in [24, 27, 55].

The next step in the development of EDS and TDS was the focus on BVPs on infinite intervals. There are some strategies to solve BVPs on an infinite interval by standard numerical techniques like finite difference methods, collocation methods or shooting methods. One possibility is to replace the boundary conditions formulated at infinity by conditions at a finite boundary point. However, it is difficult to find efficient a priori error estimates for the corresponding numerical solutions (see, e.g. [11],[17]). In some cases the BVP on an infinite interval can be replaced by a BVP on a finite interval with a free boundary (point) [16]. The next idea is to change the variables of the problem and to transform the BVP on the infinite interval into a singular BVP with an essential singularity on a finite interval [4, 5, 36]. It is shown experimentally in [4, 5] that the collocation method applied to the problem with the essential singularity provides satisfactory results. A somewhat different approach is to investigate the asymptotic behavior of the solution at infinity and to use this information to formulate asymptotic boundary conditions at a finite boundary point [37, 51, 61, 62, 63, 77]. In some cases this strategy can be useful to obtain a priori estimates (see, e.g.,[51]).

A quite different approach proposed in [56] is based on the three-point EDS on a *finite* mesh. The idea is to add an exact difference boundary condition to these difference equations and to use the resulting system as the basis for the construction of a TDS of an arbitrarily given order of accuracy. More precisely, the paper [56] deals with the BVP

$$\begin{aligned} L^{(k,q)}u &= -f(x), \quad 0 < x < \infty, \\ u(0) &= \mu_1, \quad \lim_{x\to\infty} u(x) = 0, \end{aligned} \tag{1.100}$$

under the assumptions

$$q(x) = q_0^2 + \tilde{q}(x), \tag{1.101}$$

$$k(x) \geq c_1 > 0, \quad q(x) \geq c_2 > 0, \tag{1.102}$$

$$\frac{1}{k(x)} = d^2 + \frac{p_1(x)}{x} + \frac{p_2(x)}{x^2} + \cdots + \frac{p_n(x)}{x^n} + \cdots, \quad x \to \infty,$$

$$q(x) = q_0^2 + \frac{q_1(x)}{x} + \frac{q_2(x)}{x^2} + \cdots + \frac{q_n(x)}{x^n} + \cdots, \quad x \to \infty \tag{1.103}$$

$$|p_i(x)| \le M, \quad |q_i(x)| \le M, \quad i = 1, 2, \ldots, \tag{1.104}$$

and

$$\frac{d}{dx}\left(\frac{1}{k(x)}\right), \tilde{q}(x) \in L_2(0, \infty),$$

where q_0, d and M are positive constants. Here, the solution of problem (1.100) is understood as a generalized solution from the class $W_2^1(0, \infty)$ satisfying the corresponding variational equation. For this problem the EDS as well as the TDS have been developed on the grid $\hat{\omega}_h$, where the step-sizes h_j must satisfy the conditions

$$\sum_{j=1}^{N+1} h_j = x_{N+1} \underset{N\to\infty}{\to} \infty, \quad h \overset{\text{def}}{=} \max_{1\le j\le N+1} h_j \underset{N\to\infty}{\to} 0.$$

On each interval $[x_{j-1}, x_j]$, $j = 1, 2, \ldots, N$, the three-point EDS (1.57) of the form

$$(au_{\bar{x}})_{\hat{x},j} - b(x_j)u_j = -\varphi(x_j), \quad 0 < j < N+1, \tag{1.105}$$

is used. The corresponding coefficients a, b and φ are given by

$$a(x_j) = \left[\alpha^j(0, h_j)\right]^{-1}, \quad a(x_{j+1}) = \left[\beta^j(0, h_{j+1})\right]^{-1}, \tag{1.106}$$

$$\begin{aligned} b(x_j) = {} & \frac{h_j}{\hbar_j \alpha^j(0, h_j)} \int_{-1}^{0} \alpha^j(s, h_j) q^*(s) ds \\ & + \frac{h_{j+1}}{\hbar_j \beta^j(0, h_{j+1})} \int_0^1 \beta^j(s, h_{j+1}) q^*(s) ds, \end{aligned} \tag{1.107}$$

$$\begin{aligned} \varphi(x_j) = {} & \frac{h_j}{\hbar_j \alpha^j(0, h_j)} \int_{-1}^{0} \alpha^j(s, h_j) f^*(s) ds \\ & + \frac{h_{j+1}}{\hbar_j \beta^j(0, h_{j+1})} \int_0^1 \beta^j(s, h_{j+1}) f^*(s) ds, \end{aligned} \tag{1.108}$$

where

$$q^*(s) \overset{\text{def}}{=} \begin{cases} q(x_j + sh_j), & s < 0 \\ q(x_j + sh_{j+1}), & s > 0 \end{cases}, \qquad f^*(s) \overset{\text{def}}{=} \begin{cases} f(x_j + sh_j), & s < 0 \\ f(x_j + sh_{j+1}), & s > 0 \end{cases}$$

and $\alpha^j(s, h_j)$, $\beta^j(s, h_{j+1})$ are the solutions of the IVPs

$$\frac{d}{ds}\left(k^*(s)\frac{d\alpha^j}{ds}\right) - h_j^2 q^*(s)\alpha^j = 0, \quad -1 < s < 0,$$

$$\alpha^j(-1, h_j) = 0, \quad k^*(s)\frac{d\alpha^j}{ds}\bigg|_{s=-1} = 1,$$

$$\frac{d}{ds}\left(k^*(s)\frac{d\beta^j}{ds}\right) - h_{j+1}^2 q^*(s)\beta^j = 0, \quad 0 < s < 1,$$

$$\beta^j(1, h_{j+1}) = 0, \quad k^*(s)\frac{d\beta^j}{ds}\bigg|_{s=1} = -1,$$

with

$$k^*(s) \stackrel{\text{def}}{=} \begin{cases} k(x_j + sh_j), & s < 0 \\ k(x_j + sh_{j+1}), & s > 0 \end{cases}.$$

The stencil functions $\alpha^j(s, h_j)$, $\beta^j(s, h_{j+1})$ can be represented by the series (compare that with formula (1.61))

$$\alpha^j(s, h_j) = \sum_{i=0}^{\infty} \alpha_i^j(s) h_j^{2i}, \quad \beta^j(s, h_{j+1}) = \sum_{i=0}^{\infty} \beta_i^j(s) h_{j+1}^{2i}, \tag{1.109}$$

where the coefficients $\alpha_i^j(s)$, $\beta_i^j(s)$ are given by the recurrence formulas

$$\alpha_i^j(s) = \int_{-1}^{s} \frac{1}{k^*(t)} \left[\int_{-1}^{t} q^*(\lambda) \alpha_{i-1}^j(\lambda) d\lambda \right] dt, \quad i = 1, 2, \ldots,$$

$$\beta_i^j(s) = \int_{s}^{1} \frac{1}{k^*(t)} \left[\int_{t}^{1} q^*(\lambda) \beta_{i-1}^j(\lambda) d\lambda \right] dt, \quad i = 1, 2, \ldots,$$

$$\alpha_0^j(s) = \int_{-1}^{s} \frac{1}{k^*(t)} dt, \quad \beta_0^j(s) = \int_{s}^{1} \frac{1}{k^*(t)} dt.$$

To complete the system of linear algebraic equations (1.105) we still need the exact boundary conditions at the points x_0 and x_{N+1}. For $x_0 = 0$, formula (1.100) yields $u_0 = \mu_1$. The following equation for the right boundary x_{N+1} is derived in [56]

$$u_{N+1} - \sigma_2 u_N = \sigma_3,$$

with

$$\sigma_2 \stackrel{\text{def}}{=} \tilde{v}_1(x_{N+1}),$$

$$\begin{aligned} \sigma_3 \stackrel{\text{def}}{=} & \frac{\tilde{v}_1(x_{N+1})}{D} \int_{x_N}^{x_{N+1}} [\tilde{v}_2(t) - \tilde{v}_1(t)] f(t) dt \\ & + \frac{\tilde{v}_2(x_{N+1}) - \tilde{v}_1(x_{N+1})}{D} \int_{x_{N+1}}^{\infty} [\tilde{v}_2(t) - \tilde{v}_1(t)] f(t) dt, \end{aligned} \tag{1.110}$$

$$D \stackrel{\text{def}}{=} k(x_{N+1})(\tilde{v}_1(x_{N+1})\tilde{v}_2'(x_{N+1}) - \tilde{v}_1'(x_{N+1})\tilde{v}_2(x_{N+1})),$$

where the stencil functions $\tilde{v}_1(x)$ and $\tilde{v}_2(x)$ are the solutions of the following problems:

$$L^{(k,q)} \tilde{v}_1 = 0, \quad x_N < x < \infty,$$

$$\tilde{v}_1(x_N) = 1, \quad \lim_{x \to \infty} \tilde{v}_1(x) = 0,$$

$$L^{(k,q)} \tilde{v}_2 = 0, \quad x_N < x \le x_{N+1},$$

$$\tilde{v}_2(x_N) = 1, \quad \left. \frac{d\tilde{v}_2}{dx} \right|_{x = x_N} = 0.$$

Let us set $s \stackrel{\text{def}}{=} (x - x_N)/x_N$ and $\mu \stackrel{\text{def}}{=} 1/x_N$. Then, it can be shown that the function $\tilde{v}_1(x) = \tilde{v}_1(x_N + sx_N) \stackrel{\text{def}}{=} \bar{v}_1(s)$ solves the problem

$$\frac{d\bar{v}_1}{ds} = \frac{1}{\bar{k}(s)} w, \quad \mu^2 \frac{dw}{ds} - \bar{q}(s)\bar{v}_1 = 0,$$

$$\bar{v}_1(0) = 1, \quad \lim_{s\to\infty} \bar{v}_1(s) = 0, \quad \lim_{s\to\infty} w(s) = 0,$$

where $\bar{k}(s) \stackrel{\text{def}}{=} k(x_N + sx_N)$, $\bar{q}(s) \stackrel{\text{def}}{=} q(x_N + sx_N)$ and μ is the small parameter. The assumption (1.103) implies that $1/\bar{k}(s)$ and $\bar{q}(s)$ can be represented in the form

$$\begin{aligned} \frac{1}{\bar{k}(s)} &= d^2 + \mu\bar{p}_1(s) + \mu^2\bar{p}_2(s) + \cdots + \mu^n\bar{p}_n(s) + \cdots, \\ \bar{q}(s) &= k_0^2 + \mu\bar{q}_1(s) + \mu^2\bar{q}_2(s) + \cdots + \mu^n\bar{q}_n(s) + \cdots, \end{aligned} \tag{1.111}$$

where

$$|\bar{p}_i(s)| \le M, \quad |\bar{q}_i(s)| \le M, \quad i = 1, 2, \dots . \tag{1.112}$$

Using the algorithm of the asymptotical representation with respect to μ (see [56]) for the stencil function $\tilde{v}_1(x)$ and its derivative on the interval $[x_N, \infty]$ one obtains

$$\tilde{v}_1(x) = \exp(-\kappa(x - x_N))[1 + \mu A_1(x) + \cdots + \mu^n A_n(x) + \cdots],$$

$$\frac{d\tilde{v}_1(x)}{dx} = \exp(-\kappa(x - x_N))[H + \mu B_1(x) + \cdots + \mu^n B_n(x) + \cdots],$$

where $\kappa \stackrel{\text{def}}{=} q_0 d$, $H \stackrel{\text{def}}{=} -q_0/d$ and $A_i(x)$, $B_i(x)$ are defined by the recurrence equations

$$\begin{aligned} \Pi_0\tilde{v}_1(\tau_0) &= \exp(-\kappa\tau_0), \quad \Pi_{-1}w(\tau_0) = H\exp(-\kappa\tau_0), \\ \Pi_i\tilde{v}_1(\tau_{\text{-}0}) &= \delta_1(\tau_0), \quad \Pi_{i-1}w(\tau_{\text{-}0}) = H\delta_1(\tau_0) + \delta_2(\tau_0), \quad \tau_0 = s\mu, \\ \varphi_i(\tau_0)\exp(-\kappa\tau_0) &= \bar{p}_1(\tau_0)\Pi_{i-2}w(\tau_{\text{-}0}) + \cdots + \bar{p}_i(\tau_0)\Pi_{-1}w(\tau_{\text{-}0}), \\ \psi_i(\tau_0)\exp(-\kappa\tau_0) &= \bar{q}_1(\tau_0)\Pi_{i-2}\tilde{v}_1(\tau_{\text{-}0}) + \cdots + \bar{q}_i(\tau_0)\Pi_{-1}\tilde{v}_1(\tau_{\text{-}0}), \\ \delta_2(\tau_0) &= \int_{\tau_0}^{\infty} [H\varphi_i(t) - \psi_i(t)]\exp(-\kappa(2t - \tau_0))dt, \\ \delta_1(\tau_0) &= \int_0^{\tau_0} [d^2\delta_2(t) + \varphi_i(t)\exp(-\kappa t)]\exp(-\kappa(\tau_0 - t))dt, \\ \Pi_i\tilde{v}_1(\tau_0) &= A_i(\tau_0)\exp(-\kappa\tau_0), \\ \Pi_{i-1}w(\tau_{\text{-}0}) &= B_i(\tau_0)\exp(-\kappa\tau_0), \quad i = 1, 2, \\ |A_i(\tau_0)| &\le c^1, \quad |B_i(\tau_0)| \le c^n. \end{aligned}$$

If we set $s \stackrel{\text{def}}{=} (x - x_N)/h_{N+1}$ the following problem can be derived whose solution is $\tilde{v}_2(x) = \bar{v}_2(x_N + sh_{N+1}) \stackrel{\text{def}}{=} \bar{v}_2(s)$:

$$\frac{d}{ds}\left(\bar{k}(s)\frac{d\bar{v}_2}{ds}\right) - h_{N+1}^2\bar{q}(s)\bar{v}_2(s) = 0, \quad 0 < s \le 1,$$

$$\bar{v}_2(0) = 1, \quad \bar{k}\frac{d\bar{v}_2}{ds}\bigg|_{s=0} = 0,$$

where $\bar{k}(s) \stackrel{\text{def}}{=} k(x_N + sh_{N+1})$, $\bar{q}(s) \stackrel{\text{def}}{=} q(x_N + sh_{N+1})$. This implies for the function $\bar{v}_2(s)$ the representation

$$\bar{v}_2(s) = \sum_{i=0}^{\infty} \bar{\beta}_i^j(s) h_{N+1}^{2i},$$

where

$$\bar{\beta}_0^j(s) = 1, \quad \bar{\beta}_i^j(s) = \int_0^s \frac{1}{\bar{k}(t)}\left[\int_0^t \bar{q}(\lambda)\bar{\beta}_{i-1}^j(\lambda)d\lambda\right] dt, \quad i > 0.$$

Now, in the case of problem (1.100) the equations (1.105) together with the exact boundary conditions at the points $x = 0$ and $x = x_{N+1}$ lead to the following EDS

$$(au_{\bar{x}})_{\hat{x}_j} - b(x_j)u_j = -\varphi(x_j), \quad 0 < j < N+1,$$

$$u_0 = \mu_1, \quad u_{N+1} - \sigma_2 u_N = \sigma_3,$$

where the coefficients a, b, φ are given by (1.106)–(1.108), and σ_2, σ_3 are given by (1.110).

In order to obtain a TDS for problem (1.100) we replace the stencil functions $\alpha^j(s, h_j)$, $\beta^j(s, h_{j+1})$ (see formula (1.109)) by the finite sums

$$\alpha^{(m)j}(s, h_j) \stackrel{\text{def}}{=} \sum_{i=0}^{m} \alpha_i^j(s) h_j^{2i}, \quad \beta^{(m)j}(s, h_{j+1}) \stackrel{\text{def}}{=} \sum_{i=0}^{m} \beta_i^j(s) h_{j+1}^{2i}, \tag{1.113}$$

and use in (1.110) the finite series

$$\begin{aligned} \tilde{v}_1^{(n)}(x) &\stackrel{\text{def}}{=} \exp(-\kappa(x - x_N))[1 + \mu A_1(x) + \cdots + \mu^n A_n(x)], \\ \tilde{v}_2^{(m)}(x) &\stackrel{\text{def}}{=} \sum_{i=0}^{m} \bar{\beta}_i^j(s) h_{N+1}^{2i}, \quad s = (x - x_N)/h_{N+1}, \end{aligned} \tag{1.114}$$

instead of the functions $\tilde{v}_1(x)$ and $\tilde{v}_2(x)$. This procedure leads to the following TDS:

$$\begin{aligned} (a^{(m)} y_{\bar{x}}^{(m)})_{\hat{x},i} - b_i^{(m)} y_i^{(m)} &= \varphi_i^{(m)}, \quad i = 1, 2, \ldots, N-1, \\ y_0^{(m)} = \mu_1, \quad y_{N+1}^{(m)} - \sigma_2^{(m)} y_N^{(m)} &= \sigma_3^{(m)}, \end{aligned} \tag{1.115}$$

where $a^{(m)}$, $b^{(m)}$, $\varphi^{(m)}$, $\sigma_2^{(m)}$ and $\sigma_3^{(m,n)}$ are the truncated coefficients given in (1.106), (1.110), (1.113) and (1.114). This difference scheme is sometimes called TDS of rank (m, n).

The following theorem characterizes the order of accuracy of the TDS (see [56]).

Theorem 1.14

Let the assumptions (1.114), (1.115), (1.111), (1.112) *and*

$$k(x) \in Q^1[0,\infty), \quad q(x), f(x) \in Q^0[0,\infty), \quad f \in L_t(0,\infty), \quad t \geq 1,$$

be satisfied. Then, the TDS (1.115) *of rank* (m,n) *which is defined on* $\hat{\bar{\omega}}_N$ *has the order of accuracy*

$$O(\max\{\mu^{n+1}, x_N h^{2m+2}\}), \tag{1.116}$$

where $\mu \stackrel{\text{def}}{=} 1/x_N$ *and* $m, n \geq 0$.

In order to estimate the accuracy of a TDS as a function of the number of grid points, we consider the finite equidistant grid $\bar{\omega}_N = \{x_0 = 0, x_1, \ldots, x_{N+1}\}$. If we choose $h = N^{-\varepsilon}$, $0 < \varepsilon < 1$, then it holds that

$$x_N = Nh \underset{N\to\infty}{\to} \infty, \quad \mu = 1/x_N \underset{N\to\infty}{\to} 0.$$

Let us set

$$\varepsilon = \varepsilon(m,n) = \frac{n+2}{2m+n+4}.$$

Then it can be easily shown that the two arguments of the maximum function in (1.116) are equal. Thus, if m is fixed and n satisfies $n \leq m$, the order of accuracy of the TDS (1.115) is a maximum. For $n = 2m$ we have $h = N^{-1/2}$ and we obtain for the order of accuracy $O(h^{2m+1})$.

The representation of the coefficients of the TDS by truncated series in powers of the mesh-size has the same disadvantages as the techniques mentioned in the former publications on this topic discussed above. The EDS as well as an alternative TDS in which the coefficients are computed with IVP-solvers have been proposed and justified in [25, 26] for the nonlinear BVP on the half-axis

$$\frac{d^2u}{dx^2} - m^2 u = -f(x,u), \quad x \in (0,\infty),$$

$$u(0) = \mu_1, \quad \lim_{x\to\infty} u(x) = 0.$$

Under the assumption that the coefficients of the TDS are computed by an IVP-solver of the order n, it is shown that the implementation of the EDS by the TDS has the order of accuracy $\bar{n} = 2[(n+1)/2]$, where n is a positive natural number and $[\cdot]$ denotes the entire part of the argument in these brackets.

There are other techniques to construct EDS and the corresponding TDS of a high order of accuracy which cannot be classified with the general approach described above. Without claim of completeness let us mention here some of them. In [13, 52] for the problem (1.46), (1.47) an interesting approach has been proposed to develop 3-point TDS for sufficiently smooth input data $k(x)$, $q(x)$ and $f(x)$. This approach was modified in [19] for piece-wise smooth input data with discontinuities at the grid nodes. Moreover, the former assumption $q(x) \geq 0$ is replaced in this paper by

$$|q(x)| \leq c_2. \tag{1.117}$$

In order to construct the associated TDS (i.e. to calculate their coefficients) the solution of a system of linear algebraic equations is required, the order of which depends on the desired accuracy of the approximate solution (the same is valid for the schemes proposed in [13, 52]). The recent book [3] deals with the construction and investigation of difference schemes of an arbitrarily given order of accuracy for regular and singularly perturbed BVPs for ODEs of first and second order with operator coefficients from Banach spaces. Such equations can be considered as meta-models for linear systems of ODEs as well as for linear parabolic, elliptic or hyperbolic PDEs. The authors use an EDS which is defined on two-point or three-point stencils (for the first and the second order equations, respectively) as the basis for the construction of TDS of high accuracy. They approximate the function e^{-z} (the principal part of the EDS) by Padé approximations and thus obtain difference schemes of an arbitrarily given order of accuracy p with explicit coefficients which become more and more complex with increasing p. As a second approach the authors use the so-called Taylor's decompositions (with explicitly given coefficients) on two and three points. Unfortunately, there are no numerical examples testing and comparing the robustness of both types of difference schemes.

For some time dependent PDEs, special techniques to construct EDS and the related algorithms were proposed and analysed in [50, 64, 65, 66, 69].

Chapter 2

Two-point difference schemes for systems of nonlinear BVPs

> No amount of experimentation can ever prove me right; a single experiment can prove me wrong.
>
> *Albert Einstein (1879–1955)*

This chapter deals with BVPs of the form

$$\boldsymbol{u}'(x) + A(x)\boldsymbol{u} = \boldsymbol{f}(x, \boldsymbol{u}), \quad x \in (0, 1), \quad B_0\boldsymbol{u}(0) + B_1\boldsymbol{u}(1) = \boldsymbol{\beta}, \tag{2.1}$$

where

$$A(x), B_0, B_1, \in \mathbb{R}^{d\times d}, \quad \operatorname{rank}[B_0, B_1] = d, \quad \boldsymbol{f}(x, \boldsymbol{u}), \boldsymbol{\beta}, \boldsymbol{u}(x) \in \mathbb{R}^d,$$

and $\boldsymbol{u}$ is an unknown d-dimensional vector-function. On an arbitrary closed non-equidistant grid (1.3) there exists a unique 2-point EDS (see Definition 1.8) such that its solution coincides with the grid function $\{\boldsymbol{u}(x_j)\}_{j=0}^N$, $x_j \in \hat{\omega}_h$, of the exact solution $\boldsymbol{u}(x)$. Algorithmical realizations of the EDS are the TDS (see Definition 1.9). These schemes have an order of accuracy $\mathcal{O}(h^m)$, with respect to the maximal step size h, which can arbitrarily be prescribed by the user.

Note that the EDS and TDS are very similar to the multiple shooting method [2, 35, 39, 40, 79]. Both techniques are based on the successive solution of IVPs on small subintervals and are theoretically supported by a posteriori error estimates. However, the advantage of our difference methods is that a unified theory of a priori estimates can be established.

2.1 Existence and uniqueness of the solution

Let us consider the BVP (2.1). By the linear part of the ODE in (2.1) a so-called fundamental matrix (or the evolution operator) $U(x;\xi)$ is defined.

Definition 2.1. For any $\xi \in [0,1]$ a fundamental matrix $U(x;\xi) \in \mathbb{R}^{d\times d}$ is defined as a function satisfying the matrix IVP

$$U'(x;\xi) + A(x)U(x;\xi) = 0, \quad 0 < x < 1, \qquad U(\xi;\xi) = I, \tag{2.2}$$

where $I \in \mathbb{R}^{d\times d}$ denotes the identity matrix. In (2.2) the differentiation is with respect to x, and ξ is a parameter. □

In what follows we denote by $\|\boldsymbol{u}\| \stackrel{\text{def}}{=} \sqrt{\boldsymbol{u}^T\boldsymbol{u}}$ the Euclidian norm of $\boldsymbol{u} \in \mathbb{R}^d$ and we will use the subordinate matrix norm generated by this vector norm.

Let us make the following assumptions.

Assumption 2.1.

(i) The linear homogeneous problem corresponding to (2.1) possesses only the trivial solution.

(ii) For the elements of $A(x) = [a_{ij}(x)]_{i,j=1}^{d}$ it holds that $a_{ij}(x) \in \mathbb{C}[0,1]$, $i,j = 1,2,\ldots,d$. □

The Assumption 2.1,(ii) implies the existence of a constant c_1 such that

$$\|A(x)\| \le c_1 \quad \text{for all } x \in [0,1].$$

It can easily be shown that Assumption 2.1,(i) guarantees the nonsingularity of the matrix $Q \stackrel{\text{def}}{=} B_0 + B_1U(1;0)$, i.e. the following auxiliary statement holds.

Lemma 2.1

The matrix Q is nonsingular if and only if the linear homogeneous problem corresponding to (2.1) *has only the trivial solution* $\boldsymbol{u}(x) \equiv 0$.

Proof. See e.g. [70], p.226, and [2], p.91. ■

Some sufficient conditions which guarantee that the linear homogeneous BVP corresponding to (2.1) has only the trivial solution are given in the following two lemmas.

Lemma 2.2

Let $A(x) \ge 0$ for all $x \in [0,1]$, i.e., $A(x)$ is positive semidefinite. Moreover, we assume that one of the following conditions is satisfied:

(i) B_0^{-1} *exists and* $\left\|B_0^{-1}B_1\right\| \stackrel{\text{def}}{=} \alpha < 1$*; or*

(ii) B_1^{-1} *exists and* $\|B_0^{-1}B_1\left(B_1^{-1}B_0\right)^T\| < 1$*; or*

(iii) $(B_0+B_1)^{-1}$ *exists and* $\|\left(B_0+B_1\right)^{-1}B_1\| < 1/2$.

Then, problem (2.1) *with* $\boldsymbol{f}(x,\boldsymbol{u}) \equiv \boldsymbol{0}$ *and* $\boldsymbol{\beta} = \boldsymbol{0}$ *has only the trivial solution.*

Proof. Here we show only the statement (iii). Assumption 2.1,(ii) guarantees the existence and uniqueness of the solution to the IVP (2.2). From the homogeneous ODE we derive

$$\boldsymbol{u}(x)^T\boldsymbol{u}'(x) + \boldsymbol{u}(x)^T A(x)\boldsymbol{u}(x) = 0.$$

Now the assumptions postulated in the lemma imply

$$\frac{1}{2}\frac{d}{dx}\left\|\boldsymbol{u}\right\|^2 \le 0. \tag{2.3}$$

Thus

$$\|\boldsymbol{u}(1)\| \le \|\boldsymbol{u}(0)\|. \tag{2.4}$$

The consequences of the homogeneous boundary conditions are

$$\boldsymbol{u}(0) = (B_0+B_1)^{-1}B_1\left[\boldsymbol{u}(0)-\boldsymbol{u}(1)\right], \quad \boldsymbol{u}(1) = (B_0+B_1)^{-1}B_0\left[\boldsymbol{u}(1)-\boldsymbol{u}(0)\right]. \tag{2.5}$$

Let us define

$$P \stackrel{\text{def}}{=} (B_0+B_1)^{-1}B_1, \quad \boldsymbol{v} \stackrel{\text{def}}{=} \boldsymbol{u}(1)-\boldsymbol{u}(0).$$

Then $\boldsymbol{u}(0) = -P\boldsymbol{v}$, $\boldsymbol{u}(1) = \boldsymbol{v} - P\boldsymbol{v}$ and it follows from (2.4) that

$$\|(I-P)\,\boldsymbol{v}\| \le \|P\boldsymbol{v}\|.$$

This implies

$$\|\boldsymbol{v}\| \le \|\boldsymbol{v}-P\boldsymbol{v}\| + \|P\boldsymbol{v}\| \le 2\,\|P\boldsymbol{v}\| \le 2\|P\|\,\|\boldsymbol{v}\|.$$

From the last inequality and condition (iii) we get $\|\boldsymbol{v}\| = 0$, i.e., $\boldsymbol{u}(0) = \boldsymbol{u}(1)$, which together with (2.5) implies $\boldsymbol{u}(0) = \boldsymbol{u}(1) = 0$. Now using (2.3) the claim $\boldsymbol{u}(x) \equiv \boldsymbol{0}$ is proved. ∎

Lemma 2.3

Let $A \in \mathbb{R}^{d\times d}$ *be a constant matrix such that the inverse* $[B_0 + B_1 e^{-A}]^{-1}$ *exists and the estimate*

$$\max_{0\le x\le 1}\left\{\int_0^x \|e^{-xA}(B_0+B_1e^{-A})^{-1}B_0e^{\xi A}(A-A(x))\|\,d\xi\right.$$
$$\left.+\int_x^1 \left\|e^{-xA}(B_0+B_1e^{-A})^{-1}B_1e^{-(1-\xi)A}(A-A(x))\right\|\,d\xi\right\} < 1$$

holds. Then, the matrix Q is nonsingular and the linear homogeneous BVP *corresponding to (2.1) possesses only the trivial solution.*

Proof. Let us write the homogeneous equation corresponding to (2.1) in the equivalent form

$$\boldsymbol{u}'(x) + A\boldsymbol{u} = [A - A(x)]\boldsymbol{u}, \quad x \in (0,1).$$

We get

$$\boldsymbol{u}(x) = e^{-Ax}\boldsymbol{u}(0) + \int_0^x e^{-(1-\xi)A}[A - A(\xi)]\boldsymbol{u}(\xi)d\xi. \tag{2.6}$$

Substituting this expression into the boundary condition we obtain

$$B_0\boldsymbol{u}(0) + B_1 e^{-A}\boldsymbol{u}(0) + B_1 \int_0^1 e^{-(1-\xi)A}[A - A(\xi)]\boldsymbol{u}(\xi)d\xi = \mathbf{0}.$$

Thus,

$$\boldsymbol{u}(0) = -[B_0 + B_1 e^{-A}]^{-1} B_1 \int_0^1 e^{-(1-\xi)A}[A - A(\xi)]\boldsymbol{u}(\xi)d\xi.$$

Using this in (2.6) we see that the solution of problem (2.1) satisfies the integral equation

$$\boldsymbol{u}(x) = \int_0^x \left\{ e^{-(x-\xi)A} - e^{-xA}[B_0 + B_1 e^{-A}]^{-1} B_1 e^{-(1-\xi)A} \right\} [A - A(\xi)]\boldsymbol{u}(\xi)d\xi$$

$$- \int_x^1 e^{-xA}[B_0 + B_1 e^{-A}]^{-1} B_1 e^{-(1-\xi)A}[A - A(\xi)]\boldsymbol{u}(\xi)d\xi$$

and we obtain the estimate

$$\|\boldsymbol{u}(x)\| \le \left\{ \int_0^x \|e^{-xA}[I - (B_0 + B_1 e^{-A})^{-1} B_1 e^{-A}] e^{\xi A}[A - A(\xi)]\| \, d\xi \right.$$

$$\left. + \int_x^1 \left\| e^{-xA}[B_0 + B_1 e^{-A}]^{-1} B_1 e^{-(1-\xi)A}[A - A(\xi)] \right\| d\xi \right\} \|\boldsymbol{u}\|_\infty$$

from which the assertion of the lemma follows. ∎

Let us introduce the vector-function

$$\boldsymbol{u}^{(0)}(x) \stackrel{\text{def}}{=} U(x;0)Q^{-1}\boldsymbol{\beta},$$

(which exists due to Assumption 2.1,(i) for all $x \in [0,1]$) and the set

$$\Omega(D, \beta(x)) \stackrel{\text{def}}{=} \left\{ \boldsymbol{v}(x) = (v_1(x), \dots, v_d(x))^T, \; v_j \in \mathbb{C}[0,1], \; j = 1, 2, \dots, d, \right.$$

$$\left. \left\| \boldsymbol{v}(x) - \boldsymbol{u}^{(0)}(x) \right\| \le \beta(x), \; x \in D \right\}, \tag{2.7}$$

where $D \subseteq [0,1]$ is a closed set.

Due to Assumption 2.1,(i) the problem (2.1) is equivalent to the integral equation

$$\boldsymbol{u}(x) = \int_0^1 G(x,\xi)\,\boldsymbol{f}(\xi,\boldsymbol{u}(\xi))d\xi + \boldsymbol{u}^{(0)}(x),\ x \in [0,1], \tag{2.8}$$

where $G(x,\xi)$ is the Green's function of the corresponding linear differential operator (see, e.g. [2], p.226), which can be written in the form

$$G(x,\xi) \stackrel{\text{def}}{=} \begin{cases} -U(x;0)HU(1;\xi), & 0 \le x \le \xi, \\ -U(x;0)HU(1;\xi) + U(x;\xi), & \xi \le x \le 1, \end{cases} \tag{2.9}$$

where $H \stackrel{\text{def}}{=} Q^{-1}B_1$.

Now we can formulate the next auxiliary statement.

Lemma 2.4

Let Assumption 2.1 *be satisfied. Then*

$$\|U(x;\xi)\| \le \exp[c_1(x-\xi)], \tag{2.10}$$

$$\|G(x,\xi)\| \le \begin{cases} \exp[c_1(1+x-\xi)]\,\|H\|, & 0 \le x \le \xi, \\ \exp[c_1(x-\xi)]\,[1+\|H\|\exp(c_1)], & \xi \le x \le 1. \end{cases} \tag{2.11}$$

Proof. In order to prove (2.10) let us rewrite the IVP (2.2) in the equivalent form

$$U(x;\xi) = I - \int_\xi^x A(\eta)U(\eta;\xi)d\eta.$$

Then we have

$$\|U(x;\xi)\| \le 1 + \int_\xi^x \|A(\eta)\|\,\|U(\eta;\xi)\|\,d\eta.$$

With Gronwall's Lemma (see e.g. [33], p. 24) we get (2.10). The estimate (2.11) follows from (2.9) and (2.10). ■

In addition to Assumption 2.1 we postulate

Assumption 2.2. The vector-function $\boldsymbol{f}(x,\boldsymbol{u}) = \{f_j(x,\boldsymbol{u})\}_{j=1}^d$ satisfies the conditions

(i) $f_j \in \mathbb{C}([0,1] \times \Omega([0,1], r(x)))$, $\|\boldsymbol{f}(x,\boldsymbol{u})\| \le K$ for all $\boldsymbol{u} \in \Omega([0,1], r(x))$,

(ii) $\|\boldsymbol{f}(x,\boldsymbol{u}) - \boldsymbol{f}(x,\boldsymbol{v})\| \le L\,\|\boldsymbol{u}-\boldsymbol{v}\|$ for all $x \in [0,1]$ and $\boldsymbol{u},\boldsymbol{v} \in \Omega([0,1], r(x))$, where $r(x) \stackrel{\text{def}}{=} K\exp(c_1 x)\,[x + \|H\|\exp(c_1)]$. □

Now, we discuss sufficient conditions which guarantee the existence and uniqueness of a solution of problem (2.1). Later we will use these conditions to prove the existence of an EDS and the corresponding TDS.

We begin with the following statement.

Theorem 2.1

Let Assumptions 2.1 and 2.2 be satisfied. Suppose

$$q \stackrel{\text{def}}{=} L\exp(c_1)\,[1 + \|H\|\exp(c_1)] < 1. \tag{2.12}$$

Then there exists a unique solution $\boldsymbol{u} \in \Omega([0,1], r(x))$ of problem (2.1) which can be determined using the iteration procedure

$$\boldsymbol{u}^{(k)}(x) = \int_0^1 G(x,\xi)\boldsymbol{f}(\xi, \boldsymbol{u}^{(k-1)}(\xi))d\xi + \boldsymbol{u}^{(0)}(x), \quad x \in [0,1]. \tag{2.13}$$

The corresponding error estimate is

$$\left\|\boldsymbol{u}^{(k)} - \boldsymbol{u}\right\|_{0,\infty,[0,1]} \le \frac{q^k}{1-q} r(1), \tag{2.14}$$

where $\|\boldsymbol{v}\|_{0,\infty,[0,1]} \stackrel{\text{def}}{=} \max_{x\in[0,1]} \|\boldsymbol{v}(x)\|$.

Proof. Let us show that the operator

$$\Re(x, \boldsymbol{u}(\cdot)) \stackrel{\text{def}}{=} \int_0^1 G(x,\xi)\boldsymbol{f}(\xi, \boldsymbol{u}(\xi))d\xi + \boldsymbol{u}^{(0)}(x), \quad x \in [0,1]$$

transforms the set $\Omega([0,1], r(x))$ into itself. Taking into account the assumptions of the theorem and the estimate (2.11), for all $\boldsymbol{v} \in \Omega([0,1], r(x))$ we get

$$\begin{aligned}
&\left\|\Re(x, \boldsymbol{v}(\cdot)) - \boldsymbol{u}^{(0)}(\cdot)\right\|_{0,\infty,[0,1]} \\
&\le K\left[\exp[c_1(1+x)\|H\|] \int_0^1 \exp(-c_1\xi)d\xi + \exp(c_1 x)\int_0^x \exp(-c_1\xi)d\xi\right] \\
&\le K\exp(c_1 x)\,[x + \|H\|\exp(c_1)] = r(x).
\end{aligned}$$

Moreover, for all $\boldsymbol{u}, \boldsymbol{v} \in \Omega([0,1], r(x))$ we have

$$\begin{aligned}
&\|\Re(x,\boldsymbol{u}(\cdot)) - \Re(x,\boldsymbol{v}(\cdot))\|_{0,\infty,[0,1]} \\
&\leq \exp(c_1)\left[1 + \|H\| \exp(c_1)\right] \int_0^1 \|\boldsymbol{f}(\xi,\boldsymbol{u}(\xi)) - \boldsymbol{f}(\xi,\boldsymbol{v}(\xi))\| \, d\xi \\
&\leq q \,\|\boldsymbol{u} - \boldsymbol{v}\|_{0,\infty,[0,1]} \,.
\end{aligned}$$

Due to (2.12) it holds that $q < 1$. Thus, $\Re(x, \boldsymbol{u}(\cdot))$ is a contraction operator on the set $\Omega([0,1], r(x))$. Consequently, the assumptions of Banach's Fixed Point Theorem (see Theorem 1.1) are fulfilled and we can conclude that the equation (2.8) or the problem (2.1) has a unique solution which is the fixed point of the iteration procedure (2.13). Here, we omit the standard proof of (2.14) and refer, e.g., to [68], Theorem 5.1.3. ■

In the case of scalar second-order ODEs it is possible to formulate the statement of Theorem 2.1 in a stronger version. More precisely, let us consider the problem

$$\frac{d^2 u(x)}{dx^2} - q(x)u(x) = -f(x, u(x)), \quad x \in (0,1), \quad u(0) = \mu_1,\; u(1) = \mu_2, \tag{2.15}$$

where $q(x)$ is a given piece-wise continuous function and

$$q(x) \geq 0 \quad \text{for all } x \in [0,1]. \tag{2.16}$$

Let us write problem (2.15) in the form (2.1):

$$\frac{d\boldsymbol{u}}{dx} + A(x)\boldsymbol{u} = \boldsymbol{F}(x,\boldsymbol{u}), \qquad B_0\boldsymbol{u}(0) + B_1\boldsymbol{u}(1) = \boldsymbol{\beta} \tag{2.17}$$

with

$$\boldsymbol{u} = \begin{pmatrix} u(x) \\ u'(x) \end{pmatrix}, \quad A(x) = \begin{pmatrix} 0 & -1 \\ -q(x) & 0 \end{pmatrix}, \quad \boldsymbol{F}(x,\boldsymbol{u}) = \begin{pmatrix} 0 \\ -f(x,u) \end{pmatrix},$$

$$B_0 = \begin{pmatrix} 1 & 0 \\ 0 & 0 \end{pmatrix}, \quad B_1 = \begin{pmatrix} 0 & 0 \\ 1 & 0 \end{pmatrix}, \quad \boldsymbol{\beta} = \begin{pmatrix} \mu_1 \\ \mu_2 \end{pmatrix}.$$

Note, that the often used Theorem 7.3.3.4 in [79] (for linear boundary conditions this statement was proved in [40]) cannot be applied to the problem (2.15) since the corresponding assumptions are violated. This theorem assumes that a boundary condition of the form

$$r(u(0), u(1))) = 0$$

is given and that the matrix

$$P(u,v) = D_u r(u,v) + D_v r(u,v)$$

admits for all $u, v \in \mathbb{R}^d$ a representation

$$P(u,v) = P_0(I + M(u,v)),$$

with a constant nonsingular matrix P_0. In our case this matrix agrees with the matrix B_0 which is singular.

For a matrix $M = \begin{pmatrix} m_{11} & m_{12} \\ m_{21} & m_{22} \end{pmatrix}$ and a vector-function $\boldsymbol{f}(x) = \begin{pmatrix} f_1(x) \\ f_2(x) \end{pmatrix}$ we define

$$\|M\| \stackrel{\text{def}}{=} \|M\|_1 = \max_{i=1,2} \sum_{j=1}^{2} |m_{ij}|, \quad \|\boldsymbol{f}(x)\| = \|\boldsymbol{f}(x)\|_1 = \max\{f_1(x), f_2(x)\},$$

$$\|\boldsymbol{f}\|_\infty \stackrel{\text{def}}{=} \max_{x\in[0,1]} \|\boldsymbol{f}(x)\|.$$

It is easy to check that

$$B_0 + B_1 e^{-A} = \begin{pmatrix} 1 & 0 \\ 1 & 1 \end{pmatrix}, \quad (B_0 + B_1 e^{-A})^{-1} = \begin{pmatrix} 1 & 0 \\ -1 & 1 \end{pmatrix},$$

$$e^{-xA}(B_0 + B_1 e^{-A})^{-1} B_0 e^{\xi A} = \begin{pmatrix} 1-x & -\xi(1-x) \\ -1 & \xi \end{pmatrix},$$

$$e^{-xA}(B_0 + B_1 e^{-A})^{-1} B_1 e^{-(1-\xi)A} = \begin{pmatrix} x & (1-\xi)x \\ 1 & 1-\xi \end{pmatrix}.$$

Choosing $A = \begin{pmatrix} 0 & -1 \\ 0 & 0 \end{pmatrix}$, we further get

$$\begin{aligned} &\left\|e^{-xA}(B_0 + B_1 e^{-A})^{-1} B_0 e^{\xi A}[A - A(x)]\right\| \\ &= \left\|\begin{pmatrix} 1-x & -\xi(1-x) \\ -1 & \xi \end{pmatrix}\begin{pmatrix} 0 & 0 \\ q(\xi) & 0 \end{pmatrix}\right\| = \left\|\begin{pmatrix} -\xi(1-x)q(\xi) & 0 \\ \xi q(\xi) & 0 \end{pmatrix}\right\| \le \|q\|_\infty \xi \end{aligned}$$

and

$$\begin{aligned} &\left\|e^{-xA}(B_0 + B_1 e^{-A})^{-1} B_1 e^{-(1-\xi)A}[A - A(x)]\right\| \\ &= \left\|\begin{pmatrix} x(1-\xi)q(\xi) & 0 \\ (1-\xi)q(\xi) & 0 \end{pmatrix}\right\| \le \|q\|_\infty (1-\xi). \end{aligned}$$

This yields

$$\begin{aligned} &\max_{0\le x\le 1}\left\{\int_0^x \left\|e^{-xA}(B_0 + B_1 e^{-A})^{-1} B_0 e^{\xi A}(A - A(x))\right\| d\xi \right. \\ &\left. + \int_x^1 \left\|e^{-xA}(B_0 + B_1 e^{-A})^{-1} B_1 e^{-(1-\xi)A}(A - A(x))\right\| d\xi \right\} \\ &\le \|q\|_\infty \max_{x\in[0,1]}\left\{\int_0^x \xi d\xi + \int_x^1 (1-\xi) d\xi\right\} = \|q\|_\infty / 2. \end{aligned}$$

The statement of Lemma 2.3 now implies that for a function $q(x)$, with $\|q\|_\infty \leq 2$, there exists Q^{-1}. Due to formulae (2.8) and (2.9) we get

$$\begin{aligned} \boldsymbol{u}(x) &= \int_0^1 G(x,\xi)\boldsymbol{F}(\xi,\boldsymbol{u}(\xi))d\xi + \boldsymbol{u}^{(0)}(x) \\ &= -\int_0^1 \begin{pmatrix} G_{12}(x,\xi) \\ G_{22}(x,\xi) \end{pmatrix} f(\xi,u(\xi))d\xi + \begin{pmatrix} u_1^{(0)}(x) \\ u_2^{(0)}(x) \end{pmatrix}, \end{aligned} \tag{2.18}$$

where $G_{i,j}(x,\xi)$, $i,j = 1,2$ are the components of the Green's function $G(x,\xi)$. Thus, we obtain

$$H = \frac{1}{U_{12}(1,0)} \begin{pmatrix} U_{12}(1,0) & 0 \\ -U_{11}(1,0) & 1 \end{pmatrix} \begin{pmatrix} 0 & 0 \\ 1 & 0 \end{pmatrix} = \frac{1}{U_{12}(1,0)} \begin{pmatrix} 0 & 0 \\ 1 & 0 \end{pmatrix},$$

where $U_{ij}(x,\xi)$, $i,j = 1,2$ are the elements of the matrix $U(x;\xi)$. It is not favorable to use the estimates (2.10) and (2.11) since the integral equation (2.18) contains only two components of the matrix $G(x,\xi)$. Moreover, the component-wise representation of (2.18) is

$$u(x) = -\int_0^1 G_{12}(x,\xi) f(\xi,u(\xi))d\xi + u_1^{(0)}(x),$$

$$u'(x) = -\int_0^1 G_{22}(x,\xi) f(\xi,u(\xi))d\xi + u_2^{(0)}(x).$$

The first equation agrees with the equation

$$u(x) = \int_0^1 G(x,\xi,q(\cdot)) \{f(\xi,u(\xi)) - q(\xi)\left[\mu_1 + \xi(\mu_2 - \mu_1)\right]\} d\xi + \mu_1 + x(\mu_2 - \mu_1),$$

where $G(x,\xi,q(\cdot))$ is the Green's function corresponding to the homogeneous part of the ODE (2.15). Let us now investigate this equation. It is well known that

$$0 \leq G(x,\xi,q(\cdot)) \leq G(x,\xi,0) \quad \text{and} \quad G(x,\xi,0) = \begin{cases} x(1-\xi), & 0 \leq x \leq \xi, \\ \xi(1-x), & \xi \leq x \leq 1. \end{cases}$$

We introduce the set

$$\begin{aligned} &\Omega\big([0,1],p(x),r(x)\big) \\ &\stackrel{\text{def}}{=} \big\{v \in \mathbb{C}[0,1] : \, p(x) \leq v(x) - \mu_1 - x(\mu_2 - \mu_1) \leq r(x) \text{ for all } x \in [0,1]\big\} \end{aligned} \tag{2.19}$$

and formulate an appropriate condition with respect to the right-hand side of the ODE (2.15). Let for all $u, v \in \Omega\left([0,1],p(x),r(x)\right)$ and $x \in [0,1]$ the following inequalities be fulfilled:

$$\begin{aligned} &K_1(x) \leq f(x,u(x)) \leq K_2(x), \quad K_1(x) \leq 0, \\ &|f(x,u(x)) - f(x,v(x))| \leq L(x)\,|u(x) - v(x)|\,, \end{aligned} \tag{2.20}$$

where $K_1, K_2, L \in \mathbb{C}[0,1]$ satisfy

$$\int_0^1 G(x,\xi,0)L(\xi)d\xi \le L_1 < 1, \quad x \in [0,1], \tag{2.21}$$

and

$$\int_0^1 G(x,\xi,0)\left\{K_2(\xi) - q(\xi)\left[\mu_1 + \xi(\mu_2-\mu_1)\right]^-\right\}d\xi \le r(x), \tag{2.22}$$

$$p(x) \le \int_0^1 G(x,\xi,0)\left\{K_1(\xi) - q(\xi)\left[\mu_1 + \xi(\mu_2-\mu_1)\right]^+\right\}d\xi,$$

with $[g(x)]^+ \stackrel{\text{def}}{=} \max\{0, g(x)\}$ and $[g(x)]^- \stackrel{\text{def}}{=} \min\{0, g(x)\}$.

We can now prove the following result.

Theorem 2.2

Let the conditions (2.16), (2.20), (2.21) *and* (2.22) *be satisfied. Then, the BVP* (2.15) *has a unique solution in the set* $\Omega([0,1], p(x), r(x))$ *which is defined in* (2.19). *This solution can be determined with the fixed point iteration.*

Proof. Under the assumptions formulated above the operator

$$\Re(x,u(\cdot)) = \int_0^1 G(x,\xi,q(\cdot))\left\{f(\xi,u(\xi)) - q(\xi)\left[\mu_1 + \xi(\mu_2-\mu_1)\right]\right\}d\xi + \mu_1 + x(\mu_2-\mu_1)$$

transforms the set $\Omega([0,1], p(x), r(x))$ into itself and is contractive. Banach's Fixed Point Theorem (see Theorem 1.1) yields the claim of the theorem. ∎

Let us illustrate this theorem by examples.

Example 2.1. We consider the problem (see e.g. [79], p.169)

$$\frac{d^2u(x)}{dx^2} = \frac{3}{2}u^2(x), \; x \in (0,1), \quad u(0) = 4, \; u(1) = 1. \tag{2.23}$$

Here

$$q(x) \equiv 0, \; f(x,u(x)) = -\frac{3}{2}u^2(x), \; p(x) \equiv 0, \; r(x) = 4 - 3x, \; \mu_1 = 4, \; \mu_2 = 1,$$

$$\Re(x,u(\cdot)) = -\frac{3}{2}\int_0^1 G(x,\xi,0)u^2(\xi)d\xi + 4 - 3x,$$

$$-\frac{3}{2}(4-3x)^2 \le f(x,u(x)) \le 0, \quad K_1(x) = -\frac{3}{2}(4-3x)^2, \quad K_2(x) \equiv 0,$$

$$L(x) = 3(4-3x),\ L_1 = 3\left(\frac{1}{2}x_{\max}^3 - 2x_{\max}^2 + \frac{3}{2}x_{\max}\right) < 0.95,\ x_{\max} = \frac{4-\sqrt{7}}{3}.$$

This problem can be represented in the form (2.17), with

$$A(x) = \begin{pmatrix} 0 & -1 \\ 0 & 0 \end{pmatrix}, \quad U(x;\xi) = \begin{pmatrix} 1 & x-\xi \\ 0 & 1 \end{pmatrix}, \quad H = \begin{pmatrix} 0 & 0 \\ 1 & 0 \end{pmatrix},$$

$$u^{(0)}(x) = \begin{pmatrix} 4-3x \\ -3 \end{pmatrix}, \quad G(x,\xi) = \begin{cases} -\begin{pmatrix} x & x(1-\xi) \\ 1 & 1-\xi \end{pmatrix}, & x \le \xi, \\ -\begin{pmatrix} 1-x & -\xi(1-x) \\ -1 & \xi \end{pmatrix}, & \xi < x. \end{cases}$$

Note, the estimates (2.10) and (2.11) take here the form

$$\|U(x;\xi)\| \le e^{x-\xi}, \quad \|G(x,\xi)\| \le \begin{cases} e^{1+x-\xi} & \text{if } x \le \xi, \\ e^{x-\xi}(1+e) & \text{if } \xi \le x \end{cases}.$$

Obviously, the bounds are not very sharp.

Now Theorem 2.2 says that the operator $\Re(x, u(\cdot))$ is a contractive mapping on the set $\Omega([0,1], 0, 4-3x)$ which transforms Ω into itself. Thus, problem (2.23) has a unique solution in Ω which can be determined by the fixed point iteration. □

Example 2.2. The next example goes back to B. A. Troesch (see e.g. [83]) and represents a well-known test problem for numerical BVP-software,

$$\frac{d^2u(x)}{dx^2} = \lambda \sinh(\lambda u(x)), \quad x \in (0,1), \quad \lambda > 0, \quad u(0) = 0,\ u(1) = 1. \qquad (2.24)$$

Here

$$q(x) \equiv 0,\ f(x,u(x)) = -\lambda\sinh(\lambda u(x)),\ p(x) \equiv 0,\ r(x) = x,\ \mu_1 = 0,\ \mu_2 = 1,$$

$$\Re(x, u(\cdot)) = -\lambda \int_0^1 G(x,\xi,0)\sinh(\lambda u(\xi))d\xi + x,$$

$$-\lambda\sinh(\lambda u(x)) \le f(x,u(x)) \le 0,\ K_1(x) = -\lambda\sinh(\lambda x),\ K_2(x) \equiv 0,$$

$$L(x) = 2\lambda\sinh\left(\frac{\lambda}{2}\right)\int_0^1 G(x,\xi,0)\cosh(\lambda\xi)d\xi \le L_1 < 0.81 \text{ for } \lambda \in (0, 1.9].$$

We can formulate this problem in the form (2.17) with the same $U(x;\xi)$, $A(x)$, H and $G(x,\xi)$ as given in Example 2.1 and with $\boldsymbol{u}^{(0)} = (x,1)^T$.

Using Theorem 2.2 we see that the operator $\Re(x,u(\cdot))$ is a contractive mapping on the set $\Omega([0,1],0,x)$ which transforms Ω into itself. For $\lambda \in (0, 1.9)$ problem (2.24) has a unique solution which can be determined by the fixed point iteration. Note that this restriction on λ is merely technical and is essentially the result of the actual choice of the set $\Omega([0,1],0,x)$. A numerical test with values of λ up to $\lambda = 62$ is described in Example 2.4. □

2.2 Existence of a two-point EDS

Let us consider the space of grid functions $\{\boldsymbol{u}_j\}_{j=0}^N$ defined on the grid (1.3) and equipped with the norm

$$\|\boldsymbol{u}\|_{0,\infty,\widehat{\overline{\omega}}_h} \stackrel{\text{def}}{=} \max_{0\le j\le N} \|\boldsymbol{u}_j\|.$$

Throughout this chapter M denotes a generic positive constant which does not dependent on h.

Given $\{\boldsymbol{v}_j\}_{j=0}^N$ we define the IVPs (each of the dimension d)

$$\begin{gathered}\frac{d\boldsymbol{Y}^j(x;\boldsymbol{v}_{j-1})}{dx} + A(x)\boldsymbol{Y}^j(x;\boldsymbol{v}_{j-1}) = \boldsymbol{f}(x,\boldsymbol{Y}^j(x;\boldsymbol{v}_{j-1})), \quad x\in(x_{j-1},x_j],\\ \boldsymbol{Y}^j(x_{j-1};\boldsymbol{v}_{j-1}) = \boldsymbol{v}_{j-1}, \quad j=1,2,\ldots,N.\end{gathered} \tag{2.25}$$

The existence of a unique solution of (2.25) is postulated in the following lemma.

Lemma 2.5

Let Assumptions 2.1 and 2.2 be satisfied. Suppose that the grid function $\{\boldsymbol{v}_j\}_{j=0}^N$ belongs to $\Omega\left(\widehat{\overline{\omega}}_h, r(\cdot)\right)$, where $\Omega\left(\widehat{\overline{\omega}}_h, r(\cdot)\right)$ is defined in (2.7) Then the problem (2.25) has a unique solution.

Proof. The question about the existence and uniqueness of the solution of the IVPs in (2.25) is equivalent to the same question for the integral equation

$$\boldsymbol{Y}^j(x;\boldsymbol{v}_{j-1}) = \Im(x,\boldsymbol{v}_{j-1},\boldsymbol{Y}^j), \tag{2.26}$$

where

$$\Im(x,\boldsymbol{v}_{j-1},\boldsymbol{Y}^j) \stackrel{\text{def}}{=} U(x;x_{j-1})\boldsymbol{v}_{j-1} + \int_{x_{j-1}}^{x} U(x;\xi)\boldsymbol{f}(\xi,\boldsymbol{Y}^j(\xi;\boldsymbol{v}_{j-1}))d\xi, \quad x\in\bar{e}^j.$$

We define the n-th power of the operator $\Im(x, \boldsymbol{v}_{j-1}, \boldsymbol{Y}^j)$ by

$$\Im^n(x, \boldsymbol{v}_{j-1}, \boldsymbol{Y}^j) \stackrel{\text{def}}{=} \Im(x, \boldsymbol{v}_{j-1}, \Im^{n-1}(x, \boldsymbol{v}_{j-1}, \boldsymbol{Y}^j)), \quad n = 2, 3, \ldots .$$

Let $\boldsymbol{Y}^j(x; \boldsymbol{v}_{j-1}) \in \Omega\left(\bar{e}^j, r(\cdot)\right)$ for $\{\boldsymbol{v}_j\}_{j=0}^N \in \Omega\left(\widehat{\overline{\omega}}_h, r(\cdot)\right)$. Then

$$\begin{aligned}
&\left\|\Im(x, \boldsymbol{v}_{j-1}, \boldsymbol{Y}^j) - \boldsymbol{u}^{(0)}(x)\right\| \\
&\quad \le \|U(x; x_{j-1})\| \left\|\boldsymbol{v}_{j-1} - \boldsymbol{u}^{(0)}(x_{j-1})\right\| + \int\limits_{x_{j-1}}^{x} \|U(x;\xi)\| \left\|\boldsymbol{f}(\xi, \boldsymbol{Y}^j(\xi; \boldsymbol{v}_{j-1}))\right\| d\xi \\
&\quad \le K \exp(c_1 x)\left[x_{j-1} + \|H\| \exp(c_1)\right] + K(x - x_{j-1}) \exp\left[c_1(x - x_{j-1})\right] \\
&\quad \le K \exp(c_1 x)\left[x + \|H\| \exp(c_1)\right] = r(x), \quad x \in \bar{e}^j,
\end{aligned}$$

that is, for grid functions $\{\boldsymbol{v}_j\}_{j=0}^N \in \Omega\left(\widehat{\overline{\omega}}_h, r(\cdot)\right)$ the operator $\Im(x, \boldsymbol{v}_{j-1}, \boldsymbol{Y}^j)$ transforms the set $\Omega\left(\bar{e}^j, r(\cdot)\right)$ into itself.

Moreover, for $\boldsymbol{Y}^j(x; \boldsymbol{v}_{j-1}), \widetilde{\boldsymbol{Y}}^j(x; \boldsymbol{v}_{j-1}) \in \Omega\left(\bar{e}^j, r(\cdot)\right)$ we have the estimate

$$\begin{aligned}
&\left\|\Im(x, \boldsymbol{v}_{j-1}, \boldsymbol{Y}^j) - \Im(x, \boldsymbol{v}_{j-1}, \widetilde{\boldsymbol{Y}}^j)\right\| \\
&\quad \le \int\limits_{x_{j-1}}^{x} \|U(x;\xi)\| \left\|\boldsymbol{f}(\xi, \boldsymbol{Y}^j(\xi, \boldsymbol{v}_{j-1})) - \boldsymbol{f}(\xi, \widetilde{\boldsymbol{Y}}^j(\xi, \boldsymbol{v}_{j-1}))\right\| d\xi \\
&\quad \le L \exp(c_1 h_j) \int\limits_{x_{j-1}}^{x} \left\|\boldsymbol{Y}^j(\xi, \boldsymbol{v}_{j-1}) - \widetilde{\boldsymbol{Y}}^j(\xi, \boldsymbol{v}_{j-1})\right\| d\xi \\
&\quad \le L \exp(c_1 h_j)(x - x_{j-1}) \left\|\boldsymbol{Y}^j - \widetilde{\boldsymbol{Y}}^j\right\|_{0,\infty,\bar{e}^j} .
\end{aligned}$$

Using this estimate we get

$$\begin{aligned}
&\left\|\Im^2(x, \boldsymbol{v}_{j-1}, \boldsymbol{Y}^j) - \Im^2(x, \boldsymbol{v}_{j-1}, \widetilde{\boldsymbol{Y}}^j)\right\| \\
&\quad \le L \exp(c_1 h_j) \int\limits_{x_{j-1}}^{x} \left\|\Im(x, \boldsymbol{v}_{j-1}, \boldsymbol{Y}^j) - \Im(x, \boldsymbol{v}_{j-1}, \widetilde{\boldsymbol{Y}}^j)\right\| d\xi \\
&\quad \le \frac{\left[L \exp(c_1 h_j)(x - x_{j-1})\right]^2}{2!} \left\|\boldsymbol{Y}^j - \widetilde{\boldsymbol{Y}}^j\right\|_{0,\infty,\bar{e}^j} .
\end{aligned}$$

If we continue to determine such estimates we get by induction

$$\|\Im^n(x, \boldsymbol{v}_{j-1}, \boldsymbol{Y}^j) - \Im^n(x, \boldsymbol{v}_{j-1}, \widetilde{\boldsymbol{Y}}^j)\|$$
$$\leq \frac{[L\exp(c_1h_j)(x-x_{j-1})]^n}{n!}\left\|\boldsymbol{Y}^j - \widetilde{\boldsymbol{Y}}^j\right\|_{0,\infty,\bar{e}^j}$$

and it follows that

$$\|\Im^n(\cdot, \boldsymbol{v}_{j-1}, \boldsymbol{Y}^j) - \Im^n(\cdot, \boldsymbol{v}_{j-1}, \widetilde{\boldsymbol{Y}}^j)\|_{0,\infty,[x_{j-1},x_j]}$$
$$\leq \frac{[L\exp(c_1h_j)h_j]^n}{n!}\left\|\boldsymbol{Y}^j - \widetilde{\boldsymbol{Y}}^j\right\|_{0,\infty,\bar{e}^j}.$$

In view of the fact that

$$\frac{[L\exp(c_1h_j)h_j]^n}{n!} \to 0 \text{ for } n \to \infty,$$

we can fix the number n large enough such that

$$\frac{[L\exp(c_1h_j)h_j]^n}{n!} < 1,$$

which yields that the operator $\Im^n(x, \boldsymbol{v}_{j-1}, \boldsymbol{Y}^j)$ is a contractive mapping of the set $\Omega\left(\bar{e}^j, r(\cdot)\right)$ into itself. Thus, for $\{\boldsymbol{v}_j\}_{j=0}^N \in \Omega\left(\widehat{\overline{\omega}}_h, r(x)\right)$ problem (2.26) [or problem (2.25)] has a unique solution (see, e.g. [82], pp. 392–393). ■

We are now in the position to prove the main result of this section.

Theorem 2.3

Let the assumptions of Theorem 2.1 be satisfied. Then, there exists a 2-point EDS for problem (2.1). It is of the form

$$\boldsymbol{u}_j = \boldsymbol{Y}^j(x_j; \boldsymbol{u}_{j-1}), \quad j = 1, 2, \dots N, \tag{2.27}$$

$$B_0\boldsymbol{u}_0 + B_1\boldsymbol{u}_N = \boldsymbol{\beta}. \tag{2.28}$$

Proof. It is easy to see that

$$\frac{d}{dx}\boldsymbol{Y}^j(x; \boldsymbol{u}_{j-1}) + A(x)\boldsymbol{Y}^j(x; \boldsymbol{u}_{j-1}) = \boldsymbol{f}(x, \boldsymbol{Y}^j(x; \boldsymbol{u}_{j-1})), \quad x \in (x_{j-1}, x_j],$$
$$\boldsymbol{Y}^j(x_{j-1}; \boldsymbol{u}_{j-1}) = \boldsymbol{u}_{j-1}, \quad j = 1, 2, \dots N.$$

Due to Lemma 2.5 the solvability of the last problem is equivalent to the solvability of the problem (2.1). Thus, the solution of the problem (2.1) can be represented by

$$\boldsymbol{u}(x) = \boldsymbol{Y}^j(x; \boldsymbol{u}_{j-1}), \quad x \in \bar{e}^j, \quad j = 1, 2, \dots, N. \tag{2.29}$$

Substituting $x = x_j$ into (2.29) we get the 2-point EDS (2.27),(2.28). ■

For further examination of the 2-point EDS we need the following lemma.

Lemma 2.6

Let the assumptions of Lemma 2.5 be satisfied. Then, for two grid functions $\{\boldsymbol{u}_j\}_{j=0}^N, \{\boldsymbol{v}_j\}_{j=0}^N \in \Omega(\widehat{\overline{\omega}}_h, r(\cdot))$ the following inequality is satisfied:

$$\left\|\boldsymbol{Y}^j(x;\boldsymbol{u}_{j-1}) - \boldsymbol{Y}^j(x;\boldsymbol{v}_{j-1}) - U(x;x_{j-1})(\boldsymbol{u}_{j-1} - \boldsymbol{v}_{j-1})\right\|$$

$$\leq L(x - x_{j-1}) \exp\left\{c_1(x - x_{j-1}) + L\int\limits_{x_{j-1}}^{x} \exp\left[c_1(x-\xi)\right] d\xi\right\} \|\boldsymbol{u}_{j-1} - \boldsymbol{v}_{j-1}\|. \tag{2.30}$$

Proof. When proving Lemma 2.5 it was shown that $\boldsymbol{Y}^j(x;\boldsymbol{u}_{j-1})$ and $\boldsymbol{Y}^j(x;\boldsymbol{v}_{j-1})$ belong to $\Omega\left(\bar{e}^j, r(\cdot)\right)$. Therefore it follows from (2.25) that

$$\left\|\boldsymbol{Y}^j(x;\boldsymbol{u}_{j-1}) - \boldsymbol{Y}^j(x;\boldsymbol{v}_{j-1}) - U(x;x_{j-1})(\boldsymbol{u}_{j-1} - \boldsymbol{v}_{j-1})\right\|$$

$$\leq L\int\limits_{x_{j-1}}^{x} \exp\left[c_1(x-\xi)\right]\Big\{\exp\left[c_1(\xi - x_{j-1})\right]\|\boldsymbol{u}_{j-1} - \boldsymbol{v}_{j-1}\|$$

$$+\left\|\boldsymbol{Y}^j(\xi;\boldsymbol{u}_{j-1}) - \boldsymbol{Y}^j(\xi;\boldsymbol{v}_{j-1}) - U(\xi;x_{j-1})(\boldsymbol{u}_{j-1} - \boldsymbol{v}_{j-1})\right\|\Big\}d\xi$$

$$= L\exp\left[c_1(x - x_{j-1})\right](x - x_{j-1})\|\boldsymbol{u}_{j-1} - \boldsymbol{v}_{j-1}\|$$

$$+L\int\limits_{x_{j-1}}^{x} \exp\left[c_1(x-\xi)\right]\big\|\boldsymbol{Y}^j(\xi;\boldsymbol{u}_{j-1}) - \boldsymbol{Y}^j(\xi;\boldsymbol{v}_{j-1})$$

$$-U(\xi;x_{j-1})(\boldsymbol{u}_{j-1} - \boldsymbol{v}_{j-1})\big\|\,d\xi.$$

Now, Gronwall's Lemma implies (2.30). ■

We can now prove the uniqueness of the solution of the 2-point EDS (2.27), (2.28).

Theorem 2.4

Let the assumptions of Theorem 2.1 be satisfied. Then there exists a $h_0 > 0$ such that for $h \leq h_0$ the 2-point EDS (2.27), (2.28) has a unique solution

$$\{\boldsymbol{u}_j\}_{j=0}^N \stackrel{\text{def}}{=} \{\boldsymbol{u}(x_j)\}_{j=0}^N \in \Omega(\widehat{\overline{\omega}}_h, r(\cdot))$$

which can be determined by the modified fixed point iteration

$$\boldsymbol{u}_j^{(k)} - U(x_j;x_{j-1})\boldsymbol{u}_{j-1}^{(k)} = \boldsymbol{Y}^j(x_j;\boldsymbol{u}_{j-1}^{(k-1)}) - U(x_j;x_{j-1})\boldsymbol{u}_{j-1}^{(k-1)}, \tag{2.31}$$

$$B_0\boldsymbol{u}_0^{(k)} + B_1\boldsymbol{u}_N^{(k)} = \boldsymbol{\beta}, \quad j = 1, 2, \ldots N, \quad k = 1, 2, \ldots . \tag{2.32}$$

$$\boldsymbol{u}_j^{(0)} = U(x_j;0)Q^{-1}\boldsymbol{\beta}, \quad j = 0, 1, \dots N. \tag{2.33}$$

The corresponding error estimate is

$$\left\| \boldsymbol{u}^{(k)} - \boldsymbol{u} \right\|_{0,\infty,\widehat{\omega}_h} \le \frac{q_1^k}{1 - q_1} r(1), \tag{2.34}$$

where $q_1 \stackrel{\text{def}}{=} q \exp[L\, h\, \exp(c_1\, h)] < 1$.

Proof. Taking into account (2.4), we apply successively the formula (2.27) and get

$$\begin{aligned}
\boldsymbol{u}_1 &= U(x_1;0)\boldsymbol{u}_0 + \boldsymbol{Y}^1(x_1;\boldsymbol{u}_0) - U(x_1;0)\boldsymbol{u}_0,\\
\boldsymbol{u}_2 &= U(x_2;x_1)U(x_1;0)\boldsymbol{u}_0 + U(x_2;x_1)\left[\boldsymbol{Y}^1(x_1;\boldsymbol{u}_0) - U(x_1;0)\boldsymbol{u}_0\right] + \boldsymbol{Y}^2(x_2;\boldsymbol{u}_1)\\
&\quad - U(x_2;x_1)\boldsymbol{u}_1\\
&= U(x_2;0)\boldsymbol{u}_0 + U(x_2;x_1)\left[\boldsymbol{Y}^1(x_1;\boldsymbol{u}_0) - U(x_1;0)\boldsymbol{u}_0\right] + \boldsymbol{Y}^2(x_2;\boldsymbol{u}_1)\\
&\quad - U(x_2;x_1)\boldsymbol{u}_1,\\
&\;\;\vdots
\end{aligned}$$

$$\boldsymbol{u}_j = U(x_j;0)\boldsymbol{u}_0 + \sum_{i=1}^{j} U(x_j;x_i)\left[\boldsymbol{Y}^i(x_i;\boldsymbol{u}_{i-1}) - U(x_i;x_{i-1})\boldsymbol{u}_{i-1}\right]. \tag{2.35}$$

Substituting (2.35) into the boundary condition (2.28), we obtain

$$\begin{aligned}
\left[B_0 + B_1 U(1;0)\right]\boldsymbol{u}_0 &= Q\boldsymbol{u}_0\\
&= -B_1 \sum_{i=1}^{N} U(1;x_i)\left[\boldsymbol{Y}^i(x_i;\boldsymbol{u}_{i-1}) - U(x_i;x_{i-1})\boldsymbol{u}_{i-1}\right] + \boldsymbol{\beta}.
\end{aligned}$$

Thus,

$$\begin{aligned}
\boldsymbol{u}_j = &- U(x_j;0)H \sum_{i=1}^{N} U(1;x_i)\left[\boldsymbol{Y}^i(x_i;\boldsymbol{u}_{i-1}) - U(x_i;x_{i-1})\boldsymbol{u}_{i-1}\right]\\
&+ \sum_{i=1}^{j} U(x_j;x_i)\left[\boldsymbol{Y}^i(x_i;\boldsymbol{u}_{i-1}) - U(x_i;x_{i-1})\boldsymbol{u}_{i-1}\right] + U(x_j;0)Q^{-1}\boldsymbol{\beta}
\end{aligned}$$

or

$$\boldsymbol{u}_j = \sum_{i=1}^{N} G_h(x_j,x_i)\left[\boldsymbol{Y}^i(x_i;\boldsymbol{u}_{i-1}) - U(x_i;x_{i-1})\boldsymbol{u}_{i-1}\right] + \boldsymbol{u}^{(0)}(x_j), \tag{2.36}$$

where the discrete Green's function $G_h(x,\xi)$ of problem (2.27), (2.28) is the projection of the Green's function $G(x,\xi)$ (see formula (2.9)) onto the grid $\widehat{\omega}_h$, that is

$$G(x,\xi) = G_h(x,\xi) \quad \text{for all } x, \xi \in \widehat{\omega}_h.$$

Due to

$$\boldsymbol{Y}^i(x_i;\boldsymbol{u}_{i-1}) - U(x_i;x_{i-1})\boldsymbol{u}_{i-1} = \int\limits_{x_{i-1}}^{x_i} U(x_i;\xi)\boldsymbol{f}(\xi,\boldsymbol{Y}^i(\xi;\boldsymbol{u}_{i-1}))d\xi,$$

we have

$$\Re_h(x_j,\{\boldsymbol{u}_s\}_{s=0}^N) = \sum_{i=1}^N \int\limits_{x_{i-1}}^{x_i} G(x_j,\xi)\boldsymbol{f}(\xi,\boldsymbol{Y}^i(\xi;\boldsymbol{u}_{i-1}))d\xi + \boldsymbol{u}^{(0)}(x_j). \tag{2.37}$$

Next we show that the operator (2.37) transforms the set $\Omega\left(\widehat{\overline{\omega}}_h, r(\cdot)\right)$ into itself.

Let $\{\boldsymbol{v}_j\}_{j=0}^N \in \Omega\left(\widehat{\overline{\omega}}_h, r(\cdot)\right)$, then we have (see the proof of Lemma 2.5)

$$\boldsymbol{v}(x) = \boldsymbol{Y}^j(x;\boldsymbol{v}_{j-1}) \in \Omega\left(\bar{e}^j, r(\cdot)\right), \quad j=1,2,\ldots,N,$$

and

$$\begin{aligned}
&\left\|\Re_h(x_j,\{\boldsymbol{v}_s\}_{s=0}^N) - \boldsymbol{u}^{(0)}(x_j)\right\| \\
&\le K\left[\exp[c_1(1+x_j)]\,\|H\|\sum_{i=1}^N \int\limits_{x_{i-1}}^{x_i} \exp(-c_1\xi)d\xi + \exp(c_1x_j)\sum_{i=1}^j \int\limits_{x_{i-1}}^{x_i} \exp(-c_1\xi)d\xi\right] \\
&\le K\left[\exp(c_1x_j)\sum_{i=1}^j \exp(-c_1x_{i-1})h_i + \exp[c_1(1+x_j)]\,\|H\|\sum_{i=1}^N \exp(-c_1x_{i-1})h_i\right] \\
&\le K\exp(c_1x_j)[x_j + \|H\|\exp(c_1)] = r(x_j), \quad j=0,1,\ldots,N.
\end{aligned}$$

Moreover, the operator $\Re_h(x_j,\{\boldsymbol{u}_s\}_{s=0}^N)$ is a contraction on $\Omega\left(\widehat{\overline{\omega}}_h, r(\cdot)\right)$. This can be shown as follows. Taking into account Lemma 2.6 and the estimate

$$\|G(x,\xi)\| \le \begin{cases} \exp[c_1(1+x-\xi)]\,\|H\|, & 0\le x\le\xi, \\ \exp[c_1(x-\xi)]\,[1+\|H\|\exp(c_1)], & \xi\le x\le 1, \end{cases}$$

which has been proved above, the relation (2.36) implies

$$\begin{aligned}
&\left\|\Re_h(x_j,\{\boldsymbol{u}_s\}_{s=0}^N) - \Re_h(x_j,\{\boldsymbol{v}_s\}_{s=0}^N)\right\|_{0,\infty,\widehat{\overline{\omega}}_h} \\
&\le \sum_{i=1}^N \exp\left(c_1(x_j-x_i)\right)\left(1+\|H\|\exp(c_1)\right)L(x_i-x_{i-1}) \\
&\times\exp\left\{c_1(x_j-x_{i-1}) + L\int\limits_{x_{i-1}}^{x_i} \exp\left(c_1(x_i-\xi)\right)d\xi\right\}\|\boldsymbol{u}_{j-1}-\boldsymbol{v}_{j-1}\|
\end{aligned}$$

$$\leq \exp(c_1 x_j)\left(1+\|H\|\exp(c_1)\right) L \exp\left(L\, h \exp(c_1\, h)\right) \|\boldsymbol{u}-\boldsymbol{v}\|_{0,\infty,\widehat{\omega}_h}$$

$$\leq q \exp\left(L\, h \exp(c_1\, h)\right) \|\boldsymbol{u}-\boldsymbol{v}\|_{0,\infty,\widehat{\omega}_h} = q_1 \|\boldsymbol{u}-\boldsymbol{v}\|_{0,\infty,\widehat{\omega}_h}.$$

Since (2.12) implies $q < 1$, we have $q_1 < 1$ for h_0 small enough and the operator $\Re_h(x_j, \{\boldsymbol{u}_s\}_{s=0}^N)$ is a contraction for all $\{\boldsymbol{u}_j\}_{j=0}^N, \{\boldsymbol{v}_j\}_{j=0}^N \in \Omega\left(\widehat{\omega}_h, r(\cdot)\right)$. Then Banach's Fixed Point Theorem (see Theorem 1.1) says that the 2-point EDS (2.27), (2.28) has a unique solution which can be determined by the modified fixed point iteration (2.31) with the error estimate (2.34). ∎

2.3 Implementation of the 2-point EDS

In order to get an implementable compact 2-point difference scheme from the 2-point EDS we replace (2.27), (2.28) by the so-called truncated difference scheme of rank m (m-TDS)

$$\boldsymbol{y}_j^{(m)} = Y^{(m)j}(x_j; \boldsymbol{y}_{j-1}^{(m)}), \quad j = 1, 2, \ldots N, \tag{2.38}$$

$$B_0 \boldsymbol{y}_0^{(m)} + B_1 \boldsymbol{y}_N^{(m)} = \boldsymbol{\beta}. \tag{2.39}$$

Here, m is a positive integer and $Y^{(m)j}(x_j; \boldsymbol{y}_{j-1}^{(m)})$ denotes the numerical solution of the IVP (2.25) on the interval $\bar{e}^j$ which has been obtained by some one-step method of the order m (e.g. by the Taylor series method or a Runge-Kutta method)

$$Y^{(m)j}(x_j; \boldsymbol{y}_{j-1}^{(m)}) = \boldsymbol{y}_{j-1}^{(m)} + h_j \boldsymbol{\Phi}(x_{j-1}, \boldsymbol{y}_{j-1}^{(m)}; h_j). \tag{2.40}$$

It holds that

$$\left\| \boldsymbol{Y}^{(m)j}(x_j; \boldsymbol{u}_{j-1}) - \boldsymbol{Y}^j(x_j; \boldsymbol{u}_{j-1}) \right\| \leq M h_j^{m+1}$$

and the increment function (see e.g. [31]) $\boldsymbol{\Phi}(x, \boldsymbol{u}, h)$ satisfies the consistency condition

$$\boldsymbol{\Phi}(x, \boldsymbol{u}; 0) = \boldsymbol{f}(x, \boldsymbol{u}) - A(x)\boldsymbol{u}.$$

For example, in the case of the Taylor expansion method we have

$$\boldsymbol{\Phi}(x_{j-1}, \boldsymbol{y}_{j-1}^{(m)}; h_j) = \boldsymbol{f}(x_{j-1}, \boldsymbol{y}_{j-1}^{(m)}) - A(x_{j-1})\boldsymbol{y}_{j-1}^{(m)} + \sum_{p=2}^{m} \frac{h_j{}^{p-1}}{p!} \left. \frac{d^p \boldsymbol{Y}^j(x; \boldsymbol{y}_{j-1}^{(m)})}{dx^p} \right|_{x=x_{j-1}},$$

and in the case of an explicit Runge-Kutta method we have

$$\begin{aligned} \boldsymbol{\Phi}(x_{j-1}, \boldsymbol{y}_{j-1}^{(m)}; h_j) &= \sum_{i=1}^{s} b_i\, \boldsymbol{k}_i, \\ \boldsymbol{k}_i &= \boldsymbol{g}(x_{j-1} + c_i\, h_j, \boldsymbol{y}_{j-1}^{(m)} + h_j \sum_{l=1}^{s} a_{il}\, \boldsymbol{k}_l), \quad i = 1, \ldots, s, \end{aligned} \tag{2.41}$$

where $\boldsymbol{g}(x,\boldsymbol{u}) \stackrel{\text{def}}{=} \boldsymbol{f}(x,\boldsymbol{u}) - A(x)\,\boldsymbol{u}$ and $a_{ij} = 0$ for $i \le j$.

Example 2.3. Let us consider the BVP

$$\frac{du(x)}{dx} = -u(x)^2, \quad 0 < x < 1, \quad u(0) = \frac{1}{2}u(1). \tag{2.42}$$

The corresponding exact solution is $u(x) = \dfrac{1}{x-2}$.

On the interval $[0,1]$ we use the nonuniform grid $\hat{\bar{\omega}}_h$. Due to Theorem 2.3 there exists the following EDS for problem (2.42):

$$u_j = Y^j(x_j; u_{j-1}), \quad j = 1,2,\dots N, \quad u_0 = \frac{1}{2}u_N,$$

where $Y^j(x, u_{j-1})$ is the solution of the IVP

$$\frac{dY^j(x;u_{j-1})}{dx} = -\left(Y^j(x;u_{j-1})\right)^2, \quad x \in (x_{j-1}, x_j],$$

$$Y^j(x_{j-1}; u_{j-1}) = u_{j-1}, \quad j = 1,2,\dots N.$$

Since the function

$$Y^j(x;u_{j-1}) = \left(x - x_{j-1} + \frac{1}{u_{j-1}}\right)^{-1}, \quad j = 1,2,\dots N,$$

is the exact solution of the IVP, the EDS takes the form

$$u_j = \frac{u_{j-1}}{1 + h_j u_{j-1}}, \quad j = 1,2,\dots N, \quad u_0 = \frac{1}{2}u_N.$$

Let us now construct the TDS of rank two (2-TDS)

$$y_j^{(2)} = Y^{(2)j}(x_j; y_{j-1}^{(2)}), \quad j = 1,2,\dots N, \quad y_0^{(2)} = \frac{1}{2}y_N^{(2)},$$

where $y_j^{(2)} \approx u_j$ and $Y^{(2)j}(x_j; y_{j-1}^{(2)})$ is the approximate solution of the IVP obtained by a one-step method of order 2. Using the Taylor series method as the numerical integration technique we obtain

$$Y^{(2)j}(x_j; y_{j-1}^{(2)}) = y_{j-1}^{(2)} - h_j\left(y_{j-1}^{(2)}\right)^2 + h_j^2\left(y_{j-1}^{(2)}\right)^3.$$

Therefore, the 2-TDS for which we are looking is

$$y_j^{(2)} = y_{j-1}^{(2)} - h_j\left(y_{j-1}^{(2)}\right)^2 + h_j^2\left(y_{j-1}^{(2)}\right)^3 \quad j = 1,2,\dots N, \quad y_0^{(2)} = \frac{1}{2}y_N^{(2)}.$$ □

In order to prove the existence and uniqueness of a solution of the m-TDS (2.38), (2.39) and to investigate its accuracy the following statement is required.

Lemma 2.7

Let the method (2.40) be of the order of accuracy m. Assume that the increment function $\boldsymbol{\Phi}(x,\boldsymbol{u};h)$ is sufficiently smooth, the entries $a_{ps}(x)$ of the matrix $A(x)$ belong to $\mathbb{C}^m[0,1]$ and there exists a real number $\Delta > 0$ such that the components $f_p(x,\boldsymbol{u})$ of the vector function $\boldsymbol{f}(x,\boldsymbol{u})$ belong to $\mathbb{C}^{k,m-k}([0,1]\times\Omega([0,1],r(\cdot)+\Delta))$, with $k=0,1,\ldots,m-1$. Let the matrix $U^{(1)}(x_j;x_{j-1})$ be defined by

$$U^{(1)}(x_j;x_{j-1}) \stackrel{\text{def}}{=} I - h_j A(x_{j-1}). \tag{2.43}$$

Then

$$\left\|U^{(1)}(x_j;x_{j-1}) - U(x_j;x_{j-1})\right\| \le M\,h_j^2, \tag{2.44}$$

$$\left\|\frac{1}{h_j}\left(\boldsymbol{Y}^{(m)j}(x_j;\boldsymbol{v}_{j-1}) - U^{(1)}(x_j;x_{j-1})\boldsymbol{v}_{j-1}\right)\right\| \le K + M\,h_j, \tag{2.45}$$

$$\left\|\frac{1}{h_j}\left(\boldsymbol{Y}^{(m)j}(x_j;\boldsymbol{u}_{j-1}) - \boldsymbol{Y}^{(m)j}(x_j;\boldsymbol{v}_{j-1}) - U^{(1)}(x_j;x_{j-1})(\boldsymbol{u}_{j-1}-\boldsymbol{v}_{j-1})\right)\right\|$$
$$\le (L + M\,h_j)\|\boldsymbol{u}_{j-1}-\boldsymbol{v}_{j-1}\|, \tag{2.46}$$

where $\{\boldsymbol{u}_j\}_{j=0}^N, \{\boldsymbol{v}_j\}_{j=0}^N \in \Omega(\widehat{\overline{\omega}}_h, r(\cdot)+\Delta)$.

Proof. Inserting $x = x_j$ into the Taylor expansion of the function $U(x;x_{j-1})$ at the point x_{j-1} gives

$$U(x_j;x_{j-1}) = U^{(1)}(x_j,x_{j-1}) + \int\limits_{x_{j-1}}^{x_j} (x_j - t)\,\frac{d^2U(t;x_{j-1})}{dt^2}\,dt.$$

From this equation the inequality (2.44) follows immediately.

It is easy to verify the following identities:

$$\frac{1}{h_j}\left(\boldsymbol{Y}^{(m)j}(x_j;\boldsymbol{v}_{j-1}) - U^{(1)}(x_j;x_{j-1})\boldsymbol{v}_{j-1}\right)$$
$$= \boldsymbol{\Phi}(x_{j-1},\boldsymbol{v}_{j-1};h_j) + A(x_{j-1})\boldsymbol{v}_{j-1}$$
$$= \boldsymbol{\Phi}(x_{j-1},\boldsymbol{v}_{j-1};0) + h_j\frac{\partial\boldsymbol{\Phi}(x_{j-1},\boldsymbol{v}_{j-1};\bar{h})}{\partial h} + A(x_{j-1})\boldsymbol{v}_{j-1}$$
$$= \boldsymbol{f}(x_{j-1},\boldsymbol{v}_{j-1}) + h_j\frac{\partial\boldsymbol{\Phi}(x_{j-1},\boldsymbol{v}_{j-1};\bar{h})}{\partial h},$$

$$\frac{1}{h_j}\left(\boldsymbol{Y}^{(m)j}(x_j;\boldsymbol{u}_{j-1}) - \boldsymbol{Y}^{(m)j}(x_j;\boldsymbol{v}_{j-1}) - U^{(1)}(x_j;x_{j-1})(\boldsymbol{u}_{j-1}-\boldsymbol{v}_{j-1})\right)$$

$$= \boldsymbol{\Phi}(x_{j-1}, \boldsymbol{u}_{j-1}; h_j) - \boldsymbol{\Phi}(x_{j-1}, \boldsymbol{v}_{j-1}; h_j) + A(x_{j-1})(\boldsymbol{u}_{j-1} - \boldsymbol{v}_{j-1})$$

$$= \boldsymbol{f}(x_{j-1}, \boldsymbol{u}_{j-1}) - \boldsymbol{f}(x_{j-1}, \boldsymbol{v}_{j-1}) + h_j \left(\frac{\partial \boldsymbol{\Phi}(x_{j-1}, \boldsymbol{u}_{j-1}; \bar{h})}{\partial h} - \frac{\partial \boldsymbol{\Phi}(x_{j-1}, \boldsymbol{v}_{j-1}; \bar{h})}{\partial h} \right)$$

$$= \boldsymbol{f}(x_{j-1}, \boldsymbol{u}_{j-1}) - \boldsymbol{f}(x_{j-1}, \boldsymbol{v}_{j-1})$$

$$+ h_j \int_0^1 \frac{\partial^2 \boldsymbol{\Phi}(x_{j-1}, \theta \boldsymbol{u}_{j-1} + (1-\theta)\boldsymbol{v}_{j-1}; \bar{h})}{\partial h \partial \boldsymbol{u}} d\theta \, (\boldsymbol{u}_{j-1} - \boldsymbol{v}_{j-1}),$$

where $\bar{h} \in (0, h)$. These equalities imply the formulas (2.45), (2.46) and thus the proof is complete. ■

Now we are in a position to prove the main result of this section.

Theorem 2.5

Let the assumptions of Theorem 2.1 and Lemma 2.7 be satisfied. Then, there exists a real number $h_0 > 0$ such that for $h \le h_0$ the m-TDS (2.38), (2.39) has a unique solution which can be determined by the modified fixed point iteration

$$\boldsymbol{y}_j^{(m,n)} - U^{(1)}(x_j; x_{j-1})\boldsymbol{y}_{j-1}^{(m,n)} = \boldsymbol{Y}^{(m)j}(x_j; \boldsymbol{y}_{j-1}^{(m,n)}) - U^{(1)}(x_j; x_{j-1})\boldsymbol{y}_{j-1}^{(m,n-1)},$$

$$B_0 \boldsymbol{y}_0^{(m,n)} + B_1 \boldsymbol{y}_N^{(m,n)} = \boldsymbol{\beta},$$

$$\boldsymbol{y}_j^{(m,0)} = \prod_{k=1}^{j} U^{(1)}(x_{j-k+1}; x_{j-k}) \left[B_0 + B_1 \prod_{k=1}^{N} U^{(1)}(x_{N-k+1}; x_{N-k}) \right]^{-1} \boldsymbol{\beta},$$

$$j = 1, 2, \ldots N, \quad n = 1, 2, \ldots, \tag{2.47}$$

The corresponding error estimate is

$$\left\| \boldsymbol{y}^{(m,n)} - \boldsymbol{u} \right\|_{0,\infty,\widehat{\overline{\omega}}_h} \le M \left(q_2^n + h^m \right), \tag{2.48}$$

where $q_2 \stackrel{\text{def}}{=} q + M\,h < 1$.

Proof. From the equations (2.38) we deduce successively

$$\boldsymbol{y}_1^{(m)} = U^{(1)}(x_1; x_0)\boldsymbol{y}_0^{(m)} + \boldsymbol{Y}^{(m)1}(x_1; \boldsymbol{y}_0^{(m)}) - U^{(1)}(x_1; x_0)\boldsymbol{y}_0^{(m)},$$

$$\boldsymbol{y}_2^{(m)} = U^{(1)}(x_2; x_1)\, U^{(1)}(x_1; x_0)\boldsymbol{y}_0^{(m)} + U^{(1)}(x_2; x_1)$$
$$\times \left(\boldsymbol{Y}^{(m)1}(x_1; \boldsymbol{y}_0^{(m)}) - U^{(1)}(x_1; x_0)\boldsymbol{y}_0^{(m)} \right) + \boldsymbol{Y}^{(m)2}(x_2; \boldsymbol{y}_1^{(m)})$$
$$- U^{(1)}(x_2; x_1)\boldsymbol{y}_1^{(m)}$$

$$\vdots$$

Thus

$$\boldsymbol{y}_j^{(m)} = \prod_{k=1}^{j} U^{(1)}(x_{j-k+1}; x_{j-k}) \boldsymbol{y}_0^{(m)}$$
$$+ \sum_{i=1}^{j} \prod_{k=1}^{j-i} U^{(1)}(x_{j-k+1}; x_{j-k}) \left(\boldsymbol{Y}^{(m)i}(x_i; \boldsymbol{y}_{i-1}^{(m)}) - U^{(1)}(x_i; x_{i-1}) \boldsymbol{y}_{i-1}^{(m)} \right). \tag{2.49}$$

Substituting $\boldsymbol{y}_N^{(m)}$ into the boundary conditions (2.39) we get

$$\left[B_0 + B_1 \prod_{k=1}^{N} U^{(1)}(x_{N-k+1}; x_{N-k}) \right] \boldsymbol{y}_0^{(m)}$$
$$= -B_1 \sum_{i=1}^{N} \prod_{k=1}^{N-i} U^{(1)}(x_{N-k+1}; x_{N-k}) \left(\boldsymbol{Y}^{(m)i}(x_i; \boldsymbol{y}_{i-1}^{(m)}) - U^{(1)}(x_i; x_{i-1}) \boldsymbol{y}_{i-1}^{(m)} \right) + \boldsymbol{\beta}. \tag{2.50}$$

Let us show that the matrix in square brackets on the left-hand side of (2.50) is regular.

Here and in the following we use the inequality

$$\begin{aligned} \left\| U^{(1)}(x_j; x_{j-1}) \right\| &\le \| U(x_j; x_{j-1}) \| + \left\| U^{(1)}(x_j; x_{j-1}) - U(x_j; x_{j-1}) \right\| \\ &\le \exp(c_1\, h_j) + M\, h_j^2 \end{aligned} \tag{2.51}$$

which can be easily derived using the estimate (2.44).

We have

$$\| (B_0 + B_1 \prod_{k=1}^{N} U^{(1)}(x_{N-k+1}; x_{N-k})] - [B_0 + B_1 U(1; 0)) \|$$
$$= \| B_1 (\prod_{k=1}^{N} U^{(1)}(x_{N-k+1}; x_{N-k}) - \prod_{k=1}^{N} U(x_{N-k+1}; x_{N-k})) \|$$
$$= \| B_1 \sum_{j=1}^{N} U(x_N; x_{N-j+1}) (U^{(1)}(x_{N-j+1}; x_{N-j}) - U(x_{N-j+1}; x_{N-j}))$$
$$\times \prod_{i=j+1}^{N} U^{(1)}(x_{N-i+1}; x_{N-i}) \|$$
$$\le \| B_1 \| \sum_{j=1}^{N} \exp\left((1 - x_{N-j+1}) c_1\right) M h_{N-j+1}^2 \prod_{i=j+1}^{N} \exp\left((c_1 h_{N-i+1}) + M h_{N-i+1}^2\right)$$
$$\le M\, h, \tag{2.52}$$

that is

$$\Big\|\Big(B_0+B_1\prod_{k=1}^{N}U^{(1)}(x_{N-k+1};x_{N-k})\Big)-\big(B_0+B_1U(1;0)\big)\Big\|<1 \tag{2.53}$$

for h_0 small enough. Here we have used the inequality

$$\prod_{i=j+1}^{N}\exp[(c_1h_{N-i+1})+M\,h_{N-i+1}^2]\le\exp(c_1)\left(1+M\,h^2\right)^{N-j}$$
$$\le\exp(c_1)\exp[M(N-j)h^2]\le\exp(c_1)\exp[M_1\,h]\le\exp(c_1)+M\,h.$$

Since $Q\overset{\text{def}}{=}B_0+B_1U(1;0)$ is nonsingular, it follows from (2.53) that the inverse

$$\left[B_0+B_1\prod_{k=1}^{N}U^{(m)}(x_{N-k+1};x_{N-k})\right]^{-1}$$

exists and due to (2.52) the following estimate holds:

$$\begin{aligned}
&\Big\|\Big(B_0+B_1\prod_{k=1}^{N}U^{(1)}(x_{N-k+1};x_{N-k})\Big)^{-1}B_1\Big\|\\
&\le\Big\|\Big(B_0+B_1\prod_{k=1}^{N}U^{(1)}(x_{N-k+1};x_{N-k})\Big)^{-1}\\
&\quad-\Big(B_0+B_1\prod_{k=1}^{N}U(x_{N-k+1};x_{N-k})\Big)^{-1}\Big\|\,\|B_1\|+\|Q^{-1}B_1\|\\
&\le M\,h+\|H\|\,.
\end{aligned} \tag{2.54}$$

Moreover, from (2.49) and (2.50) we have

$$\begin{aligned}
\boldsymbol{y}_j^{(m)}&=-\prod_{k=1}^{j}U^{(1)}(x_{j-k+1};x_{j-k})\Bigg(B_0+B_1\prod_{k=1}^{N}U^{(1)}(x_{N-k+1};x_{N-k})\Bigg)^{-1}\\
&\quad\times B_1\sum_{i=1}^{N}\prod_{k=1}^{N-i}U^{(1)}(x_{N-k+1};x_{N-k})\left(\boldsymbol{Y}^{(m)i}(x_i,\boldsymbol{y}_{i-1}^{(m)})-U^{(1)}(x_i;x_{i-1})\boldsymbol{y}_{i-1}^{(m)}\right)\\
&\quad+\sum_{i=1}^{j}\prod_{k=1}^{j-i}U^{(1)}(x_{j-k+1};x_{j-k})\left(\boldsymbol{Y}^{(m)i}(x_i;\boldsymbol{y}_{i-1}^{(m)})-U^{(1)}(x_i;x_{i-1})\boldsymbol{y}_{i-1}^{(m)}\right)\\
&\quad+\prod_{k=1}^{j}U^{(1)}(x_{j-k+1};x_{j-k})\left(B_0+B_1\prod_{k=1}^{N}U^{(1)}(x_{N-k+1};x_{N-k})\right)^{-1}\boldsymbol{\beta},
\end{aligned}$$

or

$$\boldsymbol{y}_j^{(m)} = \Re_h^{(m)}(x_j, \{\boldsymbol{y}_s^{(m)}\}_{s=0}^N),$$

where

$$\Re_h^{(m)}(x_j, \{\boldsymbol{y}_s^{(m)}\}_{s=0}^N)$$

$$\stackrel{\text{def}}{=} \sum_{i=1}^N G_h^{(1)}(x_j, x_i) \left(\boldsymbol{Y}^{(m)i}(x_i; \boldsymbol{y}_{i-1}^{(m)}) - U^{(1)}(x_i; x_{i-1}) \boldsymbol{y}_{i-1}^{(m)} \right) + \boldsymbol{y}_j^{(m,0)},$$

and $G_h^{(1)}(x, \xi)$ is the Green's function of the problem (2.38), (2.39) given by

$$G_h^{(1)}(x_j, x_i) = -\prod_{k=1}^j U^{(1)}(x_{j-k+1}; x_{j-k}) \left[B_0 + B_1 \prod_{k=1}^N U^{(1)}(x_{N-k+1}; x_{N-k}) \right]^{-1}$$

$$\times B_1 \prod_{k=1}^{N-i} U^{(1)}(x_{N-k+1}; x_{N-k}) + \begin{cases} 0, & i \geq j, \\ \prod_{k=1}^{j-i} U^{(1)}(x_{j-k+1}; x_{j-k}), & i < j. \end{cases}$$

The estimates (2.51) and (2.54) imply

$$\left\| G_h^{(1)}(x_j, x_i) \right\| \leq \begin{cases} \exp\left(c_1(1 + x_j - x_i)\right) \|H\| + M\,h, & i \geq j, \\ \exp\left(c_1(x_j - x_i)\right) \left(1 + \|H\| \exp\left(c_1\right)\right) + M\,h, & i < j. \end{cases} \tag{2.55}$$

Now we use Banach's Fixed Point Theorem (see Theorem 1.1). First of all we show that the operator $\Re_h^{(m)}(x_j, \{\boldsymbol{v}_k\}_{k=0}^N)$ transforms the set $\Omega(\widehat{\overline{\omega}}_h, r(x) + \Delta)$ into itself. Using (2.45), (2.55) we get for all $\{\boldsymbol{v}_k\}_{k=0}^N \in \Omega(\widehat{\overline{\omega}}_h, r(\cdot) + \Delta)$:

$$\begin{aligned}
&\left\| \Re_h^{(m)}(x_j, \{\boldsymbol{v}_k\}_{k=0}^N) - \boldsymbol{u}^{(0)}(x_j) \right\| \\
&\leq (K + M\,h) \left\{ \exp\left(c_1(1 + x_j)\right) \|H\| \sum_{i=1}^N h_i \exp\left(-c_1 x_{i-1}\right) \right. \\
&\qquad \left. + \exp\left(c_1 x_j\right) \sum_{i=1}^j h_i \exp\left(-c_1 x_{i-1}\right) + M\,h \right\} + M\,h \\
&\leq (K + M\,h) \exp\left(c_1 x_j\right) \left(x_j + \|H\| \exp\left(c_1\right) + M\,h\right) + M\,h \\
&\leq r(x_j) + M\,h \leq r(x_j) + \Delta.
\end{aligned}$$

It remains to be shown that $\Re_h^{(m)}(x_j, \{\boldsymbol{u}_s\}_{s=0}^N)$ is a contractive operator. Due to (2.46) and (2.55) we have

$$\begin{aligned}
&\left\|\mathfrak{R}_h^{(m)}(x_j,\{\boldsymbol{u}_s\}_{s=0}^N) - \mathfrak{R}_h^{(m)}(x_j,\{\boldsymbol{v}_s\}_{s=0}^N)\right\|_{0,\infty,\widehat{\omega}_h} \\
&\le \left(\exp(c_1)\left(1+\|H\|\exp(c_1)\right) + M\,h\right) \\
&\quad\times \max_{1\le j\le N}\left\|\frac{1}{h_j}\left(\boldsymbol{Y}^{(m)j}(x_j;\boldsymbol{u}_{j-1}) - \boldsymbol{Y}^{(m)j}(x_j;\boldsymbol{v}_{j-1})\right.\right. \\
&\qquad\qquad \left.\left. -U^{(1)}(x_j;x_{j-1})(\boldsymbol{u}_{j-1}-\boldsymbol{v}_{j-1})\right)\right\| \\
&\le \left(\exp(c_1)\left(1+\|H\|\exp(c_1)\right) + M\,h\right)(q+M\,h)\,\|\boldsymbol{u}-\boldsymbol{v}\|_{0,\infty,\widehat{\omega}_h} \\
&\le q_2\|\boldsymbol{u}-\boldsymbol{v}\|_{0,\infty,\widehat{\omega}_h},
\end{aligned}$$

where $\{\boldsymbol{u}_k\}_{k=0}^N, \{\boldsymbol{v}_k\}_{k=0}^N \in \Omega(\widehat{\omega}_h, r(\cdot)+\Delta)$ and $q_2 \stackrel{\text{def}}{=} [q+M\,h] < 1$ provided that h_0 is small enough. This means that $\mathfrak{R}_h^{(m)}(x_j,\{\boldsymbol{u}_s\}_{s=0}^N)$ is a contractive operator. Thus, the scheme (2.38), (2.39) has a unique solution which can be determined by the modified fixed point iteration (2.47) with the error estimate

$$\left\|\boldsymbol{y}^{(m,n)} - \boldsymbol{y}^{(m)}\right\|_{0,\infty,\widehat{\omega}_h} \le \frac{q_2^n}{1-q_2}(r(1)+\Delta). \tag{2.56}$$

The error $\boldsymbol{z}_j^{(m)} \stackrel{\text{def}}{=} \boldsymbol{y}_j^{(m)} - \boldsymbol{u}_j$ of the solution of the scheme (2.38), (2.39) satisfies

$$\begin{aligned}
&\boldsymbol{z}_j^{(m)} - U^{(1)}(x_j;x_{j-1})\boldsymbol{z}_{j-1}^{(m)} = \boldsymbol{\psi}^{(m)}(x_j;\boldsymbol{y}_{j-1}^{(m)}), \qquad j=1,2,\ldots,N, \\
&B_0\boldsymbol{z}_0^{(m)} + B_1\boldsymbol{z}_N^{(m)} = \boldsymbol{0},
\end{aligned} \tag{2.57}$$

where the residuum (the approximation error) $\boldsymbol{\psi}^{(m)}(x_j;\boldsymbol{y}_{j-1}^{(m)})$ is given by

$$\begin{aligned}
&\boldsymbol{\psi}^{(m)}(x_j;\boldsymbol{y}_{j-1}^{(m)}) \\
&= \left(\boldsymbol{Y}^{(m)j}(x_j;\boldsymbol{u}(x_{j-1})) - \boldsymbol{Y}^{j}(x_j;\boldsymbol{u}(x_{j-1}))\right) \\
&\quad + \left(\boldsymbol{Y}^{(m)j}(x_j;\boldsymbol{y}_{j-1}^{(m)}) - \boldsymbol{Y}^{(m)j}(x_j;\boldsymbol{u}(x_{j-1})) - U^{(1)}(x_j;x_{j-1})(\boldsymbol{y}_{j-1}^{(m)} - \boldsymbol{u}(x_{j-1}))\right).
\end{aligned}$$

We reformulate problem (2.57) in the equivalent form

$$\boldsymbol{z}_j^{(m)} = \sum_{i=1}^N G_h^{(1)}(x_j,x_i)\boldsymbol{\psi}^{(m)}(x_i;\boldsymbol{y}_{i-1}^{(m)}).$$

Then (2.55) and Lemma 2.7 imply

$$\begin{aligned}\left\|\boldsymbol{z}_j^{(m)}\right\| &\le (\exp(c_1)(1+\|H\|\exp(c_1)) + M\,h)\sum_{i=1}^{N}\|\boldsymbol{\psi}^{(m)}(x_i;\boldsymbol{y}_{i-1}^{(m)})\| \\ &\le (\exp(c_1)(1+\|H\|\exp(c_1)) + M\,h) \\ &\quad\times\left(M\,h^m + \sum_{i=1}^{N} h_i(L+h_iM)\|\boldsymbol{z}_{i-1}^{(m)}\|\right) \\ &\le q_2\left\|\boldsymbol{z}^{(m)}\right\|_{0,\infty,\widehat{\overline{\omega}}_h} + M\,h^m.\end{aligned}$$

The last inequality yields

$$\|\boldsymbol{z}^{(m)}\|_{0,\infty,\widehat{\overline{\omega}}_h} \le M\,h^m. \tag{2.58}$$

Now, from (2.56) and (2.58) we get the error estimate for the method (2.48):

$$\|\boldsymbol{y}^{(m,n)} - \boldsymbol{u}\|_{0,\infty,\widehat{\overline{\omega}}_h} \le \|\boldsymbol{y}^{(m,n)} - \boldsymbol{y}^{(m)}\|_{0,\infty,\widehat{\overline{\omega}}_h} + \|\boldsymbol{y}^{(m)} - \boldsymbol{u}\|_{0,\infty,\widehat{\overline{\omega}}_h} \le M\,(q_2^n + h^m)$$

which completes the proof. ∎

Remark 2.1. Using the matrix $U^{(1)}$ (see formula (2.43)) instead of the fundamental matrix U in (2.47) preserves the order of accuracy but reduces the computational costs significantly. □

Above we have shown that the system of nonlinear equations which represents the TDS can be solved by the modified fixed point iteration. In practice, however, Newton's method is used due to its higher convergence rate. Newton's method applied to the system (2.38), (2.39) has the form

$$\begin{aligned}&\nabla\,\boldsymbol{y}_j^{(m,n)} - \frac{\partial \boldsymbol{Y}^{(m)j}(x_j;\boldsymbol{y}_{j-1}^{(m,n-1)})}{\partial \boldsymbol{u}}\,\nabla\,\boldsymbol{y}_{j-1}^{(m,n)} = \boldsymbol{Y}^{(m)j}(x_j;\boldsymbol{y}_{j-1}^{(m,n-1)}) - \boldsymbol{y}_{j-1}^{(m,n-1)}, \\ &B_0\,\nabla\,\boldsymbol{y}_0^{(m,n)} + B_1\,\nabla\,\boldsymbol{y}_N^{(m,n)} = \boldsymbol{0}, \\ &\boldsymbol{y}_j^{(m,n)} = \boldsymbol{y}_j^{(m,n-1)} + \nabla\boldsymbol{y}_j^{(m,n)}, \quad j=0,1,\ldots,N, \quad n=1,2,\ldots,\end{aligned} \tag{2.59}$$

where

$$\begin{aligned}\frac{\partial \boldsymbol{Y}^{(m)j}(x_j;\boldsymbol{y}_{j-1}^{(m)})}{\partial \boldsymbol{u}} &= I + h_j\frac{\partial \boldsymbol{\Phi}(x_{j-1},\boldsymbol{y}_{j-1}^{(m)};h_j)}{\partial \boldsymbol{u}} \\ &= I + h_j\left[\frac{\partial \boldsymbol{f}(x_{j-1},\boldsymbol{y}_{j-1}^{(m)})}{\partial \boldsymbol{u}} - A(x_{j-1})\right] + O\left(h_j^2\right)\end{aligned}$$

and $\partial \boldsymbol{f}(x_{j-1}, \boldsymbol{y}_{j-1}^{(m)})/\partial \boldsymbol{u}$ is the Jacobian of the vector function $\boldsymbol{f}(x, \boldsymbol{u})$ at the point $(x_{j-1}, \boldsymbol{y}_{j-1}^{(m)})$.

Setting

$$S_j \stackrel{\text{def}}{=} \frac{\partial \boldsymbol{Y}^{(m)j}(x_j; \boldsymbol{y}_{j-1}^{(m,n-1)})}{\partial \boldsymbol{u}},$$

the system (2.59) can be written in the equivalent form

$$(B_0 + B_1 S) \bigtriangledown \boldsymbol{y}_0^{(m,n)} = -B_1 \boldsymbol{\varphi}, \qquad \boldsymbol{y}_0^{(m,n)} = \boldsymbol{y}_0^{(m,n-1)} + \bigtriangledown \boldsymbol{y}_0^{(m,n)}, \tag{2.60}$$

where

$$S \stackrel{\text{def}}{=} S_N S_{N-1} \cdots S_1, \qquad \boldsymbol{\varphi} \stackrel{\text{def}}{=} \boldsymbol{\varphi}_N, \qquad \boldsymbol{\varphi}_0 = \boldsymbol{0},$$

$$\boldsymbol{\varphi}_j = S_j \boldsymbol{\varphi}_{j-1} + \boldsymbol{Y}^{(m)j}(x_j; \boldsymbol{y}_{j-1}^{(m,n-1)}) - \boldsymbol{y}_{j-1}^{(m,n-1)}, \quad j = 1, 2, \ldots, N.$$

The coefficient matrix of the system (2.60) has the dimension $d \times d$. It can be solved by the Gaussian elimination with $O(N)$ flops since the dimension d is small in comparison with N. After that the solution of the system (2.59) is computed recursively by

$$\bigtriangledown \boldsymbol{y}_j^{(m,n)} = S_j S_{j-1} \cdots S_1 \bigtriangledown \boldsymbol{y}_0^{(m,n)} + \boldsymbol{\varphi}_j,$$

$$\boldsymbol{y}_j^{(m,n)} = \boldsymbol{y}_j^{(m,n-1)} + \bigtriangledown \boldsymbol{y}_j^{(m,n)}, \quad j = 1, 2, \ldots, N.$$

When using Newton's method or a quasi-Newton method, the problem of choosing an appropriate initial approach $\boldsymbol{y}_j^{(m,0)}$, $j = 1, 2, \ldots, N$, arises. If the original problem contains a natural parameter, say λ, and for some values of this parameter the solution is known or can be easily obtained, one can try to continue the solution along this parameter (see, e.g., [2], pp. 344–353). Thus, let us suppose that our problem can be written in the generic form

$$\boldsymbol{u}'(x) + A(x)\boldsymbol{u} = \boldsymbol{h}(x, \boldsymbol{u}; \lambda), \quad x \in (0, 1), \quad B_0 \boldsymbol{u}(0) + B_1 \boldsymbol{u}(1) = \boldsymbol{\beta}. \tag{2.61}$$

We assume that for each $\lambda \in [\lambda_0, \lambda_k]$ an isolated solution $\boldsymbol{u}(x; \lambda)$ exists and depends smoothly on λ. Moreover, let us suppose that for $\lambda = \lambda_0$ a solution of (2.61) is known or can easily be determined.

If the problem (2.61) does not contain a natural parameter, then we can introduce such a parameter λ artificially by forming the homotopy function

$$\boldsymbol{h}(x, \boldsymbol{u}; \lambda) \stackrel{\text{def}}{=} \lambda \boldsymbol{f}(x, \boldsymbol{u}) + (1 - \lambda) \boldsymbol{f}_1(x),$$

where $\boldsymbol{f}_1(x)$ is a given function for which the problem (2.61) has a (known) unique solution. Obviously, for $\lambda = 0$ we have the linear BVP

$$\boldsymbol{u}'(x) + A(x)\boldsymbol{u} = \boldsymbol{f}_1(x), \quad x \in (0, 1), \quad B_0 \boldsymbol{u}(0) + B_1 \boldsymbol{u}(1) = \boldsymbol{\beta},$$

with the known solution, and for $\lambda = 1$ we obtain our original nonlinear problem (2.61).

The m-TDS for the problem (2.61) is of the form

$$\boldsymbol{y}_j^{(m)}(\lambda) = Y^{(m)j}(x_j; \boldsymbol{y}_{j-1}^{(m)}; \lambda), \quad j = 1, 2, \ldots, N,$$

$$B_0 \boldsymbol{y}_0^{(m)}(\lambda) + B_1 \boldsymbol{y}_N^{(m)}(\lambda) = \boldsymbol{\beta}.$$

The differentiation with respect to λ leads to the BVP

$$\frac{d\boldsymbol{y}_j^{(m)}(\lambda)}{d\lambda} = \frac{\partial \boldsymbol{Y}^{(m)j}(x_j; \boldsymbol{y}_{j-1}^{(m)}; \lambda)}{\partial \lambda} + \frac{\partial \boldsymbol{Y}^{(m)j}(x_j; \boldsymbol{y}_{j-1}^{(m)}; \lambda)}{\partial \boldsymbol{u}} \frac{d\boldsymbol{y}_j^{(m)}(\lambda)}{d\lambda},$$

$$j = 1, 2, \ldots, N,$$

$$B_0 \frac{d\boldsymbol{y}_0^{(m)}(\lambda)}{d\lambda} + B_1 \frac{d\boldsymbol{y}_N^{(m)}(\lambda)}{d\lambda} = \boldsymbol{0},$$

which can be further reduced to the following system of linear algebraic equations for the unknown function $\boldsymbol{v}_0^{(m)}(\lambda) \stackrel{\text{def}}{=} d\boldsymbol{y}_0^{(m)}(\lambda)/d\lambda$:

$$[B_0 + B_1 \tilde{S}]\, \boldsymbol{v}_0^{(m)}(\lambda) = -B_1 \tilde{\boldsymbol{\varphi}},$$

where

$$\tilde{S} \stackrel{\text{def}}{=} \tilde{S}_N \tilde{S}_{N-1} \cdots \tilde{S}_1, \quad \tilde{S}_j \stackrel{\text{def}}{=} \frac{\partial \boldsymbol{Y}^{(m)j}(x_j; \boldsymbol{y}_{j-1}^{(m)}; \lambda)}{\partial \boldsymbol{u}}, \quad j = 1, 2, \ldots, N,$$

$$\tilde{\boldsymbol{\varphi}} \stackrel{\text{def}}{=} \tilde{\boldsymbol{\varphi}}_N, \quad \tilde{\boldsymbol{\varphi}}_0 = \boldsymbol{0}, \quad \tilde{\boldsymbol{\varphi}}_j = \tilde{S}_j \tilde{\boldsymbol{\varphi}}_{j-1} + \frac{\partial \boldsymbol{Y}^{(m)j}(x_j; \boldsymbol{y}_{j-1}^{(m)}; \lambda)}{\partial \lambda}, \quad j = 1, 2, \ldots, N.$$

Moreover, for $\boldsymbol{v}_j^{(m)}(\lambda) = d\boldsymbol{y}_j^{(m)}(\lambda)/d\lambda$ we have the formulas

$$\boldsymbol{v}_j^{(m)}(\lambda) = \tilde{S}_j \tilde{S}_{j-1} \cdots \tilde{S}_1 \boldsymbol{v}_0^{(m)} + \tilde{\boldsymbol{\varphi}}_j, \quad j = 1, 2, \ldots, N.$$

The initial approach for Newton's methods can now be obtained by

$$\boldsymbol{y}_j^{(m,0)}(\lambda + \triangle\lambda) = \boldsymbol{y}_j^{(m)}(\lambda) + \triangle\lambda \boldsymbol{v}_j^{(m)}(\lambda), \quad j = 0, 1, \ldots, N. \tag{2.62}$$

The following three examples illustrate the behaviour of our m-TDS.

Example 2.4. Let us apply the m-TDS to Troesch's test problem (2.24). It takes the generic form (2.61) with

$$\boldsymbol{u} = \begin{pmatrix} u_1 \\ u_2 \end{pmatrix}, \quad A = \begin{pmatrix} 0 & -1 \\ 0 & 0 \end{pmatrix}, \quad \boldsymbol{h}(x, \boldsymbol{u}; \lambda) = \begin{pmatrix} 0 \\ \lambda \sinh(\lambda u_1) \end{pmatrix},$$

$$B_0 = \begin{pmatrix} 1 & 0 \\ 0 & 0 \end{pmatrix}, \quad B_1 = \begin{pmatrix} 0 & 0 \\ 1 & 0 \end{pmatrix}, \quad \boldsymbol{\beta} = \begin{pmatrix} 0 \\ 1 \end{pmatrix}.$$

The corresponding m-TDS is

$$\boldsymbol{y}_j^{(m)} = \boldsymbol{Y}^{(m)j}(x_j; \boldsymbol{y}_{j-1}^{(m)}), \quad j = 1, 2, \dots N,$$

$$B_0 \boldsymbol{y}_0^{(m)} + B_1 \boldsymbol{y}_N^{(m)} = \boldsymbol{\beta},$$

where the following Taylor series IVP-solver has been used:

$$\boldsymbol{Y}^{(m)j}(x_j; \boldsymbol{y}_{j-1}^{(m)}) = \boldsymbol{y}_{j-1}^{(m)} + h_j \boldsymbol{g}(x_{j-1}, \boldsymbol{y}_{j-1}^{(m)}) + \sum_{p=2}^{m} \frac{h_j^p}{p!} \left. \frac{d^p \boldsymbol{Y}^j(x; \boldsymbol{y}_{j-1}^{(m)})}{dx^p} \right|_{x=x_{j-1}},$$

$$\boldsymbol{y}_j^{(m)} = \begin{pmatrix} y_{1,j}^{(m)} \\ y_{2,j}^{(m)} \end{pmatrix}, \quad \boldsymbol{g}(x, \boldsymbol{u}) = -A\boldsymbol{u} + \boldsymbol{f}(x, \boldsymbol{u}) = \begin{pmatrix} -u_2 \\ \lambda \sinh(\lambda u_1) \end{pmatrix}. \tag{2.63}$$

We now show how an algorithm for the computation of $\boldsymbol{Y}^{(m)j}(x_j; \boldsymbol{y}_{j-1}^{(m)})$ can be developed which is based on the formula in (2.63). Setting

$$Y_{1,p} \stackrel{\text{def}}{=} \frac{1}{p!} \left. \frac{d^p Y_1^j(x, \boldsymbol{y}_{j-1}^{(m)})}{dx^p} \right|_{x=x_{j-1}},$$

we get

$$\frac{1}{p!} \left. \frac{d^p \boldsymbol{Y}^j(x; \boldsymbol{y}_{j-1}^{(m)})}{dx^p} \right|_{x=x_{j-1}} = \begin{pmatrix} Y_{1,p} \\ (p+1) Y_{1,p+1} \end{pmatrix}$$

and it can be seen that in order to compute the vectors $\left. \frac{1}{p!} \frac{d^p \boldsymbol{Y}^j(x; \boldsymbol{y}_{j-1}^{(m)})}{dx^p} \right|_{x=x_{j-1}}$ it is sufficient to find $Y_{1,p}$ as the Taylor coefficients of the function $Y_1^j(x; \boldsymbol{y}_{j-1}^{(m)})$ at the point $x = x_{j-1}$. This function satisfies the IVP

$$\begin{aligned} &\frac{d^2 Y_1^j(x; \boldsymbol{y}_{j-1}^{(m)})}{dx^2} = \lambda \sinh\left(\lambda Y_1^j(x; \boldsymbol{y}_{j-1}^{(m)})\right), \\ &Y_1^j(x_{j-1}; \boldsymbol{y}_{j-1}^{(m)}) = y_{1,j-1}^{(m)}, \quad \frac{dY_1^j(x_{j-1}; \boldsymbol{y}_{j-1}^{(m)})}{dx} = y_{2,j-1}^{(m)}. \end{aligned} \tag{2.64}$$

Let

$$\tilde{r}(x) \stackrel{\text{def}}{=} \sinh\left(\lambda Y_1^j(x; \boldsymbol{y}_{j-1}^{(m)})\right) = \sum_{i=0}^{\infty} (x - x_{j-1})^i R_i.$$

Substituting this series into the ODE (2.64) we get

$$Y_{1,i+2} = \frac{\lambda R_i}{(i+1)(i+2)}.$$

Setting

$$\tilde{p}(x) \stackrel{\text{def}}{=} \lambda Y_1^j(x; \boldsymbol{y}_{j-1}^{(m)}) = \sum_{i=0}^{\infty} (x - x_{j-1})^i P_i,$$

we have

$$\tilde{r}(x) = \sinh(\tilde{p}(x)), \quad \tilde{s}(x) \stackrel{\text{def}}{=} \cosh(\tilde{p}(x)) = \sum_{i=0}^{\infty} (x - x_{j-1})^i S_i.$$

Performing the simple transformations

$$\tilde{r}' = \cosh(\tilde{p})\, \tilde{p}' = \tilde{p}' \tilde{s}, \quad \tilde{s}' = \sinh(\tilde{p})\, \tilde{p}' = \tilde{p}' \tilde{r}$$

and applying formula (8.20b) from [31], we obtain the recurrence equations

$$R_i = \frac{1}{i} \sum_{k=0}^{i-1} (i-k) S_k P_{i-k}, \quad S_i = \frac{1}{i} \sum_{k=0}^{i-1} (i-k) R_k P_{i-k}, \quad i = 1, 2, \ldots,$$

$$P_i = \lambda Y_{1,i}, \quad i = 2, 3, \ldots.$$

The corresponding initial conditions are

$$P_0 = \lambda\, y_{1,j-1}^{(m)}, \quad P_1 = \lambda\, y_{2,j-1}^{(m)}, \quad R_0 = \sinh(\lambda\, y_{1,j-1}^{(m)}), \quad S_0 = \cosh(\lambda\, y_{1,j-1}^{(m)}).$$

The Jacobian is given by

$$\frac{\partial \boldsymbol{Y}^{(m)j}(x_j; \boldsymbol{y}_{j-1}^{(m)})}{\partial \boldsymbol{u}} = I + h_j \begin{pmatrix} 0 & 1 \\ \lambda^2 \cosh(\lambda\, y_{1,j-1}^{(m)}) & 0 \end{pmatrix}$$
$$+ \sum_{p=2}^{m} \frac{h_j^p}{p!} \begin{pmatrix} Y_{1,p,u_1} & Y_{1,p,u_2} \\ (p+1) Y_{1,p+1,u_1} & (p+1) Y_{1,p+1,u_2} \end{pmatrix},$$

with

$$\boldsymbol{u} = \begin{pmatrix} u_1 \\ u_2 \end{pmatrix}, \quad Y_{1,p,u_l} \stackrel{\text{def}}{=} \frac{\partial Y_{1,p}(x_j; \boldsymbol{y}_{j-1}^{(m)})}{\partial u_l}, \quad l = 1, 2.$$

Since the functions $Y_{1,u_l}(x; \boldsymbol{y}_{j-1}^{(m)}) \stackrel{\text{def}}{=} \dfrac{\partial Y_1(x; \boldsymbol{y}_{j-1}^{(m)})}{\partial u_l}$ satisfy the ODEs

$$\frac{d^2 Y_{1,u_1}}{dx^2} = \lambda^2 \cosh(\tilde{p}(x))(1 + Y_{1,u_1}), \quad \frac{d^2 Y_{1,u_2}}{dx^2} = \lambda^2 \cosh(\tilde{p}(x))((x - x_{j-1}) + Y_{1,u_2}),$$

we get the following recursive algorithm for the computation of Y_{1,p,u_l}:

$$Y_{1,i+2,u_1} = \frac{\lambda^2}{(i+1)(i+2)}\left[S_i + \sum_{k=2}^{i} Y_{1,k,u_1} S_{i-k}\right], \quad i = 2, 3, \ldots,$$

$$Y_{1,2,u_1} = \frac{\lambda^2 S_0}{2}, \quad Y_{1,3,u_1} = \frac{\lambda^2 S_1}{6},$$

$$Y_{1,i+2,u_2} = \frac{\lambda^2}{(i+1)(i+2)}\left[S_{i-1} + \sum_{k=2}^{i} Y_{1,k,u_2} S_{i-k}\right], \quad i = 2, 3, \ldots,$$

$$Y_{1,2,u_2} = 0, \quad Y_{1,3,u_2} = \frac{\lambda^2 S_0}{6}.$$

For the vector $\partial \boldsymbol{Y}^{(m)j}(x_j; \boldsymbol{y}_{j-1}^{(m)}; \lambda)/\partial\lambda$ we have the formula

$$\frac{\partial \boldsymbol{Y}^{(m)j}(x_j; \boldsymbol{y}_{j-1}^{(m)}; \lambda)}{\partial \lambda}$$

$$= h_j \begin{pmatrix} 0 \\ \sinh(\lambda\, y_{1,j-1}^{(m)}) + \lambda\, y_{1,j-1}^{(m)} \cosh(\lambda\, y_{1,j-1}^{(m)}) \end{pmatrix} + \sum_{p=2}^{m} \frac{h_j^p}{p!} \begin{pmatrix} Y_{1,p,\lambda} \\ (p+1) Y_{1,p+1,\lambda} \end{pmatrix},$$

where

$$Y_{1,p,\lambda} \stackrel{\text{def}}{=} \frac{\partial Y_{1,p}(x_j; \boldsymbol{y}_{j-1}^{(m)}; \lambda)}{\partial \lambda}.$$

Taking into account that $Y_{1,\lambda}^j(x_j; \boldsymbol{y}_{j-1}^{(m)}; \lambda) \stackrel{\text{def}}{=} \partial Y_1^j(x_j; \boldsymbol{y}_{j-1}^{(m)}; \lambda)/\partial\lambda$ satisfies the ODE

$$\frac{d^2 Y_{1,\lambda}^j}{dx^2} = \lambda^2 \cosh(\tilde{p}(x)) Y_{1,\lambda}^j + \sinh(\tilde{p}(x)) + \tilde{p}(x)\cos(\tilde{p}(x)),$$

we obtain for $Y_{1,p,\lambda}$ the recurrence relation

$$Y_{1,i+2,\lambda} = \frac{1}{(i+1)(i+2)}\left[\lambda^2 \sum_{k=2}^{i} Y_{1,k,\lambda} S_{i-k} + R_i + \sum_{k=0}^{i} P_k S_{i-k}\right], \quad i = 2, 3, \ldots,$$

$$Y_{1,2,\lambda} = \frac{1}{2}\left(R_0 + P_0 S_0\right), \quad Y_{1,3,\lambda} = \frac{1}{6}\left(R_1 + P_0 S_1 + P_1 S_0\right).$$

Considering the behavior of the solution, we choose the grid

$$\widehat{\omega}_h = \left\{ x_j = \frac{\exp(j\alpha/N) - 1}{\exp(\alpha) - 1},\ j = 0, 1, 2, \ldots, N \right\}, \quad \alpha < 0, \tag{2.65}$$

which becomes dense for $x \to 1$. The step sizes of this grid are given by $h_1 = x_1$ and $h_{j+1} = h_j \exp(\alpha/N)$, $j = 1, 2, \ldots, N-1$. Note that the use of the formula

$h_j = x_j - x_{j-1}$, $j = 1, 2, \ldots, N$, for $j \to N$ and $|\alpha|$ large enough, leads to a large absolute roundoff error since some x_j, x_{j-1} lie very close together.

The a posteriori Runge estimator was used to arrive at the right boundary with a given tolerance ε. Thus, the tolerance was assumed to be achieved if the following inequality is fulfilled:

$$\max\left\{\left\|\frac{y_N^{(m)} - y_{2N}^{(m)}}{\max\left(\left|y_{2N}^{(m)}\right|, 10^{-5}\right)}\right\|_{0,\infty,\hat{\omega}_h}, \left\|\frac{\dfrac{dy_N^{(m)}}{dx} - \dfrac{dy_{2N}^{(m)}}{dx}}{\max\left(\left|\dfrac{dy_{2N}^{(m)}}{dx}\right|, 10^{-5}\right)}\right\|_{0,\infty,\hat{\omega}_h}\right\} \leq (2^m - 1)\,\varepsilon.$$

Otherwise a doubling of the number of the grid points has been made. Here, $y_N^{(m)}$ denotes the solution of the difference scheme of the order of accuracy m on the grid $\{x_0, \ldots, x_N\}$, and $y_{2N}^{(m)}$ denotes the solution of the same scheme on the finer grid $\{x_0, \ldots, x_{2N}\}$. The difference scheme (represented by a system of nonlinear algebraic equations) was solved by Newton's method with the stopping criterion

$$\max\left\{\left\|\frac{y^{(m,n)} - y^{(m,n-1)}}{\max\left(\left|y^{(m,n)}\right|, 10^{-5}\right)}\right\|_{0,\infty,\hat{\omega}_h}, \left\|\frac{\dfrac{dy^{(m,n)}}{dx} - \dfrac{dy^{(m,n-1)}}{dx}}{\max\left(\left|\dfrac{dy^{(m,n)}}{dx}\right|, 10^{-5}\right)}\right\|_{0,\infty,\hat{\omega}_h}\right\} \leq 0.5\,\varepsilon,$$

where n denotes the number of iterations.

Setting the unknown value of the first derivative of the exact solution of (2.24) at the left boundary $x = 0$ equal to s, this solution can be written in the form (see, for example [79])

$$u(x; s) = \frac{2}{\lambda} \operatorname{arcsinh}\left(\frac{s \cdot \operatorname{sn}(\lambda x, k)}{2 \cdot \operatorname{cn}(\lambda x, k)}\right), \quad k^2 = 1 - \frac{s^2}{4}, \tag{2.66}$$

where $\operatorname{sn}(\lambda x, k)$ and $\operatorname{cn}(\lambda x, k)$ denote the elliptic Jacobi functions. The parameter s is the solution of the equation

$$\frac{2}{\lambda} \operatorname{arcsinh}\left(\frac{s \cdot \operatorname{sn}(\lambda, k)}{2 \cdot \operatorname{cn}(\lambda, k)}\right) = 1.$$

For example, for $\lambda = 5$ one gets $s = 0.457504614063 \cdot 10^{-1}$, and for $\lambda = 10$ one gets $s = 0.35833778463 \cdot 10^{-3}$. Using the continuation method (2.62) we have computed numerical solutions of Troesch's problem (2.24) for $\lambda \in [1, 62]$ using different step sizes $\Delta\lambda$. The numerical results for $\lambda = 10, 20, 30, 40, 45, 50$ and 61 computed with the difference scheme of order of accuracy 7 on the grid (2.65) with $\alpha = -26$ are given in Table 2.1.

Here, CPU denotes the time required by the processor to solve the sequence of Troesch's problems beginning with $\lambda = 1$ and using the step size $\Delta\lambda$ until the parameter λ reaches the value given in the table.

λ	ε	N	$u'(0)$	$u(1)$	CPU (sec)
10	10^{-7}	512	$3.5833778 \cdot 10^{-4}$	1	0.02
20	10^{-7}	512	$1.6487734 \cdot 10^{-8}$	1	0.04
30	10^{-7}	512	$7.4861194 \cdot 10^{-13}$	1	0.07
40	10^{-7}	512	$3.3987988 \cdot 10^{-17}$	1	0.10
45	10^{-7}	512	$2.2902091 \cdot 10^{-19}$	1	0.11
50	10^{-7}	1024	$1.5430022 \cdot 10^{-21}$	1	0.15
61	10^{-7}	262144	$2.5770722 \cdot 10^{-26}$	1	6.10

Table 2.1: Numerical results for the 7-TDS and $\triangle\lambda = 4$

The numerical results for $\lambda = 61$ and 62 computed on the same grid as before with the difference scheme of order of accuracy 10 are given in Table 2.2.

λ	ε	N	Error	CPU (sec)
61	10^{-6}	65536	$0.860 \cdot 10^{-5}$	3.50
61	10^{-8}	131072	$0.319 \cdot 10^{-7}$	7.17
62	10^{-6}	262144	$0.232 \cdot 10^{-5}$	8.01
62	10^{-8}	262144	$0.675 \cdot 10^{-8}$	15.32

Table 2.2: Numerical results for the 10-TDS and $\triangle\lambda = 2$

The real deviation from the exact solution is measured by

$$\texttt{Error} \overset{\text{def}}{=} \max\left\{ \left\| \frac{y^{(m)} - u}{\max\left(|y^{(m)}|, 10^{-5}\right)} \right\|_{0,\infty,\hat{\omega}_h}, \left\| \frac{\dfrac{dy^{(m)}}{dx} - \dfrac{du}{dx}}{\max\left(\left|\dfrac{dy^{(m)}}{dx}\right|, 10^{-5}\right)} \right\|_{0,\infty,\hat{\omega}_h} \right\}.$$

The numerical experiments were carried out with double precision in FORTRAN. To calculate the Jacobi functions $\mathrm{sn}(x, k)$ and $\mathrm{cn}(x, k)$ for large $|x|$ the computer algebra tool MAPLE VII with Digits=80 was used. Then, the exact solution on the grid $\hat{\omega}_h$ and two approximations for the parameter s, namely $s = 0.2577072228793720338185 \cdot 10^{-25}$ satisfying $|u(1, s) - 1| < 0.17 \cdot 10^{-10}$, and $s = 0.948051891387119532089349753 \cdot 10^{-26}$ satisfying $|u(1, s) - 1| < 0.315 \cdot 10^{-15}$ were calculated.

To compare the results we have solved problem (2.24) with the multiple shooting code RWPM (see e.g. [34] or [84]). For the parameter values $\lambda = 10, 20, 30, 40$ the numerical IVP-solver used was the code `RKEX78`, an implementation of the Dormand-Prince embedded Runge-Kutta method 7(8), whereas for $\lambda = 45$ we have used the code `BGSEXP`, an implementaton of the well-known Bulirsch-Stoer-Gragg extrapolation method. In Table 2.3 we denote by m the number of the automatically determined shooting points, `NFUN` is the number of ODE calls, `it` the number of iterations and `CPU` the CPU time used.

λ	m	it	NFUN	$u'(0)$	$u(1)$	CPU (sec)
10	11	9	12641	$3.5833779 \cdot 10^{-4}$	1.0000000	0.01
20	11	13	34425	$1.6487732 \cdot 10^{-8}$	0.9999997	0.02
30	14	16	78798	$7.4860938 \cdot 10^{-13}$	1.0000008	0.05
40	15	24	172505	$3.3986834 \cdot 10^{-17}$	0.9999996	0.14
45	12	31	530085	$2.2900149 \cdot 10^{-19}$	1.0000003	0.30

Table 2.3: Numerical results for the code RWPM

One can see that the accuracy characteristics of our m-TDS method is better than that of the code RWPM. Besides, RWPM fails for values $\lambda \geq 50$. □

Example 2.5. As a next example let us consider the following BVP for a system of stiff ODEs (see [78])

$$
\begin{aligned}
&u_1' = \lambda \frac{(u_3 - u_1)u_1}{u_2}, \quad u_2' = -\lambda(u_3 - u_1),\\
&u_3' = \frac{0.9 - 10^3(u_3 - u_5) - \lambda(u_3 - u_1)u_3}{u_4},\\
&u_4' = \lambda(u_3 - u_1), \quad u_5' = -100(u_5 - u_3), \quad 0 < x < 1,\\
&u_1(0) = u_2(0) = u_3(0) = 1, \quad u_4(0) = -10, \quad u_3(1) = u_5(1).
\end{aligned}
\tag{2.67}
$$

To solve this problem numerically we have applied the 6-TDS given by

$$
\begin{aligned}
&\boldsymbol{y}_j^{(6)} = \boldsymbol{Y}^{(6)j}(x_j; \boldsymbol{y}_{j-1}^{(6)}), \quad j = 1, 2, \ldots N,\\
&B_0 \boldsymbol{y}_0^{(6)} + B_1 \boldsymbol{y}_N^{(6)} = \boldsymbol{\beta},
\end{aligned}
\tag{2.68}
$$

where $\boldsymbol{Y}^{(6)j}(x_j; \boldsymbol{y}_{j-1}^{(6)})$ is the numerical solution of the IVP (2.25) computed by a 7-stage explicit Runge-Kutta method (2.41) of order $m = 6$. The corresponding parameters are given in Table 2.4 (see also [7]).

$$
\begin{array}{c|c}
c & A \\
\hline
 & b^T
\end{array}
\quad \Longleftrightarrow \quad
\begin{array}{c|ccccccc}
0 & 0 & 0 & 0 & 0 & 0 & 0 & 0 \\
\frac{1}{2} & \frac{1}{2} & 0 & 0 & 0 & 0 & 0 & 0 \\
\frac{2}{3} & \frac{2}{9} & \frac{4}{9} & 0 & 0 & 0 & 0 & 0 \\
\frac{1}{3} & \frac{7}{36} & \frac{2}{9} & -\frac{1}{12} & 0 & 0 & 0 & 0 \\
\frac{5}{6} & -\frac{35}{144} & -\frac{55}{36} & \frac{35}{48} & \frac{15}{8} & 0 & 0 & 0 \\
\frac{1}{6} & -\frac{1}{360} & -\frac{11}{36} & -\frac{1}{8} & \frac{1}{2} & \frac{1}{10} & 0 & 0 \\
1 & -\frac{41}{260} & \frac{22}{13} & \frac{43}{156} & -\frac{118}{39} & \frac{32}{195} & \frac{80}{39} & 0 \\
\hline
 & \frac{13}{200} & 0 & \frac{11}{40} & \frac{11}{40} & \frac{4}{25} & \frac{4}{25} & \frac{13}{200}
\end{array}
$$

Table 2.4: Butcher matrix for the RK method used in Example 2.5

In Newton's method (2.59) we have approximated the matrix $\partial Y^{(6)j}(x_j, y_{j-1}^{(6)})/\partial u$ in the following way:

$$\frac{\partial \boldsymbol{Y}^{(6)j}(x_j; \boldsymbol{y}_{j-1}^{(6)})}{\partial \boldsymbol{u}} \approx I + h_j \frac{\partial \boldsymbol{g}(x_{j-1}, \boldsymbol{y}_{j-1}^{(6)})}{\partial \boldsymbol{u}}. \tag{2.69}$$

Numerical results on the equidistant grid (1.1) obtained by the 6-TDS are given in Table 2.5.

ε	NFUN	CPU (sec)
10^{-4}	24500	0.01
10^{-6}	41440	0.02
10^{-8}	77140	0.04

Table 2.5: Numerical results for the TDS with $m = 6$ ($\lambda = 100$)

We have also solved problem (2.67) by the multiple shooting code RWPM so that a comparison can be made. As an IVP-solver we used the semi-implicit extrapolation method SIMPR. A start trajectory was generated by solving the problem with

$\lambda = 0$. The numerical results are given in Table 2.6, where CPU^* denotes the cumulated time for the solution of the linear problem ($\lambda = 0$) and the solution of (2.67) with $\lambda = 100$.

□

ε	NFUN	CPU*
10^{-4}	15498	0.02
10^{-6}	31446	0.04
10^{-8}	52374	0.06

Table 2.6: Numerical results for the code RWPM ($\lambda = 100$)

Example 2.6. Our third and final example is the periodic BVP (see [70])

$$\begin{aligned} u'' = &- 0.05u' - 0.02u^2 \sin x + 0.00005 \sin x \cos^2 x \\ &- 0.05 \cos x - 0.0025 \sin x, \quad x \in (0, 2\pi), \\ u(0) &= u(2\pi), \quad u'(0) = u'(2\pi). \end{aligned} \tag{2.70}$$

It has the exact solution $u(x) = 0.05 \cos x$.

Numerical results on the equidistant grid (1.1) obtained by the 6-TDS (2.68)–(2.69) with

$$\boldsymbol{y}_j^{(6)} = \begin{pmatrix} y_{1,j}^{(6)} \\ y_{2,j}^{(6)} \end{pmatrix}, \quad B_0 = \begin{pmatrix} 1 & 0 \\ 0 & 1 \end{pmatrix}, \quad B_1 = \begin{pmatrix} -1 & 0 \\ 0 & -1 \end{pmatrix}, \quad \boldsymbol{\beta} = \begin{pmatrix} 0 \\ 0 \end{pmatrix},$$

$$\boldsymbol{g}(x, \boldsymbol{u}) = \begin{pmatrix} -u_2 \\ -0.05(u_2 + \cos x) + \sin x (0.00005 \cos^2 x - 0.02 u_1^2 - 0.0025) \end{pmatrix}$$

are given in Table 2.7.

□

2.4 A posteriori error estimation and automatic grid generation

There are various strategies to construct an algorithm which automatically generates a grid for which the norm of the error of the approximate solution is smaller than a

ε	N	NFUN	Error
10^{-4}	64	5712	$0.453 \cdot 10^{-8}$
10^{-6}	64	5712	$0.453 \cdot 10^{-8}$
10^{-8}	128	12432	$0.267 \cdot 10^{-11}$

Table 2.7: Numerical results for the TDS with $m = 6$

given tolerance ε. We will shortly discuss only the following two possibilities based on the theory developed in the previous sections.

The first possibility is the classical technique which has been proposed by Carl Runge. The estimate (2.48) via standard considerations leads to the following *a posteriori* h-$h/2$-strategy to generate an approximation with an error whose norm is smaller than a given tolerance ε (for a fixed m). Let $\boldsymbol{y}_N^{(m)}$ be the solution of the difference scheme of order of accuracy m on the grid $\{x_0, \ldots, x_N\}$, and $\boldsymbol{y}_{2N}^{(m)}$ the solution of this scheme on the grid $\{x_0, \ldots, x_{2N}\}$. Then, the inequality

$$\left\|\boldsymbol{y}_N^{(m)} - \boldsymbol{y}_{2N}^{(m)}\right\|_{0,\infty,\widehat{\omega}_h} \leq (2^m - 1)\varepsilon \tag{2.71}$$

implies $\|\boldsymbol{y}_{2N}^{(m)} - \{\boldsymbol{u}(x_j)\}_{j=0}^{2N}\|_{0,\infty,\widehat{\overline{\omega}}_h} \leq \varepsilon$. The approximation $\boldsymbol{y}_N^{(m)}$ can be further improved by applying the Richardson extrapolation formula

$$\widehat{\boldsymbol{y}}_N^{(m)}(x_j) = \boldsymbol{y}_{2N}^{(m)}(x_{2j}) + \frac{\boldsymbol{y}_{2N}^{(m)}(x_{2j}) - \boldsymbol{y}_N^{(m)}(x_j)}{2^m - 1}, \quad j = 0, 1, \ldots, N. \tag{2.72}$$

The main drawback of this strategy is that the grid for the difference scheme (2.38), (2.39) can be equidistant or quasi-uniform only.

The second approach to automatic grid generation is based on the following simple idea. Due to (2.48) the difference scheme (2.38), (2.39) has order of accuracy m which is an integer number. In order to obtain TDS of the orders m and $m+1$ by our method one should solve the IVPs (2.25) by one-step methods (2.40) of the corresponding orders. A reasonable and practical way to do that is to use two representatives of orders m and $m+1$ from a family of embedded Runge-Kutta methods (e. g. Runge-Kutta-Fehlberg, Runge-Kutta Verner or Runge-Kutta Dormand-Prince family; see [8]). An *a posteriori* error estimate for the difference scheme (2.38), (2.39) can be determined by computing an approximation $\boldsymbol{y}^{(m)}$ with the Runge-Kutta method of order m and an approximation $\boldsymbol{y}^{(m+1)}$ with the Runge-Kutta method of order $m+1$. The requested *a posteriori* error estimate is now $\|\boldsymbol{y}^{(m)} - \boldsymbol{y}^{(m+1)}\|_{0,\infty,\hat{\omega}_h}$ since it holds that

$$\left\|\boldsymbol{y}^{(m)} - \boldsymbol{y}^{(m+1)}\right\|_{0,\infty,\widehat{\overline{\omega}}_h} = \left\|\boldsymbol{y}^{(m)} - \boldsymbol{u}\right\|_{0,\infty,\widehat{\overline{\omega}}_h} + O\left(h^{m+1}\right). \tag{2.73}$$

Hence, provided that $\left\|\boldsymbol{y}^{(m)}-\boldsymbol{y}^{(m+1)}\right\|_{0,\infty,\widehat{\omega}_h}<\varepsilon$, we can compute an approximate solution of problem (2.1) by the difference scheme (2.38), (2.39) whose error is bounded by a prescribed tolerance ε.

The following algorithm generates a non-equidistant grid $\hat{\hat{\omega}}_h$ and computes an approximate solution of the IVPs (2.25) on this grid. The estimate (2.48) and the usual relation for an IVP-solver of order of accuracy m,

$$\boldsymbol{Y}^j(x_j,\boldsymbol{y}_{j-1})-\boldsymbol{Y}^{(m)j}(x_j,\boldsymbol{y}_{j-1})=\boldsymbol{\psi}^j(x_j,\boldsymbol{y}_{j-1})h_j^{m+1}+O(h_j^{m+2}), \qquad (2.74)$$

guarantee that the error of the approximate solution of the BVP (2.1) is within a given tolerance ε, provided that for each $j\in\{0,\dots,N\}$ the solutions of the IVPs (2.25) are given with tolerance $h_j\,\varepsilon$. Using the well-known idea of step-size control by embedded Runge-Kutta methods we can then construct a non-equidistant grid $\hat{\hat{\omega}}_h$ such that the IVPs (2.25) are solved with tolerance $h_j\,\varepsilon$. Having in mind the IVPs (2.25), the error of a Runge-Kutta method of order m is is given by (2.74) too, i.e., for the embedded Runge-Kutta methods of orders m and $m+1$ we have the following *a posteriori* estimate (neglecting the terms of order $O(h_j^{m+1})$)

$$\boldsymbol{Y}^{(m+1)j}(x_j,\boldsymbol{y}_{j-1})-\boldsymbol{Y}^{(m)j}(x_j,\boldsymbol{y}_{j-1})\approx\boldsymbol{\psi}^j(x_j,\boldsymbol{y}_{j-1})h_j^{m+1}. \qquad (2.75)$$

Thus if the condition $\left\|\boldsymbol{Y}^{(m)j}(x_j,\boldsymbol{y}_{j-1})-\boldsymbol{Y}^{(m+1)j}(x_j,\boldsymbol{y}_{j-1})\right\|<h_j\varepsilon$ is satisfied, then the solutions of the IVPs can be determined on the interval $[x_{j-1},x_j]$ within the prescribed tolerance. If this is not the case the actual step-size is halved.

A doubling of the step-size h_j changes the essential term of the error as follows:

$$2^{m+1}\boldsymbol{\psi}^j(x_j,\boldsymbol{y}_{j-1})h_j^{m+1}. \qquad (2.76)$$

If the approximate solutions could be determined on the previous subinterval with the prescribed error tolerance, then it would be possible to double the step-size of the next subinterval. In that case the formula (2.76) should be applied. A detailed description of the step-size control by embedded Runge-Kutta methods can be found in the book [35]. To simplify the presentation, the Runge-Kutta code which is based on a pair of Runge-Kutta methods of orders l and $l+1$ and controls the step-size by the embedding principle is denoted by $\mathtt{RK}(l)(l+1)$.

Algorithm AG ($\hat{\omega}_N$,TOL,m,h_1,RK(l)(l+1), $A(x)$,$\boldsymbol{f}$,$\boldsymbol{u}$)

Input: An error tolerance TOL, the order of accuracy m of the TDS, an initial step-size h_1, a Runge-Kutta code RK(l)($l+1$), the problem data $A(x)$, $\boldsymbol{f}$ as well as the initial values $\boldsymbol{y}_{j-1}$ for the IVPs (2.25).

Output: A non-equidistant grid $\hat{\omega}_N$ on which the numerical solution of (2.25),

$$\boldsymbol{Y}^{(l)j}(x_j, \boldsymbol{y}_{j-1}^{(l)}) \quad l = m, m+1, \quad x_j \in \hat{\omega}_N, \quad j = 1, \dots, N,$$

is determined with the given tolerance h_j TOL.

1) Set $j := 0$, $x_0 := 0$.
2) Set $j := j + 1$.
3) Solve the IVPs (2.25) on the interval $[x_{j-1}, x_j]$ using RK(m)($m+1$) and compute
$$e := \left\| \boldsymbol{Y}^{(m)j}(x_j, \boldsymbol{y}_{j-1}) - \boldsymbol{Y}^{(m+1)j}(x_j, \boldsymbol{y}_{j-1}) \right\|.$$
4) **if** $e > h_j$ TOL **then begin** $h_j := h_j/2$; go to Step 3 **end.**
5) Set $x_j := x_{j-1} + h_j$.
6) **if** $x_j < 1$ **then begin**
   if $2^{m+1}e \leq h_j$ TOL **then** $h_{j+1} := 2h_j$ **else** $h_{j+1} := h_j$;
   go to Step 2 **end.**
7) Set $N := j$.
8) **if** $x_N > 1$ **then begin** $h_N := 1 - x_{N-1}$; $x_N := 1$;
   Solve the IVPs (2.25) on the interval $[x_{N-1}, x_N]$ using RK(m)($m+1$) **end.**
9) **Stop.**

The estimation (2.48) is the basis of the following algorithm. Given the grid $\hat{\omega}_N$, it computes an approximation $\boldsymbol{y}_j^{(m+1)}$, $j = 1, \dots, N$, of the BVP (2.1) by the m-TDS (2.38), (2.39).

Algorithm A($\boldsymbol{y}^{(m+1)}$,$\hat{\hat{\omega}}_N$,TOL,m,h_1, RK(l)($l+1$),$A(x)$,$\boldsymbol{f}$)

Input: The grid $\hat{\hat{\omega}}_N$, an error tolerance TOL, the order of accuracy m of the TDS, an initial step-size h_1, a family of embedded Runge-Kutta codes RK(l)($l+1$) as well as the problem data $A(x)$ and $\boldsymbol{f}$ of problem (2.1).

Output: The solution $\boldsymbol{y}^{(m+1)}(x_j)$, $x_j \in \hat{\hat{\omega}}_N$, of the BVP (2.1) with an error smaller than the given tolerance TOL.

1) Set $\varepsilon_{it} := 0.25\,\mathtt{TOL}$.

2) Determine the starting values $\boldsymbol{u}^{(0)}(x)$.

3) Compute the grid $\hat{\hat{\omega}}_N$ and the numerical solution of (2.25) using the algorithm
   AG($\hat{\hat{\omega}}_N$,TOL,m,h_1,RK(m)($m+1$),$A(x)$,$\boldsymbol{f}$, $\boldsymbol{u}^{(0)}$).
   Set $n := 0, \quad l := m$.

4) Set $n := n+1, \quad m := l$.

5) Determine $\boldsymbol{y}_0^{(l,n)}$ by solving the system of linear algebraic equations (2.60) and compute $\boldsymbol{y}_j^{(l,n)}$, $j = 1, \ldots, N$, by the second formula in (2.60).

6) Compute $\boldsymbol{Y}^{(l)j}(x_j, \boldsymbol{y}_{j-1}^{(l)})$, $j = 1, \ldots, N$, by the Runge-Kutta code RK(l)($l+1$).

7) **If** $\left\|\boldsymbol{y}^{(l,n)} - \boldsymbol{y}^{(l,n-1)}\right\|_{0,\infty,\hat{\omega}_h} > \varepsilon_{it}$ **then go to** Step 4.

8) **If** $l = m$ **then begin** $n := 0$, $l := m+1$, **go to** Step 4 **end.**

9) **If** $\left\|\boldsymbol{y}^{(m+1)} - \boldsymbol{y}^{(m)}\right\|_{0,\infty,\hat{\omega}_h} >$ TOL **then begin**

   Find new starting values by interpolating the values $\boldsymbol{y}_i^{(m+1)}$, $i = 0, \ldots, N$, using the formula

$$\boldsymbol{u}^{(0)}(x) = \frac{1}{x_i - x_{i-1}} \left[\boldsymbol{y}_i^{(m+1)}(x - x_{i-1}) + \boldsymbol{y}_{i-1}^{(m+1)}(x_i - x)\right],$$

$$x_{j-1} \le x \le x_j, \quad i = 1, \ldots, N,$$

   go to Step 3 **end**.

10) **Stop**.

At the end of this section we will present the results of some experiments with the above algorithms. Let us consider again Troesch's test problem (see Examples 2.2 and 2.4). We have varied the value of the input variable TOL and considered the question of how a reduction of TOL influences the grid and the accuracy of the numerical solution. The corresponding results for Troesch's problem with

$\lambda = 10$ are given in Table 2.8, where `TOL` is the prescribed tolerance, `N` is the number of grid points, `NFUN` denotes the number of calls of the function $\boldsymbol{f}(x, \boldsymbol{u})$ and `Error` is the difference between the numerical and the exact solution (2.66) with $s = 0.35833778463 \cdot 10^{-3}$.

TOL	N	NFUN	Error
10^{-4}	72	11323	$0.143 \cdot 10^{-5}$
10^{-6}	124	19435	$0.538 \cdot 10^{-7}$

Table 2.8: Numerical results for the TDS using a RK7(8)

Similar results have been obtained for the periodic BVP (2.70) (see Example 2.6) and are given in Table 2.9.

TOL	N	NFUN	Error
10^{-6}	19	4940	$0.163 \cdot 10^{-7}$
10^{-8}	19	6422	$0.420 \cdot 10^{-9}$

Table 2.9: Numerical results for the TDS using a RK7(8)

Chapter 3

Three-point difference schemes for monotone second-order ODEs

Our greatest weakness lies in giving up. The most certain way to succeed is always to try just one more time.

Thomas Alva Edison (1847–1931)

In this chapter we consider nonlinear monotone ODEs with Dirichlet boundary conditions. Using a non-equidistant grid we construct an EDS on a three-point stencil and prove the existence and uniqueness of its solution. Moreover, on the basis of the EDS we develop an algorithm for the construction of a three-point TDS of rank $\bar{m} = 2\left[(m+1)/2\right]$, where $m \in \mathbb{N}$ is a given natural number and $[\cdot]$ denotes the entire part of the argument in brackets. We prove the existence and uniqueness of the solution of the TDS and determine the order of accuracy. Numerical examples are given which confirm the theoretical results.

3.1 Existence and uniqueness of a solution

Let us consider the BVP

$$\frac{d}{dx}\left[k(x)\frac{du}{dx}\right] = -f\left(x, u, \frac{du}{dx}\right), \quad x \in (0,1), \quad u(0) = \mu_1, \quad u(1) = \mu_2, \qquad (3.1)$$

where $k(x)$, $f(x, u, \xi)$ are given functions and μ_1, μ_2 are given numbers.

Note that problem (3.1) can be transformed into a system of first-order ODEs but this system is no longer monotone. The EDS for systems of first-order ODEs with a small Lipschitz constant has been constructed and analyzed in [27, 55]. In this section we develop and justify an EDS and the corresponding TDS of an arbitrary given order of accuracy for BVPs with an arbitrary Lipschitz constant.

We say that a function $u(x) \in W_2^1(0,1)$ is a weak solution of problem (3.1), if

$$u(x) - \varphi(x) \in \overset{\circ}{W}{}_2^1(0,1) = \left\{ v(x) \,\middle|\, v(x) \in W_2^1(0,1), v(0) = v(1) = 0 \right\}$$

and for all $v(x) \in \overset{\circ}{W}{}_2^1(0,1)$ the identity

$$\int_0^1 k(x) \frac{du}{dx} \frac{dv}{dx} dx = \int_0^1 f\left(x, u, \frac{du}{dx}\right) v(x) dx$$

holds, where $\varphi(x)$ is a function from $W_2^1(0,1)$ which satisfies the boundary conditions.

Sufficient conditions for the existence of a weak solution of problem (3.1) are given in the next theorem.

Theorem 3.1

Let the following assumptions be satisfied:

$$0 < c_1 \le k(x) \le c_2 \quad \textit{for all } x \in [0,1],\ k(x) \in \mathcal{Q}^1[0,1], \tag{3.2}$$

$$f_{u\xi}(x) \overset{\text{def}}{=} f(x,u,\xi) \in \mathcal{Q}^0[0,1] \quad \textit{for all } u, \xi \in \mathbb{R}, \tag{3.3}$$

$$f_x(u,\xi) \overset{\text{def}}{=} f(x,u,\xi) \in \mathbb{C}(\mathbb{R}^2) \quad \textit{for all } x \in [0,1], \tag{3.4}$$

$$|f(x,u,\xi) - f_0(x)| \le c(|u|)[g(x) + |\xi|] \quad \textit{for all } x \in [0,1],\ u, \xi \in \mathbb{R}, \tag{3.5}$$

$$[f(x,u,\xi) - f(x,v,\eta)](u-v) \le c_3 \left(|u-v|^2 + |\xi - \eta|^2\right) \quad \textit{for all } x \in [0,1],$$

$$u, v, \xi, \eta \in \mathbb{R}, \tag{3.6}$$

$$0 \le c_3 < \frac{\pi^2}{\pi^2+1} c_1. \tag{3.7}$$

Then, the boundary value problem (3.1) has a unique solution $u(x) \in W_2^1(0,1)$, *with* $u(x), k(x)\dfrac{du}{dx} \in \mathbb{C}[0,1]$.

Here $c(t)$ *is a continuous function,* $f_0(x) \in L_2(0,1)$, $g(x) \in L_1(0,1)$, $\mathcal{Q}^p[0,1]$ *is the class of functions having* p *piece-wise continuous derivatives and a finite number of discontinuity points of first kind, and* c_1, c_2, c_3 *are some real constants.*

Proof. Due to (3.3) and (3.5) the function $f(x,u,\xi)$ satisfies the Carathéodory conditions (see Definition 1.12 and belongs to the class $L_1(0,1)$ (see e.g. [18, p.113]).

For all $u(x) \in W_2^1(0,1)$ and $v(x) \in \overset{\circ}{W}{}_2^1(0,1)$ we can now define the operator

$$(A(x,u),v) \overset{\text{def}}{=} \int_0^1 k(x)\frac{du}{dx}\frac{dv}{dx}dx - \int_0^1 \tilde{f}\left(x,u,\frac{du}{dx}\right)v(x)dx,$$

where

$$\tilde{f}(x,u,\xi) \overset{\text{def}}{=} f(x,u,\xi) - f_0(x).$$

Note that the function $u(x) \in W_2^1(0,1)$ is absolutely continuous on $[0,1]$, and its generalized derivative $\dfrac{du}{dx}$ is equal to the classical derivative almost everywhere on $[0,1]$ (see e.g. [18, p.74]). Thus, $u(x) \in \mathbb{C}[0,1]$ and $\dfrac{du}{dx} \in L_2(0,1)$.

Let us show that the operator $A(x,u)$ is bounded. Actually, taking into account the Cauchy-Bunyakovsky-Schwarz inequality (see Theorem 1.8), the conditions (3.2) and (3.5), the inequality $c(|u|) \le C_2$ for all $x \in [0,1]$ as well as $\|v\|_{C[0,1]} \le C_1\|v\|_{1,2,(0,1)}$ for all $v(x) \in W_2^1(0,1)$ (see e.g. [18, p.112]) we obtain

$$\begin{aligned}
&|(A(x,u),v)| \\
&\le \left\{\int_0^1 \left[k(x)\frac{du}{dx}\right]^2 dx\right\}^{1/2} \left\{\int_0^1 \left[\frac{dv}{dx}\right]^2 dx\right\}^{1/2} + \|v\|_{C[0,1]}\int_0^1 \left|\tilde{f}\left(x,u,\frac{du}{dx}\right)\right| dx \\
&\le \left[c_2\|u\|_{1,2,(0,1)} + C_1\left\|\tilde{f}\right\|_{0,1,(0,1)}\right]\|v\|_{1,2,(0,1)} \\
&\le \left[(c_2 + C_1C_2)\|u\|_{1,2,(0,1)} + C_1C_2\|g\|_{0,1,(0,1)}\right]\|v\|_{1,2,(0,1)},
\end{aligned}$$

where

$$\|u\|_{C[0,1]} \overset{\text{def}}{=} \max_{x\in[0,1]}|u(x)|, \quad \|u\|_{0,1,(0,1)} \overset{\text{def}}{=} \int_0^1 |u(x)|dx,$$

$$\|u\|_{0,2,(0,1)} \overset{\text{def}}{=} \left[\int_0^1 (u(x))^2\, dx\right]^{1/2},$$

$$\|u\|_{1,2,(0,1)} \overset{\text{def}}{=} \left[\int_0^1 (u(x))^2\, dx + \int_0^1 \left(\frac{du}{dx}\right)^2 dx\right]^{1/2}.$$

If u_n tends to u_0 in $W_2^1(0,1)$, then (see [18, p.113]):

$$\tilde{f}\left(x,u_n,\frac{du_n}{dx}\right) \to \tilde{f}\left(x,u_0,\frac{du_0}{dx}\right), \quad k(x)\frac{du_n}{dx} \to k(x)\frac{du_0}{dx} \quad \text{in } L_1(0,1).$$

Thus, for all $v(x) \in \overset{\circ}{W}{}_2^1(0,1)$ we have

$$\lim_{n\to\infty} \left(A\left(x, u_n\right), v\right) = \lim_{n\to\infty} \left[\int_0^1 k(x) \frac{du_n}{dx}\frac{dv}{dx} dx - \int_0^1 \tilde{f}\left(x, u_n, \frac{du_n}{dx}\right) v(x) dx \right]$$

$$= \int_0^1 k(x) \frac{du_0}{dx}\frac{dv}{dx} dx - \int_0^1 \tilde{f}\left(x, u_0, \frac{du_0}{dx}\right) v(x) dx = \left(A\left(x, u_0\right), v\right),$$

i.e. the operator $A(x,u)$ is semi-continuous.

Let us show that the operator $A(x,u)$ is strongly monotone. Due to the conditions (3.2), (3.6) and the inequality

$$\left\| \frac{dv}{dx} \right\|_{0,2,(0,1)} \geq \frac{\pi}{\sqrt{1+\pi^2}} \|v\|_{1,2,(0,1)} \text{ for all } v(x) \in \overset{\circ}{W}{}_2^1(0,1),$$

we obtain

$$\left(A(x,u) - A(x,v), u - v\right)$$

$$= \int_0^1 k(x) \left[\frac{du}{dx} - \frac{dv}{dx}\right]^2 dx$$

$$- \int_0^1 \left[\tilde{f}\left(x, u(x), \frac{du}{dx}\right) - \tilde{f}\left(x, v(x), \frac{dv}{dx}\right)\right] [u(x) - v(x)]\, dx$$

$$\geq c_1 \left\| \frac{du}{dx} - \frac{dv}{dx} \right\|_{0,2,(0,1)}^2 - c_3 \|u - v\|_{1,2,(0,1)}^2$$

$$\geq \left(\frac{\pi^2}{1+\pi^2} c_1 - c_3\right) \|u - v\|_{1,2,(0,1)}^2 = c_4 \|u - v\|_{1,2,(0,1)}^2,$$

where in accordance with (3.7) we have $c_4 = \dfrac{\pi^2}{1+\pi^2} c_1 - c_3 > 0$, and $u - v \in \overset{\circ}{W}{}_2^1(0,1)$. From the strong monotonicity follows the coerciveness of $A(x,u)$.

Thus, the Browder-Minty Theorem (see Theorem 1.3) guaranties the existence of a unique solution $u \in W_2^1(0,1)$ of problem (3.1).

Since

$$k(x) \frac{du}{dx} = \int_0^x f\left(t, u, \frac{du}{dt}\right) dt + C$$

almost everywhere on [0,1] (see, e.g. [18, p.134]), i.e. the flux $k(x)\dfrac{du}{dx}$ is the undefined Lebesgue integral, this expression is absolutely continuous on [0,1] and the claim $k(x)\dfrac{du}{dx} \in \mathbb{C}[0,1]$ is shown. ∎

3.2 Existence of a three-point EDS

Let us cover the interval $(0,1)$ by a non-equidistant grid $\hat{\omega}_h$ which includes the discontinuity points (with respect to x) of the functions $k(x)$ and $f(x,u,\xi)$. We denote the set of discontinuity points by ρ and choose N so that $\rho \subseteq \hat{\omega}_h$. At the discontinuity points the solution of problem (3.1) should satisfy the continuity conditions

$$u(x_i-0)=u(x_i+0),\quad k(x)\frac{du}{dx}\bigg|_{x=x_i-0} = k(x)\frac{du}{dx}\bigg|_{x=x_i+0} \quad \text{for all } x_i\in\rho.$$

We introduce the following BVPs on (small) subintervals:

$$\begin{aligned}
&\frac{d}{dx}\left(k(x)\frac{dY_\alpha^j(x,u)}{dx}\right) = -f\left(x, Y_\alpha^j(x,u), \frac{dY_\alpha^j(x,u)}{dx}\right), \qquad x\in e_\alpha^j,\\
&Y_\alpha^j(x_{j-2+\alpha},u)=u(x_{j-2+\alpha}),\quad Y_\alpha^j(x_{j-1+\alpha},u)=u(x_{j-1+\alpha}),\\
&j=2-\alpha,3-\alpha,\ldots,N+1-\alpha,\quad \alpha=1,2.
\end{aligned} \tag{3.8}$$

Now the following auxiliary result can be proved.

Lemma 3.1

Let the assumptions of Theorem 3.1 be satisfied. Then each of the problems (3.8) has a unique solution $Y_\alpha^j(x,u)\in W_2^1(e_\alpha^j)$ with

$$Y_\alpha^j(x,u),\ k(x)\frac{dY_\alpha^j(x,u)}{dx}\in \mathbb{C}(\bar{e}_\alpha^j)$$

and for the solution $u(x)$ of problem (3.1) it holds that

$$u(x)=Y_\alpha^j(x,u),\quad x\in\bar{e}_\alpha^j,\quad j=2-\alpha,3-\alpha,\ldots,N+1-\alpha,\quad \alpha=1,2. \tag{3.9}$$

Proof. We introduce the nonlinear operator ${A^j}_\alpha\left(x,Y_\alpha^j\right)$ by the equation

$$\begin{aligned}
&\left(A_\alpha^j\left(x,Y_\alpha^j\right),v\right)\\
&=\int_{x_{j-2+\alpha}}^{x_{j-1+\alpha}} k(x)\frac{dY_\alpha^j(x,u)}{dx}\frac{dv(x)}{dx} - \int_{x_{j-2+\alpha}}^{x_{j-1+\alpha}} \tilde{f}\left(x,Y_\alpha^j(x,u),\frac{dY_\alpha^j(x,u)}{dx}\right)v(x)dx,
\end{aligned}$$

where

$$\tilde{f}(x,u,\xi)\stackrel{\text{def}}{=} f(x,u,\xi)-f_0(x).$$

This definition makes sense for all $Y_\alpha^j(x,u)\in W_2^1(e_\alpha^j)$ and $v(x)\in \overset{\circ}{W}{}_2^1(e_\alpha^j)$, where

$$\overset{\circ}{W}{}_2^1(e_\alpha^j)\stackrel{\text{def}}{=}\left\{v(x)\,\middle|\,v(x)\in W_2^1(e_\alpha^j),\ v(x_{j-2+\alpha})=0,\ v(x_{j-1+\alpha})=0,\ \alpha=1,2\right\}.$$

A function $Y_\alpha^j(x,u)$ is a weak solution of (3.8) if

$$Y_\alpha^j(x,u) - \varphi_\alpha^j(x) \in \overset{\circ}{W}{}_2^1(e_\alpha^j), \quad \left(A_\alpha^j\left(x, Y_\alpha^j\right), v\right) = (f_0, v) \quad \text{for all } v(x) \in \overset{\circ}{W}{}_2^1(e_\alpha^j),$$

where $\varphi_\alpha^j(x)$ is a given function from $W_2^1(e_\alpha^j)$ which satisfies the boundary conditions.

Let us show that the operator $A_\alpha^j\left(x, Y_\alpha^j\right)$ is bounded. Using (1.12), formulas (3.2) and (3.5), as well as the estimates

$$c(|Y_\alpha^j(x,u)|) \le C_2 \quad \text{for all } x \in \bar{e}_\alpha^j$$

and

$$\|v\|_{C[x_{j-2+\alpha}, x_{j-1+\alpha}]} \le C_1 \|v\|_{1,2,e_\alpha} \quad \text{for all } v(x) \in W_2^1(e_\alpha^j),$$

we obtain

$$\begin{aligned} A_\alpha^j\left(x, Y_\alpha^j\right) &\le \left\{ \int_{x_{j-2+\alpha}}^{x_{j-1+\alpha}} \left[k(x) \frac{dY_\alpha^j(x,u)}{dx} \right]^2 dx \right\}^{1/2} \left\{ \int_{x_{j-2+\alpha}}^{x_{j-1+\alpha}} \left[\frac{dv}{dx} \right]^2 dx \right\}^{1/2} \\ &\quad + \|v\|_{\mathbb{C}(\bar{e}_\alpha^j)} \int_{x_{j-2+\alpha}}^{x_{j-1+\alpha}} \left\| \tilde{f}\left(x, Y_\alpha^j(x,u), \frac{dY_\alpha^j(x,u)}{dx} \right) \right\| dx \\ &\le \left[c_2 \| Y_\alpha^j \|_{1,2,e_\alpha^j} + C_1 \| \tilde{f} \|_{0,1,e_\alpha^j} \right] \| v \|_{1,2,e_\alpha^j} \\ &\le \left[(c_2 + C_1 C_2) \| Y_\alpha^j \|_{1,2,e_\alpha^j} + C_1 C_2 \| g \|_{0,1,e_\alpha^j} \right] \| v \|_{1,2,e_\alpha^j} . \end{aligned}$$

The semi-continuity of $A_\alpha^j\left(x, Y_\alpha^j\right)$ follows from (3.5). More precisely, in [18, p.113]) it is shown that $Y_{\alpha n}^j(x,u) \to Y_{\alpha 0}^j(x,u)$ in $W_2^1(e_\alpha^j)$ implies

$$\begin{aligned} \tilde{f}\left(x, Y_{\alpha n}^j(x,u), \frac{dY_{\alpha n}^j(x,u)}{dx} \right) &\longrightarrow \tilde{f}\left(x, Y_{\alpha 0}^j(x,u), \frac{dY_{\alpha 0}^j(x,u)}{dx} \right), \\ k(x)\frac{dY_{\alpha n}^j(x,u)}{dx} &\longrightarrow k(x)\frac{dY_{\alpha 0}^j(x,u)}{dx} \end{aligned}$$

in the space $L_1(e_\alpha^j)$. Thus, for all $v(x) \in \overset{\circ}{W}{}_2^1(e_\alpha^j)$ it holds that

$$\lim_{n\to\infty}\left(A_\alpha^j\left(x, Y_{\alpha n}^j\right), v\right)$$

$$= \lim_{n\to\infty}\left\{\int\limits_{x_{j-2+\alpha}}^{x_{j-1+\alpha}} k(x)\frac{dY_{\alpha n}^j(x,u)}{dx}\frac{dv(x)}{dx}dx - \int\limits_{x_{j-2+\alpha}}^{x_{j-1+\alpha}} \tilde{f}\left(x, Y_{\alpha n}^j(x), \frac{dY_{\alpha n}^j(x,u)}{dx}\right)v(x)dx\right\}$$

$$= \int\limits_{x_{j-2+\alpha}}^{x_{j-1+\alpha}} k(x)\frac{dY_{\alpha 0}^j(x,u)}{dx}\frac{dv(x)}{dx}dx - \int\limits_{x_{j-2+\alpha}}^{x_{j-1+\alpha}} \tilde{f}\left(x, Y_{\alpha 0}^j(x,u), \frac{dY_{\alpha 0}^j(x,u)}{dx}\right)v(x)dx$$

$$= \left(A_\alpha^j\left(x, Y_{\alpha 0}^j\right), v\right),$$

i.e. the operator $A_\alpha^j\left(x, Y_\alpha^j\right)$ is semi-continuous.

Let us show that the operator $A_\alpha^j\left(x, Y_\alpha^j(x,u)\right)$ is also strongly monotone. Due to (3.2), (3.6) and to the inequality

$$\left\|\frac{dv}{dx}\right\|_{0,2,e_\alpha^j} \geq \frac{\pi}{\sqrt{1+\pi^2}}\|v\|_{1,2,e_\alpha^j} \quad \text{for all } v(x)\in \overset{\circ}{W}{}_2^1(e_\alpha^j),$$

we have

$$\left(A_\alpha^j\left(x, Y_\alpha^j(x,u)\right) - A_\alpha^j\left(x, \tilde{Y}_\alpha^j(x,u)\right), Y_\alpha^j(x,u) - \tilde{Y}_\alpha^j(x,u)\right)$$

$$= \int\limits_{x_{j-2+\alpha}}^{x_{j-1+\alpha}} k(x)\left(\frac{dY_\alpha^j(x,u)}{dx} - \frac{d\tilde{Y}_\alpha^j(x,u)}{dx}\right)^2 dx - \int\limits_{x_{j-2+\alpha}}^{x_{j-1+\alpha}} \left[f\left(x, Y_\alpha^j(x,u), \frac{dY_\alpha^j(x,u)}{dx}\right)\right.$$

$$\left. - f\left(x, \tilde{Y}_\alpha^j(x,u), \frac{d\tilde{Y}_\alpha^j(x,u)}{dx}\right)\right]\left[Y_\alpha^j(x,u) - \tilde{Y}_\alpha^j(x,u)\right]dx$$

$$\geq c_1\left\|\frac{dY_\alpha^j}{dx} - \frac{d\tilde{Y}_\alpha^j}{dx}\right\|_{0,2,e_\alpha^j}^2 - c_3\left\|Y_\alpha^j - \tilde{Y}_\alpha^j\right\|_{1,2,e_\alpha^j}^2$$

$$\geq \left(\frac{\pi^2}{1+\pi^2}c_1 - c_3\right)\left\|Y_\alpha^j - \tilde{Y}_\alpha^j\right\|_{1,2,e_\alpha^j}^2$$

$$= c_4\left\|Y_\alpha^j - \tilde{Y}_\alpha^j\right\|_{1,2,e_\alpha^j}^2.$$

Now, the strong monotonicity of $A_\alpha^j\left(x, Y_\alpha^j(x,u)\right)$ follows from its coerciveness.

Thus due to the Browder-Minty Theorem (see Theorem 1.3) there exist a unique solution of each of the problems (3.8).

We have

$$k(x)\frac{dY_\alpha^j(x,u)}{dx} = \int\limits_{x_{j-2+\alpha}}^{x_{j-1+\alpha}} f\left(t, Y_\alpha^j(t,u), \frac{dY_\alpha^j(t,u)}{dt}\right)dt + C,$$

which means that $k(x)\dfrac{dY_\alpha^j(x,u)}{dx} \in \mathbb{C}(\bar{e}_\alpha^j)$.

Since $Y_\alpha^j(x,u)$ is the solution of (3.8), this function is also the solution of problem (3.1) which is unique due to the assumptions of our lemma. ■

Now we are in a position to prove the next statement.

Theorem 3.2

Let the assumptions of Theorem 3.1 *be satisfied. Then there exists the following three-point EDS for problem* (3.1)*:*

$$(au_{\bar{x}})_{\hat{x}} = -\hat{T}^x\left(f\left(\xi, u(\xi), \frac{du}{d\xi}\right)\right), \quad x \in \hat{\omega}_h, \quad u(0) = \mu_1, \quad u(1) = \mu_2. \quad (3.10)$$

This EDS has a unique solution $u(x)$ for all $x \in \hat{\omega}_h$ which coincides with the solution of problem (3.1) *at the points of the grid $\hat{\omega}_h$. Here the divided differences $u_{\bar{x},j}$ and $u_{\hat{x},j}$ are defined as in Definition* 1.3*, and*

$$a(x_j) \stackrel{\text{def}}{=} \left[\frac{1}{h_j}V_1^j(x_j)\right]^{-1}, \quad V_1^j(x) \stackrel{\text{def}}{=} \int\limits_{x_{j-1}}^{x} \frac{dx}{k(x)}, \quad V_2^j(x) \stackrel{\text{def}}{=} \int\limits_{x}^{x_{j+1}} \frac{dx}{k(x)},$$

$$\hat{T}^{x_j}(w(\xi)) \stackrel{\text{def}}{=} \left(\hbar_j V_1^j(x_j)\right)^{-1} \int\limits_{x_{j-1}}^{x_j} V_1^j(\xi)w(\xi)d\xi + \left(\hbar_j V_2^j(x_j)\right)^{-1} \int\limits_{x_j}^{x_{j+1}} V_2^j(\xi)w(\xi)d\xi. \quad (3.11)$$

The function $u(x)$ on the right-hand side of (3.10) *is defined by* (3.9) *and depends only on $u(x_j)$, $j = 0, 1, \ldots, N$.*

Proof. Applying the operator $\hat{T}^{x_j}$ to both sides of equation (3.1) we obtain

$$\hat{T}^{x_j}\left(\frac{d}{d\xi}\left(k(\xi)\frac{du}{d\xi}\right)\right) = -\hat{T}^{x_j}\left(f\left(\xi, u(\xi), \frac{du}{d\xi}\right)\right),$$

where

$$\hat{T}^{x_j}\left(\frac{d}{d\xi}\left(k(\xi)\frac{du}{d\xi}\right)\right)$$

$$=\left[\hbar_j V_1^j(x_j)\right]^{-1}\int\limits_{x_{j-1}}^{x_j} V_1^j(\xi)\frac{d}{d\xi}\left[k(\xi)\frac{du}{d\xi}\right]d\xi$$

$$+\left[\hbar_j V_2^j(x_j)\right]^{-1}\int\limits_{x_j}^{x_{j+1}} V_2^j(\xi)\frac{d}{d\xi}\left[k(\xi)\frac{du}{d\xi}\right]d\xi.$$

The integration by parts implies

$$\hat{T}^{x_j}\left(\frac{d}{d\xi}\left(k(\xi)\frac{du}{d\xi}\right)\right)=(au_{\bar{x}})_{\hat{x},j},$$

which together with (3.9) proves existence of the EDS (3.10).

In order to show uniqueness of the solution of (3.10) we consider the operator

$$A_h(x,u)\stackrel{\text{def}}{=}-(au_{\bar{x}})_{\hat{x}}-\hat{T}^x\left(f\left(\xi,u(\xi),\frac{du}{d\xi}\right)\right)$$

which is defined in the finite-dimensional space $L_2(\hat{\omega}_h)$ equipped with the scalar products

$$(u,v)_{\hat{\omega}_h}\stackrel{\text{def}}{=}\sum_{\xi\in\hat{\omega}_h}\hbar(\xi)u(\xi)v(\xi),\quad (u,v)_{\hat{\omega}_h^+}\stackrel{\text{def}}{=}\sum_{\xi\in\hat{\omega}_h^+}h(\xi)u(\xi)v(\xi),\quad \hat{\omega}_h^+\stackrel{\text{def}}{=}\hat{\omega}_h\cup x_N,$$

and the norms

$$\|u\|_{0,2,\hat{\omega}_h}\stackrel{\text{def}}{=}(u,u)_{\hat{\omega}_h}^{1/2},\quad \|u\|_{0,2,\hat{\omega}_h^+}\stackrel{\text{def}}{=}(u,u)_{\hat{\omega}_h^+}^{1/2},$$

$$\|u\|_{1,2,\hat{\omega}_h}\stackrel{\text{def}}{=}\left(\|u\|_{0,2,\hat{\omega}_h}^2+\|u_{\bar{x}}\|_{0,2,\hat{\omega}_h^+}^2\right)^{1/2}.$$

Due to condition (3.5) the operator $A_h(x,u)$ is continuous. Let us show that the operator $A_h(x,u)$ is strongly monotone.

Taking into account the equality

$$\sum_{\xi\in\hat{\omega}_h}\hbar(\xi)\hat{T}^\xi(w(\eta))g(\xi)=\sum_{j=1}^{N}\int\limits_{x_{j-1}}^{x_j}\hat{g}(\eta)w(\eta)d\eta=\int\limits_0^1\hat{g}(\eta)w(\eta)d\eta,$$

where

$$\hat{g}(\eta)\stackrel{\text{def}}{=}g(x_j)\frac{V_1^j(\eta)}{V_1^j(x_j)}+g(x_{j-1})\frac{V_2^{j-1}(\eta)}{V_1^j(x_j)},\quad x_{j-1}\le\eta\le x_j,$$

we have

$$\Big(\hat{T}^x\big(f\left(\eta,u(\eta),u_\eta(\eta)\right)-f\left(\eta,v(\eta),v_\eta(\eta)\right)\big),u-v\Big)_{\hat{\omega}_h}$$
$$=\sum_{\xi\in\hat{\omega}_h}\hbar(\xi)\hat{T}^\xi\big(f\left(\eta,u(\eta),u_\eta(\eta)\right)-f\left(\eta,v(\eta),v_\eta(\eta)\right)\big)\big(u(\xi)-v(\xi)\big)$$
$$=\int_0^1\big(\hat{u}(\eta)-\hat{v}(\eta)\big)\big(f\left(\eta,u(\eta),u_\eta(\eta)\right)-f\left(\eta,v(\eta),v_\eta(\eta)\right)\big)d\eta,$$

where the functions $u(x)$ and $v(x)$ are defined by (3.9) and $z_\eta(\eta)\overset{\text{def}}{=}\dfrac{dz(\eta)}{d\eta}$.

Then using (3.6), we have

$$\Big(\hat{T}^x\big(f\left(\eta,u(\eta),u_\eta(\eta)\right)-f\left(\eta,v(\eta),v_\eta(\eta)\right)\big),u-v\Big)_{\hat{\omega}_h}$$
$$=\int_0^1\big(u(\eta)-v(\eta)\big)\big(f\left(\eta,u(\eta),u_\eta(\eta)\right)-f\left(\eta,v(\eta),v_\eta(\eta)\right)\big)d\eta$$
$$+\int_0^1\left(\hat{u}(\eta)-\hat{v}(\eta)-u(\eta)+v(\eta)\right)\left(f\left(\eta,u(\eta),u_\eta(\eta)\right)-f\left(\eta,v(\eta),v_\eta(\eta)\right)\right)d\eta$$
$$\le-\int_0^1\big(\hat{u}(\eta)-\hat{v}(\eta)\big)\frac{d}{d\eta}\left\{k(\eta)\frac{d}{d\eta}\big(u(\eta)-v(\eta)\big)\right\}d\eta$$
$$+\int_0^1\big(u(\eta)-v(\eta)\big)\frac{d}{d\eta}\left\{k(\eta)\frac{d}{d\eta}\big(u(\eta)-v(\eta)\big)\right\}d\eta+c_3\|u-v\|^2_{1,2,(0,1)}$$
$$=-\sum_{j=1}^N\int_{x_{j-1}}^{x_j}\big(\hat{u}(\eta)-\hat{v}(\eta)\big)\frac{d}{d\eta}\left\{k(\eta)\frac{d}{d\eta}\big(u(\eta)-v(\eta)\big)\right\}d\eta$$
$$-\int_0^1k(\eta)\left\{\frac{d}{d\eta}\big(u(\eta)-v(\eta)\big)\right\}^2d\eta+c_3\|u-v\|^2_{1,2,(0,1)}.$$

Since

$$\sum_{j=1}^N\int_{x_{j-1}}^{x_j}\big(\hat{u}(\eta)-\hat{v}(\eta)\big)\frac{d}{d\eta}\left\{k(\eta)\frac{d}{d\eta}\big(u(\eta)-v(\eta)\big)\right\}d\eta$$
$$=-\sum_{j=1}^N\int_{x_{j-1}}^{x_j}k(\eta)\left(\hat{u}_\eta-\hat{v}_\eta\right)\left(u_\eta-v_\eta\right)d\eta=-\big(a(u_{\bar{x}}-v_{\bar{x}}),u_{\bar{x}}-v_{\bar{x}}\big)_{\hat{\omega}_h^+},$$

we have

$$\left(\hat{T}^x\left(f\left(\eta,u,u_\eta\right)-f\left(\eta,v,v_\eta\right)\right),u-v\right)_{\hat{\omega}_h}$$
$$\leq \left(a(u_{\bar{x}}-v_{\bar{x}}),u_{\bar{x}}-v_{\bar{x}}\right)_{\hat{\omega}_h^+}-c_1\left\|\frac{du}{dx}-\frac{dv}{dx}\right\|^2_{0,2,(0,1)}+c_3\left\|u-v\right\|^2_{1,2,(0,1)}. \tag{3.12}$$

Due to the inequality $\left\|\frac{dv}{dx}\right\|_{0,2,(0,1)}\leq\frac{\pi}{\sqrt{1+\pi^2}}\|v\|_{1,2,(0,1)}$ with $v(x)\in\overset{\circ}{W}{}^1_2(0,1)$, as well as the condition (3.7), we have

$$\begin{aligned}
&\left(A_h\left(x,u\right)-A_h\left(x,v\right),u-v\right)_{\hat{\omega}_h}\\
&=\left(a(u_{\bar{x}}-v_{\bar{x}}),u_{\bar{x}}-v_{\bar{x}}\right)_{\hat{\omega}_h^+}-\left(\hat{T}^x\left(f\left(\eta,u,u_\eta\right)-f\left(\eta,v,v_\eta\right)\right),u-v\right)_{\hat{\omega}_h}\\
&\geq c_1\left\|\frac{du}{dx}-\frac{dv}{dx}\right\|^2_{0,2,(0,1)}-c_3\|u-v\|^2_{1,2,(0,1)}\\
&\geq c_4\frac{1+\pi^2}{\pi^2}\left\|\frac{du}{dx}-\frac{dv}{dx}\right\|^2_{0,2,(0,1)},
\end{aligned}$$

where $c_4\overset{\text{def}}{=}\frac{\pi^2}{1+\pi^2}c_1-c_3>0$. The inequality

$$\begin{aligned}
&\int_0^1 k(\eta)\left\{\frac{d}{d\eta}\left(\hat{u}(\eta)-\hat{v}(\eta)-u(\eta)+v(\eta)\right)\right\}^2 d\eta\\
&=-\sum_{j=1}^N h_j a(x_j)(u_{\bar{x},j}-v_{\bar{x},j})^2+\int_0^1 k(\eta)\left\{\frac{d}{d\eta}\left(u(\eta)-v(\eta)\right)\right\}^2 d\eta\\
&=-\left(a(u_{\bar{x}}-v_{\bar{x}}),u_{\bar{x}}-v_{\bar{x}}\right)_{\hat{\omega}_h^+}+\left(k\left(\frac{du}{dx}-\frac{dv}{dx}\right),\frac{du}{dx}-\frac{dv}{dx}\right)_{\hat{\omega}_h}\geq 0,
\end{aligned}$$

implies

$$\left\|\frac{du}{dx}-\frac{dv}{dx}\right\|^2_{0,2,(0,1)}\geq\frac{1}{c_2}\left(a(u_{\bar{x}}-v_{\bar{x}}),u_{\bar{x}}-v_{\bar{x}}\right)_{\hat{\omega}_h^+}.$$

Thus

$$\begin{aligned}
&\left(A_h(x,u)-A_h(x,v),u-v\right)_{\hat{\omega}_h}\\
&\geq\frac{1+\pi^2}{\pi^2}\frac{c_4}{c_2}\left(a(u_{\bar{x}}-v_{\bar{x}}),u_{\bar{x}}-v_{\bar{x}}\right)_{\hat{\omega}_h^+}\geq\frac{1+\pi^2}{\pi^2}\frac{c_4c_1}{c_2}\left\|u_{\bar{x}}-v_{\bar{x}}\right\|^2_{\hat{\omega}_h^+}\\
&\geq 8\frac{1+\pi^2}{\pi^2}\frac{c_4c_1}{c_2}\left\|u-v\right\|^2_{0,2,\hat{\omega}_h},
\end{aligned} \tag{3.13}$$

i.e. the operator $A_h(x,u)$ is strongly monotone. This yields (see e.g. [82, p.461]) the uniqueness of the solution of the equation $A_h(x,u)=0$. ■

Lemma 3.2

Let the assumptions of Theorem 3.1 be satisfied and

$$|f(x,u,\xi)-f(x,v,\eta)| \le L\{|u-v|+|\xi-\eta|\} \quad \textit{for all } x\in(0,1),\ u,v,\xi,\eta\in\mathbb{R}.$$

Then the iteration method

$$\begin{aligned} &B_h\frac{u^{(n)}-u^{(n-1)}}{\tau}+A_h\left(x,u^{(n-1)}\right)=0, \quad x\in\hat{\omega}_h,\\ &u^{(n)}(0)=\mu_1, \quad u^{(n)}(1)=\mu_2, \quad n=1,2,\ldots, \qquad (3.14)\\ &u^{(0)}(x)=\frac{V_2(x)}{V_1(1)}\mu_1+\frac{V_1(x)}{V_1(1)}\mu_2, \end{aligned}$$

$$B_hu \stackrel{\text{def}}{=} -(au_{\bar{x}})_{\hat{x}}, \quad A_h(x,u)\stackrel{\text{def}}{=} B_hu-\hat{T}^x\left(f\left(\xi,u(\xi),u_\xi(\xi)\right)\right),$$

$$\tau \stackrel{\text{def}}{=} \tau_0=\frac{1+\pi^2}{\pi^2}\frac{c_4}{c_2}\left(1+\frac{L}{c_1}\left(\frac{1}{2}+\frac{c_2^2}{c_1^2}\right)^{1/2}\left(1+\frac{\sqrt{2}L}{c_4}\right)\right)^{-2}$$

converges in the space H_{B_h} and the error estimate

$$\left\|u^{(n)}-u\right\|_{B_h}\le q^n\left\|u^{(0)}-u\right\|_{B_h} \qquad (3.15)$$

holds with

$$q\stackrel{\text{def}}{=}\sqrt{1-\frac{1+\pi^2}{\pi^2}\frac{c_4}{c_2}\tau_0},$$

where

$$c_4=\frac{\pi^2}{1+\pi^2}c_1-c_3>0 \quad \textit{and} \quad \|u\|_{B_h}=(B_hu,u)_{\hat{\omega}_h}^{1/2}.$$

Proof. It follows from (3.13) that

$$(A_h(x,u)-A_h(x,v),u-v)_{\hat{\omega}_h}\ge\frac{1+\pi^2}{\pi^2}\frac{c_4}{c_2}\|u-v\|_{B_h}^2. \qquad (3.16)$$

We have

$$\begin{aligned} &\left(A_h(x,u)-A_h(x,v),z\right)_{\hat{\omega}_h}\\ &=(B_hu-B_hv,z)_{\hat{\omega}_h}-\sum_{\xi\in\omega_h}\hbar(\xi)T^\xi\left(f\left(\eta,u(\eta),u_\eta(\eta)\right)-f\left(\eta,v(\eta),v_\eta(\eta)\right)\right)z(\xi) \end{aligned}$$

$$= (B_h u - B_h v, z)_{\hat{\omega}_h} - \int_0^1 \left(f\left(\eta, u(\eta), u_\eta(\eta)\right) - f\left(\eta, v(\eta), v_\eta(\eta)\right)\right) \hat{z}(\eta) d\eta.$$

Using the Cauchy-Bunyakovsky-Schwarz inequality (see Theorem 1.8) we can now deduce

$$\begin{aligned}
&(A_h(x,u) - A_h(x,v), z)_{\hat{\omega}_h} \\
&\le \|u-v\|_{B_h} \|z\|_{B_h} \\
&\quad + \left\{ \int_0^1 \left(f\left(\eta, u(\eta), u_\eta(\eta)\right) - f\left(\eta, v(\eta), v_\eta(\eta)\right)\right)^2 d\eta \right\}^{1/2} \left\{ \int_0^1 (\hat{z}(\eta))^2 d\eta \right\}^{1/2} \\
&\le \|u-v\|_{B_h} \|z\|_{B_h} + \sqrt{2} L \|u-v\|_{1,2,(0,1)} \|\hat{z}\|_{0,2,(0,1)}.
\end{aligned}$$

Since $V_1^j(x) \le V_1^j(x_j)$, $V_2^{j-1}(x) \le V_1^j(x_j)$ for all $x \in [x_{j-1}, x_j]$, we have

$$\begin{aligned}
\|\hat{z}\|_{0,2,(0,1)}^2 &= \sum_{j=1}^N \int_{x_{j-1}}^{x_j} \left[z_j \frac{V_1^j(x)}{V_1^j(x_j)} + z_{j-1} \frac{V_2^{j-1}(x)}{V_1^j(x_j)} \right]^2 dx \\
&\le 2 \sum_{j=1}^N \int_{x_{j-1}}^{x_j} \left\{ z_j^2 \left[\frac{V_1^j(x)}{V_1^j(x_j)} \right]^2 + z_{j-1}^2 \left[\frac{V_2^{j-1}(x)}{V_1^j(x_j)} \right]^2 \right\} dx \le 4 \|z\|_{0,2,\hat{\omega}_h}^2
\end{aligned} \tag{3.17}$$

and

$$\begin{aligned}
\left\| \frac{d\hat{z}}{dx} \right\|_{0,2,(0,1)}^2 &\le \sum_{j=1}^N \int_{x_{j-1}}^{x_j} \left[\frac{1}{k(x)} \frac{1}{V_1^j(x_j)} (z_j - z_{j-1}) \right]^2 dx \\
&\le \frac{c_2^2}{c_1^2} \sum_{j=1}^N h_j z_{\bar{x},j}^2 = \frac{c_2^2}{c_1^2} \|z_{\bar{x}}\|_{0,2,\hat{\omega}_h^+}^2 .
\end{aligned} \tag{3.18}$$

Let us now show that

$$\|u-v\|_{1,2,(0,1)} \le \left(\frac{1}{2} + \frac{c_2^2}{c_1^2} \right)^{1/2} \left(1 + \frac{\sqrt{2} L}{c_4} \right) \|u_{\bar{x}} - v_{\bar{x}}\|_{0,2,\hat{\omega}_h^+} . \tag{3.19}$$

We write $u(x) = \tilde{u}(x) + \hat{u}(x)$, $x \in \bar{e}_\alpha^j$, and reduce the problem

$$\frac{d}{dx}\left[k(x) \frac{du}{dx} \right] = -f\left(x, u(x), \frac{du}{dx} \right), \quad x \in e_\alpha^j,$$

$$u(x_{j-2+\alpha}) = u_{j-2+\alpha}, \quad u(x_{j-1+\alpha}) = u_{j-1+\alpha}, \quad \alpha = 1, 2,$$

to

$$\frac{d}{dx}\left[k(x)\frac{d\tilde{u}}{dx}\right] = -f\left(x, \tilde{u}(x) + \hat{u}(x), \frac{d\tilde{u}}{dx} + \frac{d\hat{u}}{dx}\right), \quad x \in e_\alpha^j,$$

$$\tilde{u}(x_{j-2+\alpha}) = 0, \quad \tilde{u}(x_{j-1+\alpha}) = 0, \quad \alpha = 1, 2.$$

Considering (3.2), (3.6) and using a Lipschitz condition for all $\tilde{u}(x), \tilde{v}(x) \in \overset{\circ}{W}{}_2^1(e_\alpha^j)$, we get

$$c_1 \frac{\pi^2}{1+\pi^2}\|\tilde{u} - \tilde{v}\|_{1,2,e_\alpha^j}^2 \le c_1 \left\|\frac{d\tilde{u}}{dx} - \frac{d\tilde{v}}{dx}\right\|_{0,2,e_\alpha^j}^2 \le \int\limits_{x_{j-2+\alpha}}^{x_{j-1+\alpha}} k(x)\left(\frac{d\tilde{u}}{dx} - \frac{d\tilde{v}}{dx}\right)^2 dx$$

$$= \int\limits_{x_{j-2+\alpha}}^{x_{j-1+\alpha}} \left[f\left(x, \tilde{u}(x) + \hat{u}(x), \frac{d\tilde{u}}{dx} + \frac{d\hat{u}}{dx}\right)\right.$$

$$\left. - f\left(x, \tilde{v}(x) + \hat{v}(x), \frac{d\tilde{v}}{dx} + \frac{d\hat{v}}{dx}\right)\right] (\tilde{u}(x) - \tilde{v}(x))\, dx$$

$$= \int\limits_{x_{j-2+\alpha}}^{x_{j-1+\alpha}} \left[f\left(x, \tilde{u}(x) + \hat{u}(x), \frac{d\tilde{u}}{dx} + \frac{d\hat{u}}{dx}\right)\right.$$

$$\left. - f\left(x, \tilde{v}(x) + \hat{u}(x), \frac{d\tilde{v}}{dx} + \frac{d\hat{u}}{dx}\right)\right](\tilde{u}(x) - \tilde{v}(x))dx$$

$$+ \int\limits_{x_{j-2+\alpha}}^{x_{j-1+\alpha}} \left[f\left(x, \tilde{v}(x) + \hat{u}(x), \frac{d\tilde{v}}{dx} + \frac{d\hat{u}}{dx}\right)\right.$$

$$\left. - f\left(x, \tilde{v}(x) + \hat{v}(x), \frac{d\tilde{v}}{dx} + \frac{d\hat{v}}{dx}\right)\right](\tilde{u}(x) - \tilde{v}(x))dx$$

$$\le c_3\|\tilde{u} - \tilde{v}\|_{1,2,e_\alpha^j}^2 + \left\{ \int\limits_{x_{j-2+\alpha}}^{x_{j-1+\alpha}} \left[f\left(x, \tilde{v}(x) + \hat{u}(x), \frac{d\tilde{v}}{dx} + \frac{d\hat{u}}{dx}\right)\right.\right.$$

$$\left.\left. - f\left(x, \tilde{v}(x) + \hat{v}(x), \frac{d\tilde{v}}{dx} + \frac{d\hat{v}}{dx}\right)\right]^2 dx\right\}^{1/2} \left\{ \int\limits_{x_{j-2+\alpha}}^{x_{j-1+\alpha}} (\tilde{u}(x) - \tilde{v}(x))^2 dx \right\}^{1/2}$$

$$\le \sqrt{2}L\|\hat{u} - \hat{v}\|_{1,2,e_\alpha^j}\|\tilde{u} - \tilde{v}\|_{1,2,e_\alpha^j} + c_3\|\tilde{u} - \tilde{v}\|_{1,2,e_\alpha^j}^2$$

which implies

$$\|\tilde{u} - \tilde{v}\|_{1,2,e_\alpha^j} \le \frac{\sqrt{2}L}{c_4}\|\hat{u} - \hat{v}\|_{1,2,e_\alpha^j}.$$

From the inequalities (3.17), (3.18) and $\|u\|^2_{0,2,\hat{\omega}_h} \le \frac{1}{8}\|u_{\bar{x}}\|^2_{0,2,\hat{\omega}_h^+}$ it follows that

$$\|u-v\|_{1,2,(0,1)} \le \|\tilde{u}-\tilde{v}\|_{1,2,(0,1)} + \|\hat{u}-\hat{v}\|_{1,2,(0,1)} \le \left(1+\frac{\sqrt{2}L}{c_4}\right)\|\hat{u}-\hat{v}\|_{1,2,(0,1)}$$

$$\le \left(1+\frac{\sqrt{2}L}{c_4}\right)\left(4\|u-v\|^2_{0,2,\hat{\omega}_h} + \frac{c_2^2}{c_1^2}\|u_{\bar{x}}-v_{\bar{x}}\|^2_{0,2,\hat{\omega}_h^+}\right)^{1/2}$$

$$\le \left(\frac{1}{2}+\frac{c_2^2}{c_1^2}\right)^{1/2}\left(1+\frac{\sqrt{2}L}{c_4}\right)\|u_{\bar{x}}-v_{\bar{x}}\|_{0,2,\hat{\omega}_h^+}.$$

With (3.17), (3.19) we have

$$(A_h(x,u) - A_h(x,v), z)_{\hat{\omega}_h}$$

$$\le \|u-v\|_{B_h}\|z\|_{B_h} + 2\sqrt{2}L\left(\frac{1}{2}+\frac{c_2^2}{c_1^2}\right)^{1/2}\left(1+\frac{\sqrt{2}L}{c_4}\right)\|u_{\bar{x}}-v_{\bar{x}}\|_{0,2,\hat{\omega}_h^+}\|z\|_{0,2,\hat{\omega}_h}$$

$$\le \|u-v\|_{B_h}\|z\|_{B_h} + L\left(\frac{1}{2}+\frac{c_2^2}{c_1^2}\right)^{1/2}\left(1+\frac{\sqrt{2}L}{c_4}\right)\|u_{\bar{x}}-v_{\bar{x}}\|_{0,2,\hat{\omega}_h^+}\|z_{\bar{x}}\|_{0,2,\hat{\omega}_h^+}$$

$$\le \left(1+\frac{L}{c_1}\left(\frac{1}{2}+\frac{c_2^2}{c_1^2}\right)^{1/2}\left(1+\frac{\sqrt{2}L}{c_4}\right)\right)\|u-v\|_{B_h}\|z\|_{B_h}.$$

Setting $z \stackrel{\text{def}}{=} B_h^{-1}(A_h(x,u)-A_h(x,v))$, we obtain

$$\|B_h^{-1}(A_h(x,u)-A_h(x,v))\|_{B_h} \le \left(1+\frac{L}{c_1}\left(\frac{1}{2}+\frac{c_2^2}{c_1^2}\right)^{1/2}\left(1+\frac{\sqrt{2}L}{c_4}\right)\right)\|u-v\|_{B_h} \tag{3.20}$$

and further due to formulas (3.20), (3.16) we have

$$\left(A_h(x,u) - A_h(x,v), B_h^{-1}(A_h(x,u)-A_h(x,v))\right)_{\hat{\omega}_h}$$

$$\le \left(1+\frac{L}{c_1}\left(\frac{1}{2}+\frac{c_2^2}{c_1^2}\right)^{1/2}\left(1+\frac{\sqrt{2}L}{c_4}\right)\right)^2\|u-v\|^2_{B_h}$$

$$\le \frac{\pi^2}{1+\pi^2}\frac{c_2}{c_4}\left(1+\frac{L}{c_1}\left(\frac{1}{2}+\frac{c_2^2}{c_1^2}\right)^{1/2}\left(1+\frac{\sqrt{2}L}{c_4}\right)\right)^2$$

$$\times (A_h(x,u) - A_h(x,v), u-v)_{\hat{\omega}_h}.$$

The results in [74, p.502] imply the convergence of the iteration process (3.14) in H_{B_h} as well as the estimate (3.15). ■

Note that the norms of the spaces H_{B_h} and $\overset{\circ}{W}{}^1_2(\hat{\omega}_h)$ are equivalent, i.e.,

$$\gamma_1\|u\|_{1,2,\hat{\omega}_h} \le \|u\|_{B_h} \le \gamma_2\|u\|_{1,2,\hat{\omega}_h},$$

and we can now prove the following statement.

Lemma 3.3

Let the assumptions of Lemma 3.2 be satisfied. Then the iteration process (3.14) converges and in addition to (3.15) the following estimate holds:

$$\left\| k\frac{du^{(n)}}{dx} - k\frac{du}{dx} \right\|_{0,2,\hat{\omega}_h} \le M \left\| u^{(n)} - u \right\|_{1,2,\hat{\omega}_h} \le Mq^n,$$

where

$$\|u\|_{0,2,\hat{\omega}_h} \stackrel{\text{def}}{=} \left\{ \sum_{j=1}^{N-1} \hbar_j u_j^2 + \frac{1}{2}h_1 u_0^2 + \frac{1}{2}h_N u_N^2 \right\}^{1/2} = \left\{ \frac{1}{2}\sum_{j=1}^{N} h_j (u_j^2 + u_{j-1}^2) \right\}^{1/2}.$$

Proof. Using the relations

$$\begin{aligned} k\frac{du}{dx}\bigg|_{x=x_j} &= a_j u_{\bar{x},j} + \frac{1}{V_1^j(x_j)} \int_{x_{j-1}}^{x_j} V_1^j(\xi) \frac{d}{d\xi}\left[k(\xi)\frac{du}{d\xi} \right] d\xi \\ &= a_j u_{\bar{x},j} - \frac{1}{V_1^j(x_j)} \int_{x_{j-1}}^{x_j} V_1^j(\xi) f\left(\xi, u(\xi), \frac{du}{d\xi} \right) d\xi, \end{aligned}$$

$$\begin{aligned} k\frac{du}{dx}\bigg|_{x=x_{j-1}} &= a_j u_{\bar{x},j} - \frac{1}{V_2^{j-1}(x_{j-1})} \int_{x_{j-1}}^{x_j} V_2^{j-1}(\xi) \frac{d}{d\xi}\left[k(\xi)\frac{du}{d\xi} \right] d\xi \\ &= a_j u_{\bar{x},j} + \frac{1}{V_1^j(x_j)} \int_{x_{j-1}}^{x_j} V_2^{j-1}(\xi) f\left(\xi, u(\xi), \frac{du}{d\xi} \right) d\xi, \end{aligned}$$

the inequality $(a+b)^2 \le 2(a^2+b^2)$, as well as the Cauchy-Bunyakovsky-Schwarz inequality (see Theorem 1.8) and a Lipschitz condition we obtain

$$\begin{aligned} &\left\| k\frac{du^{(n)}}{dx} - k\frac{du}{dx} \right\|_{0,2,\hat{\omega}_h} \\ &= \left\{ \frac{1}{2}\sum_{j=1}^{N} h_j \left[a_j u_{\bar{x},j}^{(n)} - a_j u_{\bar{x},j} \right.\right. \\ &\qquad \left. - \frac{1}{V_1^j(x_j)} \int_{x_{j-1}}^{x_j} V_1^j(\xi) \left(f\left(\xi, u^{(n)}(\xi), \frac{du^{(n)}}{d\xi} \right) - f\left(\xi, u(\xi), \frac{du}{d\xi} \right) \right) d\xi \right]^2 \end{aligned}$$

$$+\frac{1}{2}\sum_{j=1}^{N} h_j\left[a_j u_{\bar{x},j}^{(n)} - a_j u_{\bar{x},j}\right.$$

$$\left.\left.+\frac{1}{V_1^j(x_j)}\int_{x_{j-1}}^{x_j} V_2^{j-1}(\xi)\left(f\left(\xi,u^{(n)}(\xi),\frac{du^{(n)}}{d\xi}\right)-f\left(\xi,u(\xi),\frac{du}{d\xi}\right)\right)d\xi\right]^2\right\}^{1/2}$$

$$\leq\left\{2\sum_{j=1}^{N} h_j\left[a_j u_{\bar{x},j}^{(n)} - a_j u_{\bar{x},j}\right]^2\right.$$

$$+\sum_{j=1}^{N} h_j\left[V_1^j(x_j)\right]^{-2}\left[\int_{x_{j-1}}^{x_j} V_1^j(\xi)\left(f\left(\xi,u^{(n)}(\xi),\frac{du^{(n)}}{d\xi}\right)\right.\right.$$

$$\left.\left.-f\left(\xi,u(\xi),\frac{du}{d\xi}\right)\right)d\xi\right]^2+\sum_{j=1}^{N} h_j\left[V_1^j(x_j)\right]^{-2}$$

$$\left.\times\left[\int_{x_{j-1}}^{x_j} V_2^{j-1}(\xi)\left(f\left(\xi,u^{(n)}(\xi),\frac{du^{(n)}}{d\xi}\right)-f\left(\xi,u(\xi),\frac{du}{d\xi}\right)\right)d\xi\right]^2\right\}^{1/2}$$

$$\leq\left\{2\sum_{j=1}^{N} h_j\left[a_j u_{\bar{x},j}^{(n)} - a_j u_{\bar{x},j}\right]^2+\sum_{j=1}^{N} h_j\int_{x_{j-1}}^{x_j}\frac{\left[V_1^j(\xi)\right]^2+\left[V_2^{j-1}(\xi)\right]^2}{\left[V_1^j(\xi)+V_2^{j-1}(\xi)\right]^2}d\xi\right.$$

$$\left.\times\int_{x_{j-1}}^{x_j}\left[f\left(\xi,u^{(n)}(\xi),\frac{du^{(n)}}{d\xi}\right)-f\left(\xi,u(\xi),\frac{du}{d\xi}\right)\right]^2 d\xi\right\}^{1/2}$$

$$\leq\left\{2\sum_{j=1}^{N} h_j a_j^2\left[u_{\bar{x},j}^{(n)} - u_{\bar{x},j}\right]^2\right.$$

$$\left.+L^2\sum_{j=1}^{N} h_j\int_{x_{j-1}}^{x_j}\left[\left|u^{(n)}(\xi)-u(\xi)\right|+\left|\frac{du^{(n)}}{d\xi}-\frac{du}{d\xi}\right|\right]^2 d\xi\right\}^{1/2}$$

$$\leq\sqrt{2}c_2\left\|u_{\bar{x}}^{(n)}-u_{\bar{x}}\right\|_{0,2,\hat{\omega}_h^+}+\sqrt{2}L\left\|u^{(n)}-u\right\|_{1,2,(0,1)}.$$

Now, the inequality (3.19) and Lemma 3.1 imply

$$\left\| k\frac{du^{(n)}}{dx} - k\frac{du}{dx} \right\|_{0,2,\hat{\omega}_h}$$
$$\leq \sqrt{2}\left[c_2 + \sqrt{2}L\left(\frac{1}{2} + \frac{c_2^2}{c_1^2}\right)^{1/2}\left(1 + \frac{\sqrt{2}L}{c_4}\right)\right]\left\|u^{(n)} - u\right\|_{1,2,\hat{\omega}_h}$$
$$= M_1\left\|u^{(n)} - u\right\|_{1,2,\hat{\omega}_h} \leq Mq^n.$$

■

3.3 Implementation of the three-point EDS

First let us note that the following relation holds:

$$(-1)^{\alpha+1}\int_{x_\beta}^{x_j} V_\alpha^j(\xi) f\left(\xi, u(\xi), \frac{du}{d\xi}\right) d\xi = (-1)^\alpha V_\alpha^j(x_j) Z_\alpha^j(x_j, u) + Y_\alpha^j(x_j, u) - u_\beta,$$

where we have again used the abbreviation

$$\beta \stackrel{\text{def}}{=} j + (-1)^\alpha, \tag{3.21}$$

and $Y_\alpha^j(x,u)$ and $Z_\alpha^j(x,u)$, $\alpha = 1, 2$, are the solutions of the IVPs

$$\frac{dY_\alpha^j(x,u)}{dx} = \frac{Z_\alpha^j(x,u)}{k(x)}, \quad \frac{dZ_\alpha^j(x,u)}{dx} = -f\left(x, Y_\alpha^j(x,u), \frac{Z_\alpha^j(x,u)}{k(x)}\right), \quad x \in e_\alpha^j,$$
$$Y_\alpha^j(x_\beta, u) = u_\beta, \quad Z_\alpha^j(x_\beta, u) = k(x)\left.\frac{du}{dx}\right|_{x=x_\beta},$$
$$j = 2-\alpha,\ 3-\alpha, \ldots, N+1-\alpha, \quad \alpha = 1, 2, \tag{3.22}$$

and $\bar{V}_\alpha^j(x) = (-1)^{\alpha+1} V_\alpha^j(x)$ is the solution of the IVP

$$\frac{d\bar{V}_\alpha^j(x)}{dx} = \frac{1}{k(x)}, \quad x \in e_\alpha^j,$$
$$\bar{V}_\alpha^j(x_\beta) = 0, \quad j = 2-\alpha, 3-\alpha, \ldots, N+1-\alpha, \quad \alpha = 1, 2. \tag{3.23}$$

Obviously, the right-hand side of the three-point EDS can be represented by

$$\varphi(x_j, u) = \hat{T}^{x_j}\left(f\left(\xi, u(\xi), u_\xi(\xi)\right)\right)$$
$$= \hbar_j^{-1}\sum_{\alpha=1}^{2}(-1)^\alpha\left[Z_\alpha^j(x_j, u) + (-1)^\alpha\frac{Y_\alpha^j(x_j, u) - u_\beta}{V_\alpha^j(x_j)}\right]. \tag{3.24}$$

Therefore, there are two possibilities for computing the coefficients of the EDS (3.10) for all $x_j \in \hat{\omega}_h$. The first one is based on the solution of the BVPs (3.8), whereas the second exploits the relation (3.24) by integrating the IVPs (3.22), (3.23). These integrations can be realized by the following one-step methods:

$$Y_\alpha^{(\bar{m})j}(x_j, u) = u_\beta + (-1)^{\alpha+1} h_\gamma \Phi_1(x_\beta, u_\beta, (ku')_\beta, (-1)^{\alpha+1} h_\gamma), \tag{3.25}$$

$$Z_\alpha^{(m)j}(x_j, u) = (ku')_\beta + (-1)^{\alpha+1} h_\gamma \Phi_2(x_\beta, u_\beta, (ku')_\beta, (-1)^{\alpha+1} h_\gamma), \tag{3.26}$$

$$\bar{V}_\alpha^{(\bar{m})j}(x_j) = (-1)^{\alpha+1} h_\gamma \Phi_3(x_\beta, 0, (-1)^{\alpha+1} h_\gamma), \tag{3.27}$$

where as before

$$\gamma \stackrel{\text{def}}{=} j - 1 + \alpha, \tag{3.28}$$

$\Phi_1(x, u, v, h)$, $\Phi_2(x, u, v, h)$ and $\Phi_3(x, u, h)$ are the corresponding increment functions and

$$(ku')_\beta = k(x) \frac{du}{dx}\bigg|_{x=x_\beta}.$$

Let us assume $Z_\alpha^{(m)j}(x_j, u)$ approximates $Z_\alpha^j(x_j, u)$ with order m and $Y_\alpha^{(\bar{m})j}(x_j, u)$, $\bar{V}_\alpha^{(\bar{m})j}(x_j)$ approximate $Y_\alpha^j(x_j, u)$, $\bar{V}_\alpha^j(x_j)$, resp., with order $\bar{m} = 2[(m+1)/2]$.

If $k(x)$ and the right-hand side $f(x, u, \xi)$ are sufficiently smooth, we have

$$Y_\alpha^j(x_j, u) = Y_\alpha^{(\bar{m})j}(x_j, u) + [(-1)^{\alpha+1} h_\gamma]^{\bar{m}+1} \psi_\alpha^j(x_\beta, u) + O(h_\gamma^{\bar{m}+2}), \tag{3.29}$$

$$Z_\alpha^j(x_j, u) = Z_\alpha^{(m)j}(x_j, u) + [(-1)^{\alpha+1} h_\gamma]^{m+1} \tilde{\psi}_\alpha^j(x_\beta, u) + O(h_\gamma^{m+2}), \tag{3.30}$$

$$\bar{V}_\alpha^j(x_j) = \bar{V}_\alpha^{(\bar{m})j}(x_j) + [(-1)^{\alpha+1} h_\gamma]^{\bar{m}+1} \bar{\psi}_\alpha^j(x_\beta) + O(h_\gamma^{\bar{m}+2}). \tag{3.31}$$

For example, for the Taylor series method we have

$$\Phi_1(x_\beta, u_\beta, (ku')_\beta, (-1)^{\alpha+1} h_\gamma)$$

$$= u'_\beta - \frac{(-1)^{\alpha+1} h_\gamma}{2} \frac{f(x_\beta, u_\beta, u'_\beta)}{k_\beta} + \sum_{p=3}^{\bar{m}} \frac{[(-1)^{\alpha+1} h_\gamma]^{p-1}}{p!} \frac{d^p Y_\alpha^j(x_\beta, u)}{dx^p},$$

$$\Phi_2(x_\beta, u_\beta, (ku')_\beta, (-1)^{\alpha+1} h_\gamma)$$

$$= -f(x_\beta, u_\beta, u'_\beta) + \sum_{p=2}^{m} \frac{[(-1)^{\alpha+1} h_\gamma]^{p-1}}{p!} \frac{d^p Z_\alpha^j(x_\beta, u)}{dx^p},$$

$$\Phi_3(x_\beta, 0, (-1)^{\alpha+1} h_\gamma) = \frac{1}{k_\beta} + \sum_{p=2}^{\bar{m}} \frac{[(-1)^{\alpha+1} h_\gamma]^{p-1}}{p!} \left[\frac{d^{p-1}}{dx^{p-1}} \left(\frac{1}{k(x)} \right) \right] \bigg|_{x=x_\beta},$$

and for explicit Runge-Kutta methods we have

$$\Phi_1(x_\beta, u_\beta, (ku')_\beta, (-1)^{\alpha+1} h_\gamma) = b_1 g_1 + b_2 g_2 + \cdots + b_s g_s,$$

$$\Phi_2(x_\beta, u_\beta, (ku')_\beta, (-1)^{\alpha+1}h_\gamma) = b_1\bar{g}_1 + b_2\bar{g}_2 + \cdots + b_s\bar{g}_s,$$
$$\Phi_3(x_\beta, 0, (-1)^{\alpha+1}h_\gamma) = b_1\tilde{g}_1 + b_2\tilde{g}_2 + \cdots + b_s\tilde{g}_s,$$

where

$$\tilde{g}_i \stackrel{\text{def}}{=} \frac{1}{k(x_\beta + c_i(-1)^{\alpha+1}h_\gamma)},$$

$$g_i \stackrel{\text{def}}{=} \left[(ku')_\beta + (-1)^{\alpha+1}h_\gamma \sum_{p=1}^{i-1} a_{ip}\bar{g}_p\right]\tilde{g}_i,$$

$$\bar{g}_i \stackrel{\text{def}}{=} -f\left(x_\beta + c_i(-1)^{\alpha+1}h_\gamma,\ u_\beta + (-1)^{\alpha+1}h_\gamma \sum_{p=1}^{i-1} a_{ip}g_p, g_i\right), \quad i = 1, 2, \ldots, s.$$

The following lemma shows the order of the error when the exact solutions of the IVPs are replaced by their numerical approximations.

Lemma 3.4

Let $0 < c_1 \le k(x) \le c_2$ for all $x \in [0,1]$, $k(x) \in \mathcal{Q}^{m+1}[0,1]$, and

$$f(x, u, \xi) \in \bigcup_{j=1}^{N} \mathbb{C}^m\left([x_{j-1}, x_j] \times \mathbb{R}^2\right).$$

Suppose that for the numerical one-step methods (3.25)–(3.27) the expansions (3.29)–(3.31) hold. Then

$$Y_\alpha^j(x_j, u) = Y_\alpha^{(\bar{m})j}(x_j, u) + (-1)^{\alpha+1}h_\gamma^{\bar{m}+1}\psi_1^\gamma(x_\beta, u) + O(h_\gamma^{\bar{m}+2}), \tag{3.32}$$

$$Z_\alpha^j(x_j, u) = Z_\alpha^{(m)j}(x_j, u) + \left((-1)^{\alpha+1}h_\gamma\right)^{m+1}\tilde{\psi}_1^\gamma(x_\beta, u) + O(h_\gamma^{m+2}), \tag{3.33}$$

$$V_\alpha^j(x_j) = V_\alpha^{(\bar{m})j}(x_j) + h_\gamma^{\bar{m}+1}\bar{\psi}_1^\gamma(x_\beta) + O(h_\gamma^{\bar{m}+2}), \tag{3.34}$$

$$j = 2-\alpha,\ 3-\alpha, \ldots, N+1-\alpha, \quad \alpha = 1, 2. \tag{3.35}$$

Proof. The expansions (3.29), (3.30) imply

$$\begin{aligned}
&Y_1^j(x_j, u) - Y_1^{(\bar{m})j}(x_j, u)\\
&= Y_1^j(x_j, u) - u(x_j - h_j) - h_j\Phi_1(x_j - h_j, u(x_j - h_j), ku'|_{x=x_j-h_j}, h_j) \qquad (3.36)\\
&= h_j^{\bar{m}+1}\psi_1^j(x_j - h_j, u) + O(h_j^{\bar{m}+2}),
\end{aligned}$$

$$\begin{aligned}
&Z_1^j(x_j, u) - Z_1^{(m)j}(x_j, u)\\
&= Z_1^j(x_j, u) - ku'|_{x=x_j-h_j} - h_j\Phi_2(x_j - h_j, u(x_j - h_j), ku'|_{x=x_j-h_j}, h_j) \qquad (3.37)
\end{aligned}$$

$$= h_j^{m+1}\tilde{\psi}_1^j(x_j - h_j, u) + O(h_j^{m+2}).$$

Moreover, we have

$$Y_2^j(x_j, u) - Y_2^{(\bar{m})j}(x_j, u)$$
$$= Y_2^j(x_j, u) - u_{j+1} + h_{j+1}\Phi_1(x_{j+1}, u_{j+1}, (ku')_{j+1}, -h_{j+1}),$$
$$Z_2^j(x_j, u) - Z_2^{(m)j}(x_j, u)$$
$$= Z_2^j(x_j, u) - (ku')_{j+1} + h_{j+1}\Phi_2(x_{j+1}, u_{j+1}, (ku')_{j+1}, -h_{j+1}).$$

If we replace h_j by $-h_{j+1}$ in (3.36) and (3.37), and take into account the fact that $Y_1^j(x_j, u) = Y_2^j(x_j, u)$ and $Z_1^j(x_j, u) = Z_2^j(x_j, u)$, we obtain

$$Y_2^j(x_j, u) - u_{j+1} + h_{j+1}\Phi_1(x_{j+1}, u_{j+1}, (ku')_{j+1}, -h_{j+1})$$
$$= -h_{j+1}^{\bar{m}+1}\psi_1^{j+1}(x_{j+1}, u) + O(h_{j+1}^{\bar{m}+2}),$$
$$Z_2^j(x_j, u) - (ku')_{j+1} + h_{j+1}\Phi_2(x_{j+1}, u_{j+1}, (ku')_{j+1}, -h_{j+1})$$
$$= (-1)^{m+1}h_{j+1}^{m+1}\tilde{\psi}_1^{j+1}(x_{j+1}, u) + O(h_{j+1}^{m+2}).$$

From this the relations (3.32) and (3.33) follow.

Analogously to (3.32) one can prove the relation

$$\bar{V}_\alpha^j(x_j) = \bar{V}_\alpha^{(\bar{m})j}(x_j) + (-1)^{\alpha+1}h_\gamma^{\bar{m}+1}\bar{\psi}_1^\gamma(x_\beta) + O(h_\gamma^{\bar{m}+2})$$

which due to $\bar{V}_\alpha^j(x) = (-1)^{\alpha+1}V_\alpha^j(x)$ implies (3.34). ■

If in the EDS (3.10), (3.11), (3.21) the exact solutions of the corresponding IVPs are approximated by numerical integration methods, the following truncated difference scheme of rank $\bar{m}$ (hereinafter referred to as $\bar{m}$-TDS) is obtained:

$$\begin{aligned} &(a^{(\bar{m})}y_{\bar{x}}^{(\bar{m})})_{\hat{x},j} = -\varphi^{(\bar{m})}(x_j, y^{(\bar{m})}), \quad j = 1, 2, \dots N-1, \\ &y_0^{(\bar{m})} = \mu_1, \quad y_N^{(\bar{m})} = \mu_2, \end{aligned} \tag{3.38}$$

where

$$a^{(\bar{m})}(x_j) \stackrel{\text{def}}{=} \left(\frac{1}{h_j}V_1^{(\bar{m})}(x_j)\right)^{-1},$$

$$\varphi^{(\bar{m})}(x_j, u) \stackrel{\text{def}}{=} \hbar_j^{-1}\sum_{\alpha=1}^{2}(-1)^\alpha\left(Z_\alpha^{(m)j}(x_j, u) + (-1)^\alpha\frac{Y_\alpha^{(\bar{m})j}(x_j, u) - u_\beta}{V_\alpha^{(\bar{m})j}(x_j)}\right).$$

The following statement establishes the accuracy of the approximation of the input data of this difference scheme.

Lemma 3.5

Under the hypotheses of Lemma 3.4 the following three conditions are satisfied:

$$\left|a^{(\bar{m})}(x_j) - a(x_j)\right| \leq M\, h^{\bar{m}}, \tag{3.39}$$

$$\begin{aligned} &\varphi^{(\bar{m})}(x_j,u) - \varphi(x_j,u) \\ &= \left\{ h_j^{m+1} \left[k(x) \left(\psi_1^j(x,u) - \bar{\psi}_1^j(x) k(x) \frac{du}{dx} \right) - \tilde{\psi}_1^j(x,u) \right]_{x=x_j+0} \right\}_{\hat{x}} \\ &\quad + O\left(\frac{h_j^{m+2} + h_{j+1}^{m+2}}{\hbar_j} \right) \end{aligned} \tag{3.40}$$

for m odd, and

$$\begin{aligned} &\varphi^{(\bar{m})}(x_j,u) - \varphi(x_j,u) \\ &= \left\{ h_j^{m} \left[k(x) \left(\psi_1^j(x,u) - \bar{\psi}_1^j(x) k(x) \frac{du}{dx} \right) \right]_{x=x_j+0} \right\}_{\hat{x}} \\ &\quad + O\left(\frac{h_j^{m+1} + h_{j+1}^{m+1}}{\hbar_j} \right) \end{aligned} \tag{3.41}$$

for m even.

Proof. The estimate (3.39) follows from (3.34) since

$$a^{(\bar{m})}(x_j) - a(x_j) = \frac{h_j[V_1^j(x_j) - V_1^{(\bar{m})j}(x_j)]}{V_1^j(x_j)V_1^{(\bar{m})j}(x_j)} = O(h_j^{\bar{m}}).$$

In order to prove (3.40), (3.41) we first want to point out that

$$\begin{aligned} &\varphi^{(\bar{m})}(x_j,u) - \varphi(x_j,u) \\ &= \hbar_j^{-1} \sum_{\alpha=1}^{2} (-1)^\alpha \left\{ Z_\alpha^{(m)j}(x_j,u) - Z_\alpha^j(x_j,u) \right. \\ &\quad \left. + (-1)^\alpha \left(\frac{Y_\alpha^{(\bar{m})j}(x_j,u) - u_\beta}{V_\alpha^{(\bar{m})j}(x_j)} - \frac{Y_\alpha^j(x_j,u) - u_\beta}{V_\alpha^j(x_j)} \right) \right\}, \end{aligned} \tag{3.42}$$

Lemma 3.4 and the relations

$$V_\alpha^j(x_j) = \frac{h_\gamma}{k_\beta} + O(h_\gamma^2), \qquad Y_\alpha^j(x_j,u) - u_\beta = (-1)^{\alpha+1} h_\gamma \left.\frac{du}{dx}\right|_{x=x_\beta} + O(h_\gamma^2)$$

imply

$$Z_{\alpha}^{(m)j}(x_j,u) - Z_{\alpha}^{j}(x_j,u) = -\left((-1)^{\alpha+1}h_{\gamma}\right)^{m+1}\tilde{\psi}_1^{\gamma}(x_{\beta},u) + O(h_{\gamma}^{m+2})$$

and

$$\frac{Y_{\alpha}^{(\bar{m})j}(x_j,u) - u_{\beta}}{V_{\alpha}^{(\bar{m})j}(x_j)} - \frac{Y_{\alpha}^{j}(x_j,u) - u_{\beta}}{V_{\alpha}^{j}(x_j)}$$

$$= -(-1)^{\alpha+1}h_{\gamma}^{\bar{m}}\left[k(x)\left(\psi_1^{\gamma}(x,u) - \bar{\psi}_1^{\gamma}(x)k(x)\frac{du}{dx}\right)\right]_{x=x_{\beta}} + O(h_{\gamma}^{\bar{m}+1}).$$

Now the equation (3.42) takes the form

$$\varphi^{(\bar{m})}(x_j,u) - \varphi(x_j,u)$$

$$= \frac{1}{\hbar_j}\left\{h_{j+1}^{m+1}\left[k(x)\left(\psi_1^{j+1}(x,u) - \bar{\psi}_1^{j+1}(x)k(x)\frac{du}{dx}\right) - \tilde{\psi}_1^{j+1}(x,u)\right]_{x=x_{j+1}}\right.$$

$$\left. - h_j^{m+1}\left[k(x)\left(\psi_1^{j}(x,u) - \bar{\psi}_1^{j}(x)k(x)\frac{du}{dx}\right) - \tilde{\psi}_1^{j}(x,u)\right]_{x=x_{j-1}}\right\} \tag{3.43}$$

$$+ O\left(\frac{h_j^{m+2} + h_{j+1}^{m+2}}{\hbar_j}\right),$$

provided that m is odd, and

$$\varphi^{(\bar{m})}(x_j,u) - \varphi(x_j,u)$$

$$= \frac{1}{\hbar_j}\left\{h_{j+1}^{m}\left[k(x)\left(\psi_1^{j+1}(x,u) - \bar{\psi}_1^{j+1}(x)k(x)\frac{du}{dx}\right)\right]_{x=x_{j+1}}\right. \tag{3.44}$$

$$\left. - h_j^{m}\left[k(x)\left(\psi_1^{j}(x,u) - \bar{\psi}_1^{j}(x)k(x)\frac{du}{dx}\right)\right]_{x=x_{j-1}}\right\} + O\left(\frac{h_j^{m+1} + h_{j+1}^{m+1}}{\hbar_j}\right),$$

provided that m is even. Due to

$$\left[k(x)\left(\psi_1^{j}(x,u) - \bar{\psi}_1^{j}(x)k(x)\frac{du}{dx}\right)\right]_{x=x_{j-1}}$$

$$= \left[k(x)\left(\psi_1^{j}(x,u) - \bar{\psi}_1^{j}(x)k(x)\frac{du}{dx}\right)\right]_{x=x_j} + O(h_j),$$

$$\tilde{\psi}_1^{j}(x_{j-1},u) = \tilde{\psi}_1^{j}(x_j,u) + O(h_j),$$

the estimates (3.40), (3.41) follow from (3.43), (3.44). ∎

We can now prove the following statement about the accuracy of the $\bar{m}$-TDS.

Theorem 3.3

Let the assumptions of Theorem 3.1 and Lemma 3.4 be satisfied. Then there exists an $h_0 > 0$ such that for all $\{h_j\}_{j=1}^N$ with $h \overset{\text{def}}{=} \max_{1\le j\le N} h_j \le h_0$ the $\bar{m}$-TDS (3.38) has a unique solution $y^{(\bar{m})}$. Moreover, the error of this solution can be estimated as

$$\left\|y^{(\bar{m})} - u\right\|_{1,2,\hat{\omega}_h}^{*} \overset{\text{def}}{=} \left[\left\|y^{(\bar{m})} - u\right\|_{0,2,\hat{\omega}_h}^2 + \left\|k\frac{dy^{(\bar{m})}}{dx} - k\frac{du}{dx}\right\|_{0,2,\hat{\omega}_h}^2\right]^{1/2} \le M\,h^{\bar{m}},$$

where

$$k(x)\frac{dy^{(\bar{m})}}{dx}\bigg|_{x=x_0} = Z_2^{(m)0}\left(x_0, y^{(\bar{m})}\right) + \frac{Y_2^{(\bar{m})0}\left(x_0, y^{(\bar{m})}\right) - y_0^{(\bar{m})}}{V_1^{(\bar{m})1}(x_1)},$$

$$k(x)\frac{dy^{(\bar{m})}}{dx}\bigg|_{x=x_j} = Z_1^{(m)j}\left(x_j, y^{(\bar{m})}\right) + \frac{y_j^{(\bar{m})} - Y_1^{(\bar{m})j}\left(x_j, y^{(\bar{m})}\right)}{V_1^{(\bar{m})j}(x_j)}, \quad j=1,2,\ldots,N,$$

and the constant M is independent of h.

Proof. Let us consider the operator

$$A_h^{(\bar{m})}(x,u) \overset{\text{def}}{=} B_h^{(\bar{m})} - \varphi^{(\bar{m})}(x,u)),$$

where we have used the abbreviation $B_h^{(\bar{m})} \overset{\text{def}}{=} -(a^{(\bar{m})}u_{\bar{x}})_{\hat{x}}$. The equations (3.40), (3.41) yield

$$\begin{aligned}
&\left(A_h^{(\bar{m})}(x,u) - A_h^{(\bar{m})}(x,v), u - v\right)_{\hat{\omega}_h} \\
&= \left(a^{(\bar{m})}\left(u_{\bar{x}} - v_{\bar{x}}\right), u_{\bar{x}} - v_{\bar{x}}\right)_{\hat{\omega}_h} - \left(\varphi^{(\bar{m})}(x,u) - \varphi^{(\bar{m})}(x,v), u - v\right)_{\hat{\omega}_h} \\
&= (A_h(x,u) - A_h(x,v), u - v)_{\hat{\omega}_h} + O\left(h^{\bar{m}}\right).
\end{aligned}$$

Due to (3.13) there exists a $h_0 > 0$ such that for all $\{h_j\}_{j=1}^N$ with $h \le h_0$ the following estimation holds:

$$\left(A_h^{(\bar{m})}(x,u) - A_h^{(\bar{m})}(x,v), u - v\right)_{\hat{\omega}_h} \ge c\,\|u-v\|_{B_h^{(\bar{m})}}^2 \ge 8c\tilde{c}_1\,\|u-v\|_{0,2,\hat{\omega}_h}^2, \tag{3.45}$$

where $0 < c < 1$ and $0 < \tilde{c}_1 \le a^{(\bar{m})}(x)$ are constants.

Therefore the operator $A_h^{(\bar{m})}(x,u)$ is strongly monotone and for $h \le h_0$ the $\bar{m}$-TDS (3.38) has a unique solution $y^{(\bar{m})}(x)$, $x \in \hat{\omega}_h$ (see [82, p.461]).

The error $z(x) \overset{\text{def}}{=} y^{(\bar{m})}(x) - u(x)$, $x \in \hat{\omega}_h$, of the difference scheme (3.38) satisfies the problem

$$\begin{aligned} &\left[a^{(\bar{m})}(x) z_{\bar{x}}(x)\right]_{\hat{x}} + \varphi^{(\bar{m})}(x, y^{(\bar{m})}) - \varphi^{(\bar{m})}(x, u) \\ &= \varphi(x, u) - \varphi^{(\bar{m})}(x, u) + \left[\left(a(x) - a^{(\bar{m})}(x)\right) u_{\bar{x}}(x)\right]_{\hat{x}}, \\ &z(0) = z(1) = 0. \end{aligned} \tag{3.46}$$

Using (3.46) we obtain

$$\begin{aligned} &\left(A_h^{(\bar{m})}(x, u) - A_h^{(\bar{m})}(x, y^{(\bar{m})}), z\right)_{\hat{\omega}_h} \\ &= \left(\varphi(x, u) - \varphi^{(\bar{m})}(x, u), z\right)_{\hat{\omega}_h} + \left(\left(a^{(\bar{m})} - a\right) u_{\bar{x}}, z_{\bar{x}}\right)_{\hat{\omega}_h^+}, \end{aligned} \tag{3.47}$$

which together with (3.45) implies the estimation

$$\left(A_h^{(\bar{m})}(x, u) - A_h^{(\bar{m})}(x, y^{(\bar{m})}), z\right)_{\hat{\omega}_h} \geq c\, \|z\|^2_{B_h^{(\bar{m})}}. \tag{3.48}$$

Using the Cauchy-Bunyakovsky-Schwarz inequality (see Theorem 1.8) we obtain from (3.40) and (3.41) the following estimates for the summands on the right-hand side of (3.47):

$$\begin{aligned} &\left(\left(a^{(\bar{m})} - a\right) u_{\bar{x}}, z_{\bar{x}}\right)_{\hat{\omega}_h^+} \\ &\leq \left\|a^{(\bar{m})} - a\right\|_{0,2,\hat{\omega}_h} \|u_{\bar{x}}\|_{0,2,\hat{\omega}_h^+} \|z_{\bar{x}}\|_{0,2,\hat{\omega}_h^+} \leq M\, h^{\bar{m}} \|z_{\bar{x}}\|_{0,2,\hat{\omega}_h^+} \\ &\leq \frac{M\, h^{\bar{m}}}{\tilde{c}_1} \|z\|_{B_h^{(\bar{m})}}, \end{aligned} \tag{3.49}$$

and

$$\left(\varphi(x, u) - \varphi^{(\bar{m})}(x, u), z\right)_{\hat{\omega}_h} \leq M\, h^{m+1} \|z_{\bar{x}}\|_{0,2,\hat{\omega}_h^+} \leq \frac{M\, h^{m+1}}{\tilde{c}_1} \|z\|_{B_h^{(\bar{m})}} \tag{3.50}$$

if m is odd, and

$$\left(\varphi(x, u) - \varphi^{(\bar{m})}(x, u), z\right)_{\hat{\omega}_h} \leq M\, h^{m} \|z_{\bar{x}}\|_{0,2,\hat{\omega}_h^+} \leq \frac{M\, h^{m}}{\tilde{c}_1} \|z\|_{B_h^{(\bar{m})}} \tag{3.51}$$

if m is even.

The estimates (3.48)–(3.51) yield $\|z\|_{B_h^{(\bar{m})}} \leq M\, h^{\bar{m}}$. Furthermore due to the equivalence of the norms $\|\cdot\|_{1,2,\hat{\omega}_h}$ and $\|\cdot\|_{B_h^{(\bar{m})}}$ we obtain $\|z\|_{1,2,\hat{\omega}_h} \leq M\, h^{\bar{m}}$.

Since $y_0^{(\bar{m})} = Y_2^0(x_0, y^{(\bar{m})})$, $y_j^{(\bar{m})} = Y_1^j(x_j, y^{(\bar{m})})$ and the relations (3.32)–(3.34) hold, we have

$$\begin{aligned}\left|\left[k\frac{dz}{dx}\right]_{x=x_0}\right| \le &\left|Z_2^{(m)0}(x_0, y^{(\bar{m})}) - Z_2^0(x_0, y^{(\bar{m})})\right| + \left|Z_2^0(x_0, y^{(\bar{m})}) - Z_2^0(x_0, u)\right| \\ &+ \frac{1}{\left|V_2^{(\bar{m})0}(x_0)\right|}\left|Y_2^{(\bar{m})0}(x_0, y^{(\bar{m})}) - Y_2^0(x_0, y^{(\bar{m})})\right| \\ \le &M_1\, h^{\bar{m}} + \left|\frac{\partial}{\partial u} Z_2^0(x_0, u)\right|_{u=\tilde{u}} \|z\|_{0,2,\hat{\omega}_h} \le M_2\, h^{\bar{m}},\end{aligned}$$

$$\begin{aligned}\left|\left[k\frac{dz}{dx}\right]_{x=x_j}\right| \le &\left|Z_1^{(m)j}(x_j, y^{(\bar{m})}) - Z_1^j(x_j, y^{(\bar{m})})\right| + \left|Z_1^j(x_j, y^{(\bar{m})}) - Z_1^j(x_j, u)\right| \\ &+ \frac{1}{\left|V_1^{(\bar{m})j}(x_j)\right|}\left|Y_1^j(x_j, y^{(\bar{m})}) - Y_1^{(\bar{m})j}(x_j, y^{(\bar{m})})\right| \\ \le &M_3\, h^{\bar{m}} + \left|\frac{\partial}{\partial u} Z_1^j(x_j, u)\right|_{u=\tilde{u}} \|z\|_{0,2,\hat{\omega}_h} \le M_4\, h^{\bar{m}}, \quad j = 1, 2, \ldots, N,\end{aligned}$$

$$\left\|k\frac{dz}{dx}\right\|_{0,2,\hat{\omega}_h} \le \max_{j=0,1,\ldots,N} \left|\left[k\frac{dz}{dx}\right]_{x=x_j}\right| \le M\, h^{\bar{m}}.$$

Now, using (3.39)–(3.41), we get the claim $\|z\|_{1,2,\hat{\omega}_h}^* \le M\, h^{\bar{m}}$. ∎

In order to solve the nonlinear equations of the $\bar{m}$-TDS (3.38), we use an iteration method which is characterized in the next theorem.

Theorem 3.4

Let the assumptions of Theorem 3.3 be satisfied. Then

$$\left|\varphi^{(\bar{m})}(x, u) - \varphi^{(\bar{m})}(x, v)\right| \le \tilde{L}\, |u - v|,$$

and there exists a $h_0 > 0$ such that for all $\{h_j\}_{j=1}^N$ with $h \le h_0$ the following two relations are satisfied:

$$a^{(\bar{m})}(x) \ge \tilde{c}_1 > 0, \quad \tilde{c}_1 \in \mathbb{R},$$

$$\left(A_h^{(\bar{m})}(x, u) - A_h^{(\bar{m})}(x, v), u - v\right)_{\hat{\omega}_h} \ge c\|u - v\|_{B_h^{(\bar{m})}}^2, \quad 0 < c < 1.$$

Moreover, the iteration method

$$\begin{aligned}&B_h^{(\bar{m})} \frac{y^{(\bar{m},n)} - y^{(\bar{m},n-1)}}{\tau} + A_h^{(\bar{m})}(x, y^{(\bar{m},n-1)}) = 0, \quad x \in \hat{\omega}_h, \\ &y^{(\bar{m},n)}(0) = \mu_1, \quad y^{(\bar{m},n)}(1) = \mu_2, \quad n = 1, 2, \ldots, \qquad (3.52) \\ &y^{(\bar{m},0)}(x) = \frac{V_2(x)}{V_1(1)}\mu_1 + \frac{V_1(x)}{V_1(1)}\mu_2\end{aligned}$$

converges and for the corresponding error we have

$$\left\| y^{(m,n)} - u \right\|^{*}_{1,2,\hat{\omega}_h} \leq M(h^{\bar{m}} + q^n), \tag{3.53}$$

where

$$\tau = \tau_0 \overset{\text{def}}{=} c\left(1 + \frac{\tilde{L}}{8\tilde{c}_1}\right)^{-2}, \quad q \overset{\text{def}}{=} \sqrt{1 - c\tau_0},$$

$$k(x)\frac{dy^{(\bar{m},n)}}{dx}\bigg|_{x=x_0} = Z_2^{(m)0}\left(x_0, y^{(\bar{m},n)}\right) + \frac{Y_2^{(\bar{m})0}\left(x_0, y^{(\bar{m},n)}\right) - y_0^{(\bar{m},n)}}{V_1^{(\bar{m})1}(x_1)},$$

$$k(x)\frac{dy^{(\bar{m},n)}}{dx}\bigg|_{x=x_j} = Z_1^{(m)j}\left(x_j, y^{(\bar{m},n)}\right) + \frac{y_j^{(\bar{m},n)} - Y_1^{(\bar{m})j}\left(x_j, y^{(\bar{m},n)}\right)}{V_1^{(\bar{m})j}(x_j)},$$

and the constant M does not depend on h, m and n.

Proof. Theorem 3.3 implies

$$\begin{aligned} \left\| y^{(m,n)} - u \right\|^{*}_{1,2,\hat{\omega}_h} &\leq \left\| y^{(m)} - u \right\|^{*}_{1,2,\hat{\omega}_h} + \left\| y^{(\bar{m},n)} - y^{(\bar{m})} \right\|^{*}_{1,2,\hat{\omega}_h} \\ &\leq M\, h^{\bar{m}} + \left\| y^{(\bar{m},n)} - y^{(\bar{m})} \right\|^{*}_{1,2,\hat{\omega}_h}. \end{aligned} \tag{3.54}$$

Since $f(x,u,\xi) \in \bigcup_{j=1}^{N} \mathbb{C}^{\bar{m}}([x_{j-1}, x_i] \times \mathbb{R}^2)$, we have

$$\left| \varphi^{(\bar{m})}(x,u) - \varphi^{(\bar{m})}(x,v) \right| \leq \tilde{L}|u - v|.$$

By using the Cauchy-Bunyakovsky-Schwarz inequality (see Theorem 1.8),

$$\begin{aligned} &\left(A_h^{(\bar{m})}(x,u) - A_h^{(\bar{m})}(x,v), w\right)_{\hat{\omega}_h} \\ &\leq \|u - v\|_{B_h^{(\bar{m})}} \|w\|_{B_h^{(\bar{m})}} + \left\| \varphi^{(\bar{m})}(x,u) - \varphi^{(\bar{m})}(x,v) \right\|_{0,2,\hat{\omega}_h} \|w\|_{0,2,\hat{\omega}_h} \\ &\leq \|u - v\|_{B_h^{(\bar{m})}} \|w\|_{B_h^{(\bar{m})}} + \tilde{L}\, \|u - v\|_{0,2,\hat{\omega}_h} \|w\|_{0,2,\hat{\omega}_h} \\ &\leq \|u - v\|_{B_h^{(\bar{m})}} \|w\|_{B_h^{(\bar{m})}} + \frac{\tilde{L}}{8}\, \|u_{\bar{x}} - v_{\bar{x}}\|_{0,2,\hat{\omega}_h^+} \|w_{\bar{x}}\|_{0,2,\hat{\omega}_h^+} \\ &\leq \left(1 + \frac{\tilde{L}}{8\tilde{c}_1}\right) \|u - v\|_{B_h^{(\bar{m})}} \|w\|_{B_h^{(\bar{m})}}. \end{aligned}$$

Setting $w \stackrel{\text{def}}{=} \left(B_h^{(\bar{m})}\right)^{-1}\left(A_h^{(\bar{m})}(x,u) - A_h^{(\bar{m})}(x,v)\right)$, we obtain

$$\left\|\left(B_h^{(\bar{m})}\right)^{-1}\left(A_h^{(\bar{m})}(x,u) - A_h^{(\bar{m})}(x,v)\right)\right\|_{B_h^{(\bar{m})}} \le \left(1 + \frac{\tilde{L}}{8\tilde{c}_1}\right)\|u-v\|_{B_h^{(\bar{m})}}. \quad (3.55)$$

It follows from (3.46), (3.55) that

$$\left(A_h^{(\bar{m})}(x,u) - A_h^{(\bar{m})}(x,v), \left(B_h^{(\bar{m})}\right)^{-1}\left(A_h^{(\bar{m})}(x,u) - A_h^{(\bar{m})}(x,v)\right)\right)_{\hat{\omega}_h}$$

$$\le \left(1 + \frac{\tilde{L}}{8\tilde{c}_1}\right)^2 \|u-v\|^2_{B_h^{(\bar{m})}} \le \frac{1}{c}\left(1 + \frac{\tilde{L}}{8\tilde{c}_1}\right)^2\left(A_h^{(\bar{m})}(x,u) - A_h^{(\bar{m})}(x,v), u-v\right)_{\hat{\omega}_h}.$$

Using the results of [74, p.502] we conclude that the iteration method (3.52) converges in the space $H_{B_h^{(\bar{m})}}$. Its norm is equivalent to the norm of the space $\overset{\circ}{W}{}_2^1(\hat{\omega}_h)$, and so we get

$$\left\|y^{(\bar{m},n)} - y^{(\bar{m})}\right\|_{1,2,\hat{\omega}_h} \le M_1 q^n.$$

Besides we have

$$\left|\left[k\frac{dy^{(\bar{m},n)}}{dx}\right]_{x=x_0} - \left[k\frac{dy^{(\bar{m})}}{dx}\right]_{x=x_0}\right|$$

$$\le \left|Z_2^{(m)0}(x_0, y^{(\bar{m},n)}) - Z_2^{(m)0}(x_0, y^{(\bar{m})})\right|$$

$$+ \frac{1}{\left|V_2^{(\bar{m})0}(x_0)\right|}\left|Y_2^{(\bar{m})0}(x_0, y^{(\bar{m},n)}) - Y_2^{(\bar{m})0}(x_0, y^{(\bar{m})})\right|$$

$$\le \left[\left|\frac{\partial}{\partial u}Z_2^{(m)0}(x_0,u)\right|_{u=\tilde{y}} + \frac{1}{\left|V_2^{(\bar{m})0}(x_0)\right|}\left|\frac{\partial}{\partial u}Y_2^{(\bar{m})0}(x_0,u)\right|_{u=\bar{y}}\right]\left\|y^{(\bar{m},n)} - y^{(\bar{m})}\right\|_{0,2,\hat{\omega}_h}$$

$$\le M_1\left\|y^{(\bar{m},n)} - y^{(\bar{m})}\right\|_{0,2,\hat{\omega}_h},$$

$$\left|\left[k\frac{dy^{(\bar{m},n)}}{dx}\right]_{x=x_j} - \left[k\frac{dy^{(\bar{m})}}{dx}\right]_{x=x_j}\right|$$

$$\le \left|Z_1^{(m)j}(x_j, y^{(\bar{m},n)}) - Z_1^{(m)j}(x_j, y^{(\bar{m})})\right|$$

$$+ \frac{1}{\left|V_1^{(\bar{m})j}(x_j)\right|}\left|Y_1^{(\bar{m})j}(x_j, y^{(\bar{m},n)}) - Y_1^{(\bar{m})j}(x_j, y^{(\bar{m})})\right|$$

$$\leq \left[\left| \frac{\partial}{\partial u} Z_1^{(m)j}(x_j, u) \right|_{u=\tilde{y}} + \frac{1}{\left| V_1^{(\bar{m})j}(x_j) \right|} \left| \frac{\partial}{\partial u} Y_1^{(\bar{m})j}(x_j, u) \right|_{u=\tilde{y}} \right] \left\| y^{(\bar{m},n)} - y^{(\bar{m})} \right\|_{0,2,\hat{\omega}_h}$$

$$\leq M_2 \left\| y^{(\bar{m},n)} - y^{(\bar{m})} \right\|_{0,2,\hat{\omega}_h}, \quad j = 1, 2, \ldots, N,$$

$$\left\| k \frac{dy^{(\bar{m},n)}}{dx} - k \frac{dy^{(\bar{m})}}{dx} \right\|_{0,2,\hat{\hat{\omega}}_h}$$

$$\leq \max_{j=0,1,\ldots,N} \left| \left[k \frac{dy^{(\bar{m},n)}}{dx} \right]_{x=x_j} - \left[k \frac{dy^{(\bar{m})}}{dx} \right]_{x=x_j} \right|$$

$$\leq M \left\| y^{(\bar{m},n)} - y^{(\bar{m})} \right\|_{1,2,\hat{\omega}_h},$$

from which we obtain

$$\left\| y^{(\bar{m},n)} - y^{(\bar{m})} \right\|^*_{1,2,\hat{\omega}_h} \leq M q^n. \tag{3.56}$$

The inequalities (3.54), (3.56) yield the estimate (3.53). ■

It is preferable to compute the solution of (3.38) by a slightly modified version of the well-known Newton's method which was justified in [44]. It is

$$\varphi^{(\bar{m})}(x_j, u) = \hbar_j^{-1} \sum_{\alpha=1}^{2} (-1)^\alpha \Big\{ (ku')_\beta + (-1)^{\alpha+1} h_\gamma \Phi_2 \left(x_\beta, u_\beta, (ku')_\beta, (-1)^{\alpha+1} h_\gamma \right)$$
$$- \frac{\Phi_1 \left(x_\beta, u_\beta, (ku')_\beta, (-1)^{\alpha+1} h_\gamma \right)}{\Phi_3 \left(x_\beta, 0, (-1)^{\alpha+1} h_\gamma \right)} \Big\},$$

where we have used as before the abbreviations

$$\beta \stackrel{\text{def}}{=} j + (-1)^\alpha, \quad \gamma \stackrel{\text{def}}{=} j - 1 + \alpha.$$

From (3.25)–(3.27) we have the relations

$$\Phi_1(x, u, \xi, 0) = \frac{\xi}{k(x)}, \quad \frac{\partial \Phi_1(x, u, \xi, 0)}{\partial h} = -\frac{f(x, u, \xi)}{2k(x)},$$

$$\Phi_2(x, u, \xi, 0) = -f(x, u, \xi), \quad \Phi_3(x, u, 0) = \frac{1}{k(x)}.$$

Thus we can write

$$\varphi^{(\bar{m})}(x_j, y^{(\bar{m})}) = f\Big(x_j, y_j^{(\bar{m})}, \frac{dy^{(\bar{m})}}{dx}\Big|_{x=x_j} \Big) + O\Big(\frac{h_\gamma^2}{\hbar_j} \Big),$$

$$\left.\frac{dy^{(\bar m)}}{dx}\right|_{x=x_j} = y^{(\bar m)}_{\bar x,j} + O\left(\frac{h_\gamma^2}{\hbar_j}\right).$$

The modified Newton's method now reads:

Provide starting values $\nabla y_j^{(\bar m,0)}$, $j = 0, 1, \dots, N$;

For $n = 1, 2, \dots$

1) compute the Newton correction $\nabla y_j^{(\bar m,n)}$, $j = 0, 1, \dots, N$, from the system of linear equations

$$\left(a^{(\bar m)}\nabla y_{\bar x}^{(\bar m,n)}\right)_{\hat x,j} + \frac{\partial f\left(x_j, y_j^{(\bar m,n-1)}, \left.\frac{dy^{(\bar m,n-1)}}{dx}\right|_{x=x_j}\right)}{\partial u}\nabla y_j^{(\bar m,n)}$$

$$+ \frac{\partial f\left(x_j, y_j^{(\bar m,n-1)}, \left.\frac{dy^{(\bar m,n-1)}}{dx}\right|_{x=x_j}\right)}{\partial \xi}\nabla y_{\bar x,j}^{(\bar m,n)}$$

$$= -\varphi^{(\bar m)}\left(x_j, y^{(\bar m,n-1)}\right) - \left(a^{(\bar m)} y_{\bar x}^{(\bar m,n-1)}\right)_{\hat x,j},$$

$$\nabla y_0^{(\bar m,n)} = 0, \quad \nabla y_N^{(\bar m,n)} = 0, \quad j = 1, 2, \dots, N-1,$$

2) compute the new iterate

$$y_j^{(\bar m,n)} = y_j^{(\bar m,n-1)} + \nabla y_j^{(\bar m,n)}, \quad j = 0, 1, \dots, N.$$

3.4 Boundary conditions of 3rd kind

Let us now consider the BVP

$$\frac{d}{dx}\left[k(x)\frac{du}{dx}\right] = -f\left(x, u, \frac{du}{dx}\right), \quad x \in (0,1), \tag{3.57}$$

$$k(0)\frac{du(0)}{dx} - \beta_1 u(0) = -\mu_1, \quad k(1)\frac{du(1)}{dx} + \beta_2 u(1) = \mu_2 \tag{3.58}$$

and derive an exact difference boundary condition at $x = 0$ which is satisfied by the exact solution of problem (3.57), (3.58).

To achieve this, we represent the solution of problem (3.57), (3.58) on the interval $[x_0, x_1]$ in the form

$$u(x) = Y_2^0(x, u), \tag{3.59}$$

where the function $Y_2^0(x,u)$ is the solution of the IVP

$$\frac{dY_2^0(x,u)}{dx} = \frac{Z_2^0(x,u)}{k(x)}, \quad \frac{dZ_2^0(x,u)}{dx} = -f\left(x, Y_2^0(x,u), \frac{Z_2^0(x,u)}{k(x)}\right),$$
$$Y_2^0(x_1,u) = u_1, \quad Z_2^0(x_1,u) = k(x)\frac{du}{dx}\bigg|_{x=x_1}, \quad 0 = x_0 < x < x_1. \tag{3.60}$$

We require that this function satisfies the condition

$$k(0)\frac{du(0)}{dx} - \beta_1 u(0) = -\mu_1. \tag{3.61}$$

Substituting (3.59) into (3.61) we obtain

$$a_1 u_{x,0} - \beta_1 u_0 = -\mu_1 - h_1\varphi(x_0,u), \tag{3.62}$$

where

$$a_1 \stackrel{\text{def}}{=} a(x_1) - \left[\frac{1}{h_1}V_2^0(x_0)\right]^{-1}, \quad \varphi(x_0,u) \stackrel{\text{def}}{=} \frac{1}{h_1}\left[Z_2^0(x_0,u) + \frac{Y_2^0(x_0,u) - u_1}{V_2^0(x_0)}\right].$$

The function $V_2^0(x)$ is the solution of the IVP

$$\frac{dV_2^0(x)}{dx} = \frac{1}{k(x)}, \quad x_0 < x < x_1, \quad V_2^0(x_1) = 0, \tag{3.63}$$

and can be represented explicitly in the form

$$V_2^0(x) = \int_0^{x_1} \frac{1}{k(\xi)}d\xi. \tag{3.64}$$

Analogously the boundary condition

$$k(1)\frac{du(1)}{dx} + \beta_2 u(1) = \mu_2$$

corresponds to the exact difference condition

$$a_N u_{\bar{x},N} + \beta_2 u_N = \mu_2 + h_N\varphi(x_N,u), \tag{3.65}$$

where

$$a_N \stackrel{\text{def}}{=} a(x_N) = \left[\frac{1}{h_N}V_1^N(x_N)\right]^{-1},$$

$$\varphi(x_N,u) \stackrel{\text{def}}{=} \frac{1}{h_N}\left[Z_1^N(x_N,u) - \frac{Y_1^N(x_N,u) - u_{N-1}}{V_1^N(x_N)}\right].$$

The functions $Y_1^N(x,u)$ and $Z_1^N(x,u)$ are the solutions of the IVPs

$$\frac{dY_1^N(x,u)}{dx} = \frac{Z_1^N(x,u)}{k(x)}, \quad \frac{dZ_1^N(x,u)}{dx} = -f\left(x, Y_1^N(x,u), \frac{Z_1^N(x,u)}{k(x)}\right),$$

$$Y_1^N(x_{N-1},u) = u_{N-1}, \quad Z_1^N(x_{N-1},u) = k(x)\frac{du}{dx}\bigg|_{x=x_{N-1}}, \quad x_{N-1} < x < x_N,$$

and the function $V_1^N(x)$ is the solution of the IVP

$$\frac{dV_1^N(x)}{dx} = \frac{1}{k(x)}, \quad x_{N-1} < x < x_N, \quad V_1^N(x_{N-1}) = 0.$$

Let us construct an approximation of the exact boundary condition (3.62). To determine numerically the solution of the IVPs (3.60), (3.63) we use the one-step method

$$\begin{aligned}
Y_2^{(\bar m)0}(x_0,u) &= u_1 - h_1\Phi_1(x_1,u_1,(ku')_1,-h_1),\\
Z_2^{(m)0}(x_0,u) &= (ku')_1 - h_1\Phi_2(x_1,u_1,(ku')_1,-h_1),\\
V_2^{(\bar m)0}(x_0) &= h_1\Phi_3(x_1,0,-h_1).
\end{aligned}$$

Instead of (3.62) one can use a difference rank-$\bar m$-approximation of the boundary condition (3.61):

$$a_1^{(\bar m)}y_{x,0}^{(\bar m)} - \beta_1 y_0^{(\bar m)} = -\mu_1 - h_1\varphi^{(\bar m)}(x_0,y^{(\bar m)}), \quad y_N^{(\bar m)} = \mu_2,$$

where

$$a_1^{(\bar m)} \stackrel{\text{def}}{=} a^{(\bar m)}(x_1) = \left[\frac{1}{h_1}V_2^{(\bar m)0}(x_0)\right]^{-1},$$

$$\varphi^{(\bar m)}(x_0,y^{(\bar m)}) \stackrel{\text{def}}{=} \frac{1}{h_1}\left[Z_2^{(m)0}(x_0,y^{(\bar m)}) + \frac{Y_2^{(\bar m)0}(x_0,y^{(\bar m)}) - y_1^{(\bar m)}}{V_2^{(\bar m)0}(x_0)}\right].$$

Note that the relation $V_2^0(x_0) = \dfrac{h_1}{k_1} + O(h_1^2)$ and Lemma 3.4 imply

$$\begin{aligned}
&a^{(\bar m)}(x_1) - a(x_1)\\
&\quad= \frac{h_1[V_2^0(x_0) - V_2^{(\bar m)0}(x_0)]}{V_2^0(x_0)V_2^{(\bar m)0}(x_0)} = O(h_1^{\bar m}),
\end{aligned}$$

$$\begin{aligned}
&\varphi^{(\bar m)}(x_0,u) - \varphi(x_0,u)\\
&\quad= \frac{1}{h_1}\left\{Z_2^{(m)0}(x_0,u) - Z_2^0(x_0,u) + \frac{Y_2^{(\bar m)0}(x_0,u) - u_1}{V_2^{(\bar m)0}(x_0)} - \frac{Y_2^0(x_0,u) - u_1}{V_2^0(x_0)}\right\}\\
&\quad= O(h_1^{\bar m}).
\end{aligned}$$

This means that the difference scheme

$$\begin{aligned} &(a^{(\bar{m})} y_{\bar{x}}^{(\bar{m})})_{\hat{x},j} = -\varphi^{(\bar{m})}(x_j, y^{(\bar{m})}), \quad j = 1, 2, \dots N-1, \\ &a_1^{(\bar{m})} y_{x,0}^{(\bar{m})} - \beta_1 y_0^{(\bar{m})} = -\mu_1 - h_1 \varphi^{(\bar{m})}(x_0, y^{(\bar{m})}), \quad y_N^{(\bar{m})} = \mu_2 \end{aligned} \tag{3.66}$$

approximates problem (3.57), (3.58) with order of accuracy $\bar{m}$.

Analogously, the approximation of the boundary condition (3.65) is of the form

$$a_N^{(\bar{m})} y_{\bar{x},N} + \beta_2 y_N^{(\bar{m})} = \mu_2 + h_N \varphi^{(\bar{m})}(x_N, u), \quad a_N^{(\bar{m})} = \left[\frac{1}{h_N} V_1^{(\bar{m})N}(x_N)\right]^{-1},$$

$$\varphi^{(\bar{m})}(x_N, y^{(\bar{m})}) = \frac{1}{h_N}\left[Z_1^{(m)N}(x_N, y^{(\bar{m})}) - \frac{Y_1^{(\bar{m})N}(x_N, y^{(\bar{m})}) - y_{N-1}^{(\bar{m})}}{V_1^{(\bar{m})N}(x_N)}\right].$$

3.5 Numerical examples

Example 3.1. Let us consider the BVP

$$u'' = \exp(u), \quad u(0) = 0, \quad u(1) = -\ln\cos^2\left(1/\sqrt{2}\right), \tag{3.67}$$

with the exact solution $u(x) = -\ln\cos^2\left(x/\sqrt{2}\right)$.

Let $f(x, u, \xi) = -\exp(u)$, $f_0(x) \equiv 0$, then the condition (3.5) is fulfilled if we use $c(t) = \exp(t)$, $g(x) \equiv 1$. Besides we have

$$\left[f(x, u, \xi) - f(x, v, \eta)\right](u - v) = -\exp(\theta u + (1-\theta)v)(u-v)^2 \le 0, \quad 0 < \theta < 1.$$

Thus, due to Theorem 3.1 the problem has a unique solution.

In order to solve (3.67) numerically we use the following 6-TDS:

$$y_{\bar{x}\hat{x},j}^{(6)} = -\varphi^{(6)}\left(x_j, y^{(6)}\right), \quad j = 1, 2, \dots, N-1, \quad y_0^{(6)} = \mu_1, \quad y_N^{(6)} = \mu_2, \tag{3.68}$$

where

$$\varphi^{(6)}(x_j, u) = \hbar_j^{-1} \sum_{\alpha=1}^{2} (-1)^\alpha \left[Z_\alpha^{(6)j}(x_j, u) + \frac{(-1)^\alpha}{h_\gamma}\left(Y_\alpha^{(6)j}(x_j, u) - u_\beta\right)\right],$$

$$\mu_1 = 0, \quad \mu_2 = -\ln\cos^2\left(1/\sqrt{2}\right),$$

and $Y_\alpha^{(6)j}(x, u)$, $Z_\alpha^{(6)j}(x, u)$ are the numerical solutions of the IVPs

$$\frac{dY_\alpha^j(x, u)}{dx} = Z_\alpha^j(x, u), \quad \frac{dZ_\alpha^j(x, u)}{dx} = -f\left(x, Y_\alpha^j(x, u), Z_\alpha^j(x, u)\right), \quad x \in e_\alpha^j,$$

$$Y_\alpha^j(x_\beta, u) = u_\beta, \quad Z_\alpha^j(x_\beta, u) = \left.\frac{du}{dx}\right|_{x = x_\beta}, \tag{3.69}$$

$$j = 2-\alpha, 3-\alpha, \ldots, N+1-\alpha, \quad \alpha = 1,2.$$

We have solved the IVPs (3.69) with the explicit 7-stage Runge-Kutta method of order 6 which is characterized by the Butcher tableau given in Table 2.4.

To determine the solution of the difference scheme (3.68) the following fixed point iteration can be used:

- starting values

$$y_j^{(6,0)} = (1-x_j)\mu_1 + x_j\mu_2, \quad j = 0,1,\ldots,N;$$

- iteration procedure ($n = 1,2,\ldots$)

$$\begin{aligned} y_{\bar{x}\hat{x},j}^{(6,n)} &= -\varphi^{(6)}\left(x_j, y^{(6,n-1)}\right), \quad j = 1,2,\ldots,N-1, \\ y_0^{(6,n)} &= \mu_1, \quad y_N^{(6,n)} = \mu_2. \end{aligned} \tag{3.70}$$

The equations (3.70) represent a system of linear algebraic equations for the unknowns $y^{(6,n)}(x)$, $x \in \hat{\omega}_h$. We solved this system with a special Gaussian elimination technique for linear systems with a tridiagonal matrix (see e.g. [30]).

Numerical results for the equidistant grid ω_h are given in Table 3.1. Here, we have used the formulas

$$\mathtt{er} \stackrel{\text{def}}{=} \left\|z^{(6)}\right\|_{1,2,\omega_h}^{*} = \left\|y^{(6)} - u\right\|_{1,2,\omega_h}^{*} \quad \text{and} \quad \mathtt{p} \stackrel{\text{def}}{=} \log_2 \frac{\left\|z^{(6)}\right\|_{1,2,\omega_h}^{*}}{\left\|z^{(6)}\right\|_{1,2,\omega_{h/2}}^{*}} \tag{3.71}$$

to measure the error and the order of convergence, respectively. Obviously, p has been determined by the well-known Runge technique. This and the next two examples should elucidate that the theoretical order of accuracy is actually achieved.

N	er	p
8	$0.1987\,E-8$	
16	$0.3168\,E-10$	6.0
32	$0.4894\,E-12$	6.0

Table 3.1: Numerical results for problem (3.67)

Example 3.2. Let us consider the problem

$$\frac{d^2u}{dx^2} = \frac{1}{2}\cos^2\left(\frac{du}{dx}\right), \quad u(0) = 0, \quad u(1) = \arctan\left(\frac{1}{2}\right) - \ln\left(\frac{5}{4}\right) \tag{3.72}$$

with the exact solution $u(x) = x \arctan\left(\frac{1}{2}x\right) - \ln\left(1 + \frac{1}{4}x^2\right)$.

Here we have $f(x, u, \xi) = -\frac{1}{2}\cos^2 \xi$ and further

$$[f(x,u,\xi) - f(x,v,\eta)]\,(u-v) = -\frac{1}{2}\cos^2\left(\theta\xi + (1-\theta)\eta\right)(\xi-\eta)\,(u-v)$$

$$\leq \frac{1}{2}\cos^2\left(\theta\xi + (1-\theta)\eta\right)\left[\frac{1}{2}(u-v)^2 + \frac{1}{2}(\xi-\eta)^2\right] \leq \frac{1}{4}\left(|u-v|^2 + |\xi-\eta|^2\right),$$

which means that the BVP (3.72) is monotone. The numerical results for the method with order of accuracy 6 ($m = 6$) on the equidistant grid ω_h are given in Table 3.2.

N	er	p
4	$0.6972\,E-8$	
8	$0.1099\,E-9$	6.0
16	$0.1723\,E-11$	6.0

Table 3.2: Numerical results for problem (3.72)

Example 3.3. Let us consider the BVP

$$\frac{d^2u}{dx^2} = u\frac{du}{dx}, \quad \frac{du(0)}{dx} = -0.5/\cosh^2(1/4), \quad u(1) = -\tanh(1/4). \qquad (3.73)$$

The exact solution is $u(x) = \tanh((1-2x)/4)$.

The numerical results which have been obtained for the difference scheme (3.63) of order of accuracy 6 ($m = 6$) on the equidistant grid ω_h are given in Table 3.3.

N	er	p
4	$0.2241\,E-7$	
8	$0.3148\,E-9$	6.2
16	$0.4787\,E-11$	6.0
32	$0.7677\,E-13$	6.0

Table 3.3: Numerical results for problem (3.73)

In the following three examples the implementation of the TDS uses the h-$h/2$-*a posteriori* estimation to achieve a given accuracy `EPS`. The comparison with the true error `Error` shows that this accuracy is actually achieved.

Example 3.4. The next example

$$\frac{d^2u}{dx^2} = Au\frac{du}{dx}, \quad u(0) = \tanh\left(\frac{A}{4}\right), \quad u(1) = -\tanh\left(\frac{A}{4}\right), \tag{3.74}$$

has the exact solution $u(x) = \tanh\left(\frac{A(1-2x)}{4}\right)$. Table 3.4 presents the numerical results for the parameter value $A = 7$.

EPS	N	Error
$1.0\,E-4$	16	$0.937\,E-4$
$1.0\,E-6$	256	$0.234\,E-6$
$1.0\,E-8$	2048	$0.242\,E-8$

Table 3.4: Numerical results for problem (3.74)

Example 3.5. The BVP

$$\frac{d^2u}{dx^2} = \frac{3}{2}u^2, \quad u(0) = 4, \quad u(1) = 1 \tag{3.75}$$

has the exact solution $u(x) = \frac{4}{(1+x)^2}$. The numerical results on the equidistant grid ω_h are given in Table 3.5.

EPS	N	Error
$1.0\,E-4$	32	$0.879\,E-5$
$1.0\,E-6$	64	$0.847\,E-7$
$1.0\,E-8$	512	$0.777\,E-9$

Table 3.5: Numerical results for problem (3.75)

Example 3.6. Let us consider the BVP

$$\frac{d^2u}{dx^2} = \left(\frac{du}{dx}\right)^2, \quad u(0) = 1, \quad u(1) = 0, \tag{3.76}$$

with the exact solution $u(x) = -\ln\left(x + e^{-1}(1-x)\right)$. Numerical results on an equidistant grid ω_h are presented in Table 3.6.

EPS	N	Error
$1.0\,E-4$	128	$0.4899\,E-5$
$1.0\,E-6$	1024	$0.8035\,E-7$
$1.0\,E-8$	8192	$0.1047\,E-8$

Table 3.6: Numerical results for problem (3.76)

The conclusion of this section is that the numerical results which have been obtained for the examples 3.1 – 3.6 confirm our theoretical statements on the order of accuracy of the proposed difference schemes.

Chapter 4

Three-point difference schemes for systems of monotone second-order ODEs

What do I like most about your system? It is as hard to understand as the world.

Franz Grillparzer (1791–1872)

This chapter deals with a generalization of the results from the previous chapter to the case of systems of second-order ODEs with a monotone operator.

For such systems with Dirichlet boundary conditions an exact 3-point EDS on an irregular grid is constructed, and the existence and uniqueness of its solution is shown. The EDS forms the basis for the associated TDS on a 3-point stencil that can be constructed such that it has a required order of accuracy $\bar{m}$. Considering the nonlinearity of the ODE we use the method of monotone operators (see, e.g. [18]) as the standard tool for our theoretical investigations .

4.1 Problem setting

Let us consider the following nonlinear BVP for a system of second-order ODEs

$$\frac{d}{dx}\left[K(x)\frac{d\boldsymbol{u}}{dx}\right] = -\boldsymbol{f}\left(x, \boldsymbol{u}, \frac{d\boldsymbol{u}}{dx}\right), \quad x \in (0,1), \quad \boldsymbol{u}(0) = \boldsymbol{\mu}_1, \quad \boldsymbol{u}(1) = \boldsymbol{\mu}_2, \tag{4.1}$$

where $K(x) \in \mathbb{R}^{n\times n}$ and $\boldsymbol{f}(x,\boldsymbol{u},\boldsymbol{\xi})$, $\boldsymbol{\mu}_1$, $\boldsymbol{\mu}_2 \in \mathbb{R}^n$ are given, and $\boldsymbol{u}(x) \in \mathbb{R}^n$ is the unknown vector.

Let $(\boldsymbol{u},\boldsymbol{v})$ be a scalar product in $\mathbb{R}^n$, $\|\boldsymbol{u}\| \stackrel{\text{def}}{=} (\boldsymbol{u},\boldsymbol{u})^{1/2}$ be the associated norm, and $Q^p[0,1]$ be the space of functions having piece-wise continuous derivatives up to order p with a finite number of discontinuity points of first kind. Moreover, let C_r be a non-negative, continuous function, $f_{0r}(x) \in L_2(0,1)$, $g_r(x) \in L_1(0,1)$, and c_1, c_2, c_3 be some constants.

The following theorem states sufficient conditions for the existence and uniqueness of a solution of problem (4.1).

Theorem 4.1

Let the matrix $K(x) \stackrel{\text{def}}{=} [k_{rs}(x)]_{r,s=1}^n$ *and the function* $\boldsymbol{f}(x,\boldsymbol{u},\boldsymbol{\xi}) \stackrel{\text{def}}{=} \{f_r(x,\boldsymbol{u},\boldsymbol{\xi})\}_{r=1}^n$ *satisfy the conditions*

$$K(x) = K^*(x), \quad k_{rs}(x) \in Q^1[0,1],$$

$$c_1\|\boldsymbol{u}\|^2 \le (K(x)\boldsymbol{u},\boldsymbol{u}) \le c_2\|\boldsymbol{u}\|^2 \quad \text{for all } x \in [0,1],\ \boldsymbol{u} \in \mathbb{R}^n, \quad c_1 > 0, \tag{4.2}$$

$$f_{r\boldsymbol{u}\boldsymbol{\xi}}(x) \stackrel{\text{def}}{=} f_r(x,\boldsymbol{u},\boldsymbol{\xi}) \in Q^0[0,1] \quad \text{for all } \boldsymbol{u},\boldsymbol{\xi} \in \mathbb{R}^n,$$

$$f_{rx}(\boldsymbol{u},\boldsymbol{\xi}) \stackrel{\text{def}}{=} f_r(x,\boldsymbol{u},\boldsymbol{\xi}) \in \mathbb{C}(\mathbb{R}^{2n}) \quad \text{for all } x \in [0,1], \tag{4.3}$$

$$|f_r(x,\boldsymbol{u},\boldsymbol{\xi})| \le C_r\left(\sum_{p=1}^n |u_p|\right)(g_r(x) + |\xi_r|) \quad \text{for all } x \in [0,1],\ \boldsymbol{u},\boldsymbol{\xi} \in \mathbb{R}^n, \tag{4.4}$$

$$\begin{aligned}&(\boldsymbol{f}(x,\boldsymbol{u},\boldsymbol{\xi}) - \boldsymbol{f}(x,\boldsymbol{v},\boldsymbol{\eta}),\boldsymbol{u}-\boldsymbol{v})\\ &\le c_3\left(\|\boldsymbol{u}-\boldsymbol{v}\|^2 + \|\boldsymbol{\xi}-\boldsymbol{\eta}\|^2\right) \quad \text{for all } x \in [0,1],\ \boldsymbol{u},\boldsymbol{v},\boldsymbol{\xi},\boldsymbol{\eta} \in \mathbb{R}^n,\end{aligned} \tag{4.5}$$

$$0 \le c_3 < \frac{\pi^2}{\pi^2+1}c_1. \tag{4.6}$$

Then problem (4.1) has a unique solution $\boldsymbol{u}(x) = \{u_r(x)\}_{r=1}^n$, $u_r(x) \in W_2^1(0,1)$, *with* $u_r(x)$, $\sum_{s=1}^n k_{rs}(x)\frac{du_s}{dx} \in C[0,1]$, $r = 1,2,\ldots,n$.

Proof. Due to (4.3) the components $f_r(x,\boldsymbol{u},\boldsymbol{\xi})$, $r = 1,2,\ldots,n$, of the vector-valued function $\boldsymbol{f}(x,\boldsymbol{u},\boldsymbol{\xi})$ have the Carathéodory property (see Definition 1.12) and belong to $L_1[0,1]$. Therefore we can define the operator $A(x,\boldsymbol{u})$ by the variational equality

$$(A(x,\boldsymbol{u}),\boldsymbol{v}) = \int_0^1 \left(K(x)\frac{d\boldsymbol{u}}{dx},\frac{d\boldsymbol{v}}{dx}\right)dx - \int_0^1 \left(\tilde{\boldsymbol{f}}\left(x,\boldsymbol{u}(x),\frac{d\boldsymbol{u}}{dx}\right),\boldsymbol{v}(x)\right)dx,$$

where

$$\tilde{\boldsymbol{f}}\left(x, \boldsymbol{u}(x), \frac{d\boldsymbol{u}}{dx}\right) \stackrel{\text{def}}{=} \boldsymbol{f}\left(x, \boldsymbol{u}(x), \frac{d\boldsymbol{u}}{dx}\right) - \boldsymbol{f}_0(x), \quad \boldsymbol{f}_0(x) = \{f_{0r}(x)\}_{r=1}^n,$$

which holds for all $\boldsymbol{u}(x)$, $u_r(x) \in W_2^1(0,1)$, and for all $\boldsymbol{v}(x)$, $v_r(x) \in \overset{\circ}{W}{}_2^1(0,1)$. Here we have used

$$\overset{\circ}{W}{}_2^1(0,1) \stackrel{\text{def}}{=} \{v_r(x) : v_r(x) \in W_2^1(0,1), \quad v_r(0) = v_r(1) = 0\}, \quad r = 1, \dots, n.$$

A function $\boldsymbol{u}(x)$ is a weak solution of (4.1), if

$$u_r(x) - \varphi_r(x) \in \overset{\circ}{W}{}_2^1(0,1), \quad r = 1, 2, \dots, n,$$

$$(A(x, \boldsymbol{u}), \boldsymbol{v}) = (\boldsymbol{f}_0(x), \boldsymbol{v}) \quad \text{for all } \boldsymbol{v}(x) \text{ with } v_r(x) \in \overset{\circ}{W}{}_2^1(0,1), \quad r = 1, 2, \dots, n,$$

where $\boldsymbol{\varphi}(x)$ is a vector-valued function with components from $W_2^1(0,1)$, which satisfies the boundary conditions of problem (4.1).
Since $u_r(x) \in W_2^1(0,1)$, $r = 1, 2, \dots, n$, we can deduce $u_r(x) \in \mathbb{C}[0,1]$ and the derivative $\dfrac{du_r}{dx} \in L_2(0,1)$, $r = 1, 2, \dots, n$.

Let us show that the operator $A(x, \boldsymbol{u})$ is bounded. Using the Cauchy-Bunyakovsky-Schwarz inequality (see Theorem 1.8) and taking into account the conditions (4.2) and (4.4) with

$$C_r\left(\sum_{p=1}^n |u_p|\right) \le \tilde{C}_2$$

as well as the inequality (see, e.g. [18], p.112)

$$\|\boldsymbol{v}\|_{C[0,1]} \le \tilde{C}_1 \|\boldsymbol{v}\|_{1,2,(0,1)} \quad \text{for all } \boldsymbol{v}(x) \text{ with } v_r(x) \in W_2^1(0,1),$$

we obtain

$$\begin{aligned}
|(A(x, \boldsymbol{u}), \boldsymbol{v})| &\le \left(\int_0^1 \left\| K(x)\frac{d\boldsymbol{u}}{dx} \right\|^2 dx\right)^{1/2} \left(\int_0^1 \left\| \frac{d\boldsymbol{v}}{dx} \right\|^2 dx\right)^{1/2} \\
&\quad + \|\boldsymbol{v}\|_{C[0,1]} \left(\int_0^1 \left\| \tilde{\boldsymbol{f}}\left(x, \boldsymbol{u}(x), \frac{d\boldsymbol{u}}{dx}\right) \right\|^2 dx\right)^{1/2} \\
&\le \left[c_2 \|\boldsymbol{u}\|_{1,2,(0,1)} + \tilde{C}_1 \|\tilde{\boldsymbol{f}}\|_{0,1,(0,1)} \right] \|\boldsymbol{v}\|_{1,2,(0,1)} \\
&\le \left[(c_2 + \tilde{C}_1\tilde{C}_2) \|\boldsymbol{u}\|_{1,2,(0,1)} + \tilde{C}_1\tilde{C}_2 \|\boldsymbol{g}\|_{0,1,(0,1)} \right] \|\boldsymbol{v}\|_{1,2,(0,1)},
\end{aligned}$$

where

$$\|\boldsymbol{u}\|_{C[0,1]} \stackrel{\text{def}}{=} \max_{x \in [0,1]} \|\boldsymbol{u}(x)\|, \quad \|\boldsymbol{u}\|_{0,1,(0,1)} \stackrel{\text{def}}{=} \int_0^1 \|\boldsymbol{u}(x)\| dx,$$

$$\|\boldsymbol{u}\|_{0,2,(0,1)} \overset{\text{def}}{=} \left[\int_0^1 \|\boldsymbol{u}(x)\|^2 dx \right]^{1/2},$$

$$\|\boldsymbol{u}\|_{0,2,(0,1)} \overset{\text{def}}{=} \left[\|\boldsymbol{u}\|_{0,2,(0,1)}^2 + \left\| \frac{d\boldsymbol{u}}{dx} \right\|_{0,2,(0,1)}^2 \right]^{1/2}.$$

The semi-continuity of the operator $A(x, \boldsymbol{u})$ follows from (4.3). Actually, if $\boldsymbol{u}_n \to \boldsymbol{u}_0$ component-wise in the space $W_2^1(0,1)$, then

$$\boldsymbol{f}\left(x, \boldsymbol{u}_n, \frac{d\boldsymbol{u}_n}{dx}\right) \longrightarrow \boldsymbol{f}\left(x, \boldsymbol{u}_0, \frac{d\boldsymbol{u}_0}{dx}\right), \quad K(x)\frac{d\boldsymbol{u}_n}{dx} \longrightarrow K(x)\frac{d\boldsymbol{u}_0}{dx}$$

component-wise in the space $L_2(0,1)$.

Therefore, for all $\boldsymbol{v}(x)$, $v_r(x) \in \overset{\circ}{W}{}_2^1(0,1)$, $r = 1, 2, .., n$, it holds that

$$\begin{aligned} &\lim_{n\to\infty} (A(x, \boldsymbol{u}_n), \boldsymbol{v}) \\ &= \lim_{n\to\infty} \left[\int_0^1 \left(K(x)\frac{d\boldsymbol{u}_n}{dx}, \frac{d\boldsymbol{v}}{dx} \right) dx - \int_0^1 \left(\boldsymbol{f}\left(x, \boldsymbol{u}_n(x), \frac{d\boldsymbol{u}_n}{dx} \right), \boldsymbol{v}(x) \right) dx \right] \\ &= \int_0^1 \left(K(x)\frac{d\boldsymbol{u}_0}{dx}, \frac{d\boldsymbol{v}}{dx} \right) dx - \int_0^1 \left(\boldsymbol{f}\left(x, \boldsymbol{u}_0(x), \frac{d\boldsymbol{u}_0}{dx} \right), \boldsymbol{v}(x) \right) dx \\ &= (A(x, \boldsymbol{u}_0), \boldsymbol{v}), \end{aligned}$$

i.e the operator $A(x, \boldsymbol{u})$ is semi-continuous.

The operator $A(x, \boldsymbol{u})$ is strongly monotone, since (4.2), (4.5) and the inequality

$$\left\| \frac{d\boldsymbol{v}}{dx} \right\|_{0,2,(0,1)} \geq \frac{\pi}{\sqrt{1+\pi^2}} \|\boldsymbol{v}\|_{1,2,(0,1)} \quad \text{for all } \boldsymbol{v}(x) \text{ with } v_r(x) \in \overset{\circ}{W}{}_2^1(0,1),$$

$r = 1, \ldots, n$, imply

$$\begin{aligned} &(A(x, \boldsymbol{u}) - A(x, \boldsymbol{v}), \boldsymbol{u} - \boldsymbol{v}) \\ &= \int_0^1 \left(K(x) \left[\frac{d\boldsymbol{u}}{dx} - \frac{d\boldsymbol{v}}{dx} \right], \frac{d\boldsymbol{u}}{dx} - \frac{d\boldsymbol{v}}{dx} \right) dx \\ &\quad - \int_0^1 \left(\boldsymbol{f}\left(x, \boldsymbol{u}(x), \frac{d\boldsymbol{u}}{dx} \right) - \boldsymbol{f}\left(x, \boldsymbol{v}(x), \frac{d\boldsymbol{v}}{dx} \right), \boldsymbol{u}(x) - \boldsymbol{v}(x) \right) dx \end{aligned}$$

$$\geq c_1 \left\| \frac{d\boldsymbol{u}}{dx} - \frac{d\boldsymbol{v}}{dx} \right\|^2_{0,2,(0,1)} - c_3 \left\| \boldsymbol{u} - \boldsymbol{v} \right\|^2_{1,2,(0,1)}$$

$$\geq c_4 \left\| \boldsymbol{u} - \boldsymbol{v} \right\|^2_{1,2,(0,1)},$$

where in accordance with (4.6) we have $c_4 = \frac{\pi^2}{\pi^2+1} c_1 - c_3 > 0$, and the components of the vector $\boldsymbol{u} - \boldsymbol{v}$ belong to the space $\overset{\circ}{W}{}^1_2(0,1)$. The strong monotonicity implies the coercivity of the operator $A(x, \boldsymbol{u})$.

Now, The Browder-Minty Theorem (see Theorem 1.3) yields the existence and uniqueness of a solution of (4.1). Due to

$$K(x)\frac{d\boldsymbol{u}}{dx} = \int_0^x \boldsymbol{f}\left(\xi, \boldsymbol{u}(\xi), \frac{d\boldsymbol{u}}{d\xi}\right) d\xi + C,$$

the vector $K(x)\frac{d\boldsymbol{u}}{dx}$ is the indefinite Lebesgue integral which yields that

$$\sum_{s=1}^{n} k_{rs}(x) \frac{du_s}{dx} \in C[0,1], \quad r = 1, 2, \ldots, n.$$

■

4.2 Existence of a three-point EDS

We choose the irregular grid $\hat{\omega}_h$ (see (1.4,a)) so that the set ρ of the discontinuity points of the matrix $K(x)$ and of the vector-valued function $\boldsymbol{f}(x, \boldsymbol{u}, \boldsymbol{\xi})$ is part of $\hat{\omega}_h$, i.e. the number N is large enough to assure $\rho \subseteq \hat{\omega}_h$. For the solution of problem (4.1) at each point $x_i \in \rho$ we demand that the following continuity conditions hold:

$$\boldsymbol{u}(x_i - 0) = \boldsymbol{u}(x_i + 0), \quad K(x)\frac{d\boldsymbol{u}}{dx}\bigg|_{x=x_i-0} = K(x)\frac{d\boldsymbol{u}}{dx}\bigg|_{x=x_i+0} \quad \text{for all } x_i \in \rho.$$

Let us consider the following BVPs on *small* intervals:

$$\begin{aligned} &\frac{d}{dx}\left(K(x)\frac{d\boldsymbol{Y}^j_\alpha(x,\boldsymbol{u})}{dx}\right) = -\boldsymbol{f}\left(x, \boldsymbol{Y}^j_\alpha(x,\boldsymbol{u}), \frac{d\boldsymbol{Y}^j_\alpha(x,\boldsymbol{u})}{dx}\right), \quad x \in e^j_\alpha, \\ &\boldsymbol{Y}^j_\alpha(x_{j-2+\alpha}, \boldsymbol{u}) = \boldsymbol{u}(x_{j-2+\alpha}), \quad \boldsymbol{Y}^j_\alpha(x_{j-1+\alpha}, \boldsymbol{u}) = \boldsymbol{u}(x_{j-1+\alpha}), \\ &j = 2-\alpha, 3-\alpha, \ldots, N+1-\alpha, \quad \alpha = 1, 2. \end{aligned} \tag{4.7}$$

The relationship between the solutions of the BVPs (4.1) and (4.7) is shown in the next lemma.

Lemma 4.1

Let the assumptions of Theorem 4.1 be satisfied. Then, each of the problems (4.7) has a unique solution $\boldsymbol{Y}_\alpha^j(x,\boldsymbol{u})$, $Y_{\alpha,r}^j(x,\boldsymbol{u}) \in W_2^1(e_\alpha^j)$, *where*

$$Y_{\alpha,r}^j(x,\boldsymbol{u}),\ \sum_{s=1}^n k_{rs}(x)\frac{dY_{\alpha,s}^j}{dx} \in \mathbb{C}(\bar{e}_\alpha^j), \quad r=1,2,\ldots,n.$$

These solutions are associated with the solution of problem (4.1) *by the equations*

$$\boldsymbol{u}(x) = \boldsymbol{Y}_\alpha^j(x,\boldsymbol{u}), \quad x \in \bar{e}_\alpha^j,$$
$$j = 2-\alpha, 3-\alpha, \ldots, N+1-\alpha, \quad \alpha = 1,2. \tag{4.8}$$

Proof. Since the function $\boldsymbol{f}(x,\boldsymbol{u},\boldsymbol{\xi})$ possesses the Carathéodory property (see Definition 1.12) and belongs to the space $L_2[0,1]$, we can introduce the nonlinear operator $A_\alpha^j\left(x,\boldsymbol{Y}_\alpha^j\right)$ by the relation

$$\left(A_\alpha^j\left(x,\boldsymbol{Y}_\alpha^j\right),\boldsymbol{v}\right) = \int\limits_{x_{j-2+\alpha}}^{x_{j-1+\alpha}} \left(K(x)\frac{d\boldsymbol{Y}_\alpha^j(x,\boldsymbol{u})}{dx}, \frac{d\boldsymbol{v}(x)}{dx}\right)dx$$
$$- \int\limits_{x_{j-2+\alpha}}^{x_{j-1+\alpha}} \left(\tilde{\boldsymbol{f}}\left(x,\boldsymbol{Y}_\alpha^j(x,\boldsymbol{u}),\frac{d\boldsymbol{Y}_\alpha^j(x,\boldsymbol{u})}{dx}\right),\boldsymbol{v}(x)\right)dx,$$

where

$$\tilde{\boldsymbol{f}}\left(x,\boldsymbol{Y}_\alpha^j(x,\boldsymbol{u}),\frac{d\boldsymbol{Y}_\alpha^j(x,\boldsymbol{u})}{dx}\right) \stackrel{\text{def}}{=} \boldsymbol{f}\left(x,\boldsymbol{Y}_\alpha^j(x,\boldsymbol{u}),\frac{d\boldsymbol{Y}_\alpha^j(x,\boldsymbol{u})}{dx}\right) - \boldsymbol{f}_0(x),$$
$$\boldsymbol{f}_0(x) \stackrel{\text{def}}{=} \{f_{0r}(x)\}_{r=1}^n.$$

This relation is valid for all $\boldsymbol{Y}_\alpha^j(x,\boldsymbol{u})$ with $Y_{\alpha r}^j \in W_2^1(e_\alpha)$, and all functions $\boldsymbol{v}(x)$ with $v_r(x) \in \overset{\circ}{W}{}_2^1(e_\alpha)$, $r=1,2,\ldots,n$. Here we have used the notation

$$\overset{\circ}{W}{}_2^1(e_\alpha) \stackrel{\text{def}}{=} \left\{v(x):\ v(x) \in W_2^1(e_\alpha),\ v(x_{j-2+\alpha}) = v(x_{j-1+\alpha}) = 0\right\}, \quad \alpha = 1,2.$$

The functions $\boldsymbol{Y}_\alpha^j(x,\boldsymbol{u})$, $\alpha = 1,2$, are the weak solutions of problem (4.7), provided that

$$\left(A_\alpha^j\left(x,\boldsymbol{Y}_\alpha^j\right),\boldsymbol{v}\right) = 0, \quad \text{for all } \boldsymbol{v}(x) \text{ with } v_r(x) \in \overset{\circ}{W}{}_2^1(e_\alpha), \quad r=1,2,\ldots,n,$$
$$Y_{\alpha r}^j(x,\boldsymbol{u}) - \varphi_{\alpha r}(x) \in \overset{\circ}{W}{}_2^1(e_\alpha), \quad r=1,2,\ldots,n,$$

where $\varphi_\alpha(x)$ is a given function from $W_2^1(e_\alpha)$ satisfying the boundary conditions (4.7).

The operator $A_\alpha^j\left(x, \boldsymbol{Y}_\alpha^j\right)$ is bounded since the Cauchy-Bunyakovsky-Schwarz inequality (see Theorem 1.8), Minkowski's inequality (see Theorem 1.9) and the formulas (4.2), (4.4) imply

$$\begin{aligned}
\left|\left(A_\alpha^j\left(x, \boldsymbol{Y}_\alpha^j\right), \boldsymbol{v}\right)\right| &\leq \left(\int\limits_{x_{j-2+\alpha}}^{x_{j-1+\alpha}} \left\|K(x)\frac{d\boldsymbol{Y}_\alpha^j}{dx}\right\|^2 dx\right)^{1/2} \left(\int\limits_{x_{j-2+\alpha}}^{x_{j-1+\alpha}} \left\|\frac{d\boldsymbol{v}}{dx}\right\|^2 dx\right)^{1/2} \\
&+ \|\boldsymbol{v}\|_{\mathbb{C}(\bar{e}_\alpha^j)} \left(\int\limits_{x_{j-2+\alpha}}^{x_{j-1+\alpha}} \left\|\boldsymbol{f}\left(x, \boldsymbol{Y}_\alpha^j, \frac{d\boldsymbol{Y}_\alpha^j}{dx}\right)\right\|^2 dx\right)^{1/2} \\
&\leq \left[\left(c_2 + \tilde{C}_1\tilde{C}_2\right)\left\|\boldsymbol{Y}_\alpha^j\right\|_{1,2,e_\alpha} + \tilde{C}_1\tilde{C}_2\|\boldsymbol{g}\|_{0,1,e_\alpha}\right]\|\boldsymbol{v}\|_{1,2,e_\alpha}
\end{aligned}$$

Due to (4.4) the operator $A_\alpha^j(x, \boldsymbol{Y}_\alpha^j)$ is also semi-continuous. Actually, if

$$\boldsymbol{Y}_{\alpha n}^j(x, \boldsymbol{u}) \to \boldsymbol{Y}_{\alpha 0}^j(x, \boldsymbol{u})$$

component-wise in $W_2^1(0,1)$, then

$$\boldsymbol{f}\left(x, \boldsymbol{Y}_{\alpha n}^j, \frac{d\boldsymbol{Y}_{\alpha n}^j}{dx}\right) \longrightarrow \boldsymbol{f}\left(x, \boldsymbol{Y}_{\alpha 0}^j, \frac{d\boldsymbol{Y}_{\alpha 0}^j}{dx}\right), \quad K(x)\frac{d\boldsymbol{Y}_{\alpha n}^j}{dx} \longrightarrow K(x)\frac{d\boldsymbol{Y}_{\alpha 0}^j}{dx}$$

in $L_2(0,1)$ (see, [18], p.113).

Therefore, for all $\boldsymbol{v}$ with $v_r(x) \in \overset{\circ}{W}{}_2^1(e_\alpha)$, $r = 1, 2, \ldots, n$, it holds that

$$\begin{aligned}
&\lim_{n\to\infty}\left(A_\alpha^j(x, \boldsymbol{Y}_{\alpha n}^j), \boldsymbol{v}\right) \\
&= \lim_{n\to\infty}\left[\int\limits_0^1\left(K(x)\frac{d\boldsymbol{Y}_{\alpha n}^j}{dx}, \frac{d\boldsymbol{v}}{dx}\right)dx - \int\limits_0^1\left(\boldsymbol{f}\left(x, \boldsymbol{Y}_{\alpha n}^j, \frac{d\boldsymbol{Y}_{\alpha n}^j}{dx}\right), \boldsymbol{v}(x)\right)dx\right] \\
&= \int\limits_0^1\left(K(x)\frac{d\boldsymbol{Y}_{\alpha 0}^j}{dx}, \frac{d\boldsymbol{v}}{dx}\right)dx - \int\limits_0^1\left(\boldsymbol{f}\left(x, \boldsymbol{Y}_{\alpha 0}^j, \frac{d\boldsymbol{Y}_{\alpha 0}^j}{dx}\right), \boldsymbol{v}(x)\right)dx \\
&= \left(A_\alpha^j(x, \boldsymbol{Y}_{\alpha 0}^j), \boldsymbol{v}\right),
\end{aligned}$$

i.e. the operator $A_\alpha^j\left(x, \boldsymbol{Y}_\alpha^j\right)$ is semi-continuous.

Let us show that $A^j_\alpha\left(x, \boldsymbol{Y}^j_\alpha\right)$ is strongly monotone. We have

$$\left(A^j_\alpha\left(x, \boldsymbol{Y}^j_\alpha\right) - A^j_\alpha\left(x, \tilde{\boldsymbol{Y}}^j_\alpha\right), \boldsymbol{Z}^j_\alpha(x, \boldsymbol{u})\right)$$

$$= \int\limits_{x_{j-2+\alpha}}^{x_{j-1+\alpha}} \left(K(x)\frac{d\boldsymbol{Z}^j_\alpha(x, \boldsymbol{u})}{dx}, \frac{d\boldsymbol{Z}^j_\alpha(x, \boldsymbol{u})}{dx}\right) dx$$

$$- \int\limits_{x_{j-2+\alpha}}^{x_{j-1+\alpha}} \left(\boldsymbol{f}\left(x, \boldsymbol{Y}^j_\alpha(x, \boldsymbol{u}), \frac{d\boldsymbol{Y}^j_\alpha}{dx}\right) - \boldsymbol{f}\left(x, \tilde{\boldsymbol{Y}}^j_\alpha(x, \boldsymbol{u}), \frac{d\tilde{\boldsymbol{Y}}^j_\alpha}{dx}\right), \boldsymbol{Z}^j_\alpha(x, \boldsymbol{u})\right) dx.$$

Due to (4.3), (4.5), and to the inequality

$$\left\|\frac{d\boldsymbol{v}}{dx}\right\|_{0,2,e_\alpha} \geq \frac{\pi}{\sqrt{1+\pi^2}} \|\boldsymbol{v}\|_{1,2,e_\alpha} \quad \text{for all } \boldsymbol{v}(x) \text{ with } v_r(x) \in \overset{\circ}{W}{}^1_2(0,1),$$

$r = 1, \ldots, n$, it follows that

$$\left(A^j_\alpha\left(x, \boldsymbol{Y}^j_\alpha\right) - A^j_\alpha\left(x, \tilde{\boldsymbol{Y}}^j_\alpha\right), \boldsymbol{Z}^j_\alpha(x, \boldsymbol{u})\right)$$

$$\geq c_1 \int\limits_{x_{j-2+\alpha}}^{x_{j-1+\alpha}} \left\|\frac{d\boldsymbol{Z}^j_\alpha}{dx}\right\|^2 dx - c_3 \int\limits_{x_{j-2+\alpha}}^{x_{j-1+\alpha}} \left(\left\|\boldsymbol{Z}^j_\alpha\right\|^2 + \left\|\frac{d\boldsymbol{Z}^j_\alpha}{dx}\right\|^2\right) dx$$

$$\geq c_1 \left\|\frac{d\boldsymbol{Z}^j_\alpha}{dx}\right\|^2_{0,2,e_\alpha} - c_3 \left\|\boldsymbol{Z}^j_\alpha\right\|^2_{1,2,e_\alpha}$$

$$\geq c_4 \left\|\boldsymbol{Z}^j_\alpha\right\|^2_{1,2,e_\alpha},$$

where

$$\boldsymbol{Z}^j_\alpha(x, \boldsymbol{u}) \stackrel{\text{def}}{=} \boldsymbol{Y}^j_\alpha(x, \boldsymbol{u}) - \tilde{\boldsymbol{Y}}^j_\alpha(x, \boldsymbol{u}), \quad c_4 \stackrel{\text{def}}{=} \frac{\pi^2}{1+\pi^2} c_1 - c_3 > 0,$$

and the components of the vector $\boldsymbol{Z}^j_\alpha(x, \boldsymbol{u})$ belong to $\overset{\circ}{W}{}^1_2(e_\alpha)$. The strong monotonicity yields the coercivity of $A^j_\alpha(x, \boldsymbol{Y}^j_\alpha)$.

The Browder-Minty Theorem (see Theorem 1.3) implies the existence of a unique solution of (4.7). Since $\boldsymbol{Y}^j_\alpha(x, \boldsymbol{u})$ is the solution of (4.7), this function is also the solution of problem (4.1), which in accordance with the assumptions of the lemma is unique. ■

Lemma 4.1 is the basis for the next theorem which guarantees the existence of the following EDS.

Theorem 4.2

Let the assumptions of Theorem 4.1 be satisfied. Then there exists the EDS

$$(A\boldsymbol{u}_{\bar{x}})_{\hat{x}} = -\hat{T}^{x_j}\left(\boldsymbol{f}\left(\xi, \boldsymbol{u}(\xi), \frac{d\boldsymbol{u}}{d\xi}\right)\right), \quad x \subset \hat{\omega}_h, \quad \boldsymbol{u}(0) = \boldsymbol{\mu}_1, \quad \boldsymbol{u}(1) = \boldsymbol{\mu}_2, \tag{4.9}$$

where

$$A(x_j) \stackrel{\text{def}}{=} h_j \left[V_1^j(x_j)\right]^{-1},$$

$$\hat{T}^{x_j}(\mathrm{w}(\xi)) \stackrel{\text{def}}{=} \frac{1}{\hbar_j}[V_1^j(x_j)]^{-1} \int\limits_{x_{j-1}}^{x_j} V_1^j(\xi)\mathrm{w}(\xi)\,d\xi + \frac{1}{\hbar_j}[V_2^j(x_j)]^{-1} \int\limits_{x_j}^{x_{j+1}} V_2^j(\xi)\mathrm{w}(\xi)\,d\xi,$$

$$V_1^j(x) \stackrel{\text{def}}{=} \int\limits_{x_{j-1}}^{x} K^{-1}(t)\,dt, \quad V_2^j(x) \stackrel{\text{def}}{=} \int\limits_{x}^{x_{j+1}} K^{-1}(t)\,dt.$$

The EDS (4.9) *has a unique solution $\boldsymbol{u}(x)$, $x \in \hat{\omega}_h$, which coincides with the solution of problem* (4.1) *at the nodes of $\hat{\omega}_h$. The function $\boldsymbol{u}(x)$ in the right-hand side of* (4.9) *is defined by* (4.8) *and depends only on the values $\boldsymbol{u}(x_j)$, $j = 0, 1, \ldots, N$.*

Proof. The application of the operator $\hat{T}^{x_j}$ to equation (4.1) gives

$$\hat{T}^{x_j}\left(\frac{d}{d\xi}\left(K(\xi)\frac{d\boldsymbol{u}}{d\xi}\right)\right) = -\hat{T}^{x_j}\left(\boldsymbol{f}\left(\xi, \boldsymbol{u}(\xi), \frac{d\boldsymbol{u}}{d\xi}\right)\right),$$

where

$$\hat{T}^{x_j}\left(\frac{d}{d\xi}\left(K(\xi)\frac{d\boldsymbol{u}}{d\xi}\right)\right) = \frac{1}{\hbar_j}\left[V_1^j(x_j)\right]^{-1} \int\limits_{x_{j-1}}^{x_j} V_1^j(\xi)\frac{d}{d\xi}\left[K(\xi)\frac{d\boldsymbol{u}}{d\xi}\right] d\xi$$

$$+ \frac{1}{\hbar_j}\left[V_2^j(x_j)\right]^{-1} \int\limits_{x_j}^{x_{j+1}} V_2^j(\xi)\frac{d}{d\xi}\left[K(\xi)\frac{d\boldsymbol{u}}{d\xi}\right] d\xi$$

$$= \frac{1}{\hbar_j}[V_1^j(x_j)]^{-1}\left[V_1^j(x_j)\,K(x)\frac{d\boldsymbol{u}}{d\xi}\bigg|_{\xi=x_j} - \int\limits_{x_{j-1}}^{x_j} \frac{d\boldsymbol{u}}{d\xi}d\xi\right]$$

$$+ \frac{1}{\hbar_j}[V_2^j(x_j)]^{-1}\left[-V_2^j(x_j)\,K(\xi)\frac{d\boldsymbol{u}}{d\xi}\bigg|_{\xi=x_j} + \int\limits_{x_j}^{x_{j+1}} \frac{d\boldsymbol{u}}{d\xi}d\xi\right]$$

$$= \frac{h_{j+1}}{\hbar_j}\left[V_1^{j+1}(x_{j+1})\right]^{-1} \boldsymbol{u}_{x,j} - \frac{h_j}{\hbar_j}\left[V_1^j(x_j)\right]^{-1} \boldsymbol{u}_{\bar{x},j}$$

$$= (A\boldsymbol{u}_{\bar{x}})_{\hat{x},j}\,.$$

This together with (4.8) implies the existence of the two-point EDS (4.9). Let us note that the difference scheme (4.9) is self-adjoint (see, e.g. [58]).

In order to prove the uniqueness of the solution of the EDS (4.9) we introduce the operator

$$A_h(x,\boldsymbol{u}) \stackrel{\text{def}}{=} -(A\boldsymbol{u}_{\bar{x}})_{\hat{x},j} - \hat{T}^{x_j}\left(\boldsymbol{f}\left(\xi, \boldsymbol{u}(\xi), \frac{d\boldsymbol{u}}{d\xi}\right)\right),$$

of which the components are defined on the grid spaces H_h accompanied with the scalar products

$$(\boldsymbol{u},\boldsymbol{v})_{\hat{\omega}_h} \stackrel{\text{def}}{=} \sum_{\xi\in\hat{\omega}_h} \hbar(\xi)(\boldsymbol{u}(\xi),\boldsymbol{v}(\xi)), \quad (\boldsymbol{u},\boldsymbol{v})_{\hat{\omega}_h^+} \stackrel{\text{def}}{=} \sum_{\xi\in\hat{\omega}_h^+} h(\xi)(\boldsymbol{u}(\xi),\boldsymbol{v}(\xi))$$

and the norms

$$\|\boldsymbol{u}\|_{0,2,\hat{\omega}_h} \stackrel{\text{def}}{=} (\boldsymbol{u},\boldsymbol{u})_{\hat{\omega}_h}^{1/2}, \quad \|\boldsymbol{u}\|_{0,2,\hat{\omega}_h^+} \stackrel{\text{def}}{=} (\boldsymbol{u},\boldsymbol{u})_{\hat{\omega}_h^+}^{1/2},$$

$$\|\boldsymbol{u}\|_{1,2,\hat{\omega}_h} \stackrel{\text{def}}{=} \left(\|\boldsymbol{u}\|_{0,2,\hat{\omega}_h}^2 + \|\boldsymbol{u}_{\bar{x}}\|_{0,2,\hat{\omega}_h^+}^2\right)^{1/2}.$$

Due to (4.4) the operator $A_h(x,\boldsymbol{u})$ is continuous. Let us show that $A_h(x,\boldsymbol{u})$ is strongly monotone. Taking into account the relation

$$\sum_{\xi\in\hat{\omega}_h} \hbar(\xi)\left(\hat{T}^{\xi}(\mathrm{w}(\eta)), \mathrm{g}(\xi)\right) = \sum_{j=1}^{N}\int_{x_{j-1}}^{x_j} (\mathrm{w}(\eta), \hat{\mathrm{g}}(\eta))d\eta = \int_0^1 (\mathrm{w}(\eta), \hat{\mathrm{g}}(\eta))\,d\eta,$$

where

$$\hat{\mathrm{g}}(\eta) \stackrel{\text{def}}{=} V_1^j(\eta)\left[V_1^j(x_j)\right]^{-1}\mathrm{g}(x_j) + V_2^{j-1}(\eta)\left[V_1^j(x_j)\right]^{-1}\mathrm{g}(x_{j-1}), \quad x_{j-1}\le\eta\le x_j,$$

we have

$$\left(\hat{T}^{x}\left(\boldsymbol{f}\left(\eta,\boldsymbol{u}(\eta),\frac{d\boldsymbol{u}}{d\eta}\right) - \boldsymbol{f}\left(\eta,\boldsymbol{v}(\eta),\frac{d\boldsymbol{v}}{d\eta}\right)\right), \boldsymbol{u}-\boldsymbol{v}\right)_{\hat{\omega}_h}$$

$$= \sum_{\xi\in\hat{\omega}_h} \hbar(\xi)\left(\hat{T}^{\xi}\left(\boldsymbol{f}\left(\eta,\boldsymbol{u}(\eta),\frac{d\boldsymbol{u}}{d\eta}\right) - \boldsymbol{f}\left(\eta,\boldsymbol{v}(\eta),\frac{d\boldsymbol{v}}{d\eta}\right)\right), \boldsymbol{u}(\xi)-\boldsymbol{v}(\xi)\right)$$

$$= \int_0^1 \left(\hat{\boldsymbol{u}}(\eta)-\hat{\boldsymbol{v}}(\eta), \boldsymbol{f}\left(\eta,\boldsymbol{u}(\eta),\frac{d\boldsymbol{u}}{d\eta}\right) - \boldsymbol{f}\left(\eta,\boldsymbol{v}(\eta),\frac{d\boldsymbol{v}}{d\eta}\right)\right)d\eta.$$

The functions $\boldsymbol{u}(x)$ and $\boldsymbol{v}(x)$ are defined by formulas of the form (4.8). Now, using (4.5), we obtain

$$\left(\hat{T}^x\left(\boldsymbol{f}\left(\eta, \boldsymbol{u}(\eta), \frac{d\boldsymbol{u}}{d\eta}\right) - \boldsymbol{f}\left(\eta, \boldsymbol{v}(\eta), \frac{d\boldsymbol{v}}{d\eta}\right)\right), \boldsymbol{u} - \boldsymbol{v}\right)_{\hat{\omega}_h}$$

$$= \int_0^1 \left(\boldsymbol{u}(\eta) - \boldsymbol{v}(\eta), \boldsymbol{f}\left(\eta, \boldsymbol{u}(\eta), \frac{d\boldsymbol{u}}{d\eta}\right) - \boldsymbol{f}\left(\eta, \boldsymbol{v}(\eta), \frac{d\boldsymbol{v}}{d\eta}\right)\right) d\eta$$

$$+ \int_0^1 \left(\hat{\boldsymbol{u}}(\eta) - \hat{\boldsymbol{v}}(\eta) - \boldsymbol{u}(\eta) + \boldsymbol{v}(\eta), \boldsymbol{f}\left(\eta, \boldsymbol{u}(\eta), \frac{d\boldsymbol{u}}{d\eta}\right) - \boldsymbol{f}\left(\eta, \boldsymbol{v}(\eta), \frac{d\boldsymbol{v}}{d\eta}\right)\right) d\eta$$

$$\leq c_3 \|\boldsymbol{u} - \boldsymbol{v}\|^2_{1,2,(0,1)} - \int_0^1 \left(\hat{\boldsymbol{u}}(\eta) - \hat{\boldsymbol{v}}(\eta), \frac{d}{d\eta}\left\{K(\eta)\left[\frac{d\boldsymbol{u}}{d\eta} - \frac{d\boldsymbol{v}}{d\eta}\right]\right\}\right) d\eta$$

$$+ \int_0^1 \left(\boldsymbol{u}(\eta) - \boldsymbol{v}(\eta), \frac{d}{d\eta}\left\{K(\eta)\left[\frac{d\boldsymbol{u}}{d\eta} - \frac{d\boldsymbol{v}}{d\eta}\right]\right\}\right) d\eta$$

$$= -\sum_{j=1}^{N} \int_{x_{j-1}}^{x_j} \left(\hat{\boldsymbol{u}}(\eta) - \hat{\boldsymbol{v}}(\eta), \frac{d}{d\eta}\left\{K(\eta)\left[\frac{d\boldsymbol{u}}{d\eta} - \frac{d\boldsymbol{v}}{d\eta}\right]\right\}\right) d\eta$$

$$- \int_0^1 \left(K(\eta)\left[\frac{d\boldsymbol{u}}{d\eta} - \frac{d\boldsymbol{v}}{d\eta}\right], \frac{d\boldsymbol{u}}{d\eta} - \frac{d\boldsymbol{v}}{d\eta}\right) d\eta + c_3 \|\boldsymbol{u} - \boldsymbol{v}\|^2_{1,2,(0,1)}.$$

Since

$$\sum_{j=1}^{N} \int_{x_{j-1}}^{x_j} \left(\hat{\boldsymbol{u}}(\eta) - \hat{\boldsymbol{v}}(\eta), \frac{d}{d\eta}\left\{K(\eta)\left[\frac{d\boldsymbol{u}}{d\eta} - \frac{d\boldsymbol{v}}{d\eta}\right]\right\}\right) d\eta$$

$$= -\sum_{j=1}^{N} \int_{x_{j-1}}^{x_j} \left(K(\eta)\left[\frac{d\hat{\boldsymbol{u}}}{d\eta} - \frac{d\hat{\boldsymbol{v}}}{d\eta}\right], \frac{d\boldsymbol{u}}{d\eta} - \frac{d\boldsymbol{v}}{d\eta}\right) d\eta$$

$$= -\left(A\left(\boldsymbol{u}_{\bar{x}} - \boldsymbol{v}_{\bar{x}}\right), \boldsymbol{u}_{\bar{x}} - \boldsymbol{v}_{\bar{x}}\right)_{\hat{\omega}_h^+}.$$

it follows that

$$\left(\hat{T}^x\left(\boldsymbol{f}\left(\eta, \boldsymbol{u}(\eta), \frac{d\boldsymbol{u}}{d\eta}\right) - \boldsymbol{f}\left(\eta, \boldsymbol{v}(\eta), \frac{d\boldsymbol{v}}{d\eta}\right)\right), \boldsymbol{u} - \boldsymbol{v}\right)_{\hat{\omega}_h}$$

$$\leq \left(A\left(\boldsymbol{u}_{\bar{x}} - \boldsymbol{v}_{\bar{x}}\right), \boldsymbol{u}_{\bar{x}} - \boldsymbol{v}_{\bar{x}}\right)_{\hat{\omega}_h^+} - c_1 \left\|\frac{d\boldsymbol{u}}{dx} - \frac{d\boldsymbol{v}}{dx}\right\|^2_{0,2,(0,1)} + c_3 \left\|\boldsymbol{u} - \boldsymbol{v}\right\|^2_{1,2,(0,1)}. \tag{4.10}$$

Taking into account (4.3) and the inequality

$$\left\|\frac{d\boldsymbol{v}}{dx}\right\|_{0,2,(0,1)} \leq \frac{\pi}{\sqrt{1+\pi^2}} \|\boldsymbol{v}\|_{1,2,(0,1)}, \quad v_r(x) \in \overset{\circ}{W}{}^1_2(0,1),$$

we obtain

$$\begin{aligned}
&\left(A_h(x, \boldsymbol{u}) - A_h(x, \boldsymbol{v}), \boldsymbol{u} - \boldsymbol{v}\right)_{\hat{\omega}_h} \\
&= (A(\boldsymbol{u}_{\bar{x}} - \boldsymbol{v}_{\bar{x}}), \boldsymbol{u}_{\bar{x}} - \boldsymbol{v}_{\bar{x}})_{\hat{\omega}_h^+} \\
&\quad - \left(\hat{T}^x\left(\boldsymbol{f}\left(\eta, \boldsymbol{u}(\eta), \frac{d\boldsymbol{u}}{d\eta}\right) - \boldsymbol{f}\left(\eta, \boldsymbol{v}(\eta), \frac{d\boldsymbol{v}}{d\eta}\right)\right), \boldsymbol{u} - \boldsymbol{v}\right)_{\hat{\omega}_h} \\
&\geq c_1 \left\|\frac{d\boldsymbol{u}}{dx} - \frac{d\boldsymbol{v}}{dx}\right\|^2_{0,2,(0,1)} - c_3 \|\boldsymbol{u} - \boldsymbol{v}\|^2_{1,2,(0,1)} \\
&\geq c_4 \frac{1+\pi^2}{\pi^2} \left\|\frac{d\boldsymbol{u}}{dx} - \frac{d\boldsymbol{v}}{dx}\right\|^2_{0,2,(0,1)},
\end{aligned}$$

where $c_4 \overset{\text{def}}{=} \dfrac{\pi^2}{1+\pi^2} c_1 - c_3 > 0$. Since

$$\begin{aligned}
&\int_0^1 \left(K(\eta)\left[\frac{d\hat{\boldsymbol{u}}}{d\eta} - \frac{d\hat{\boldsymbol{v}}}{d\eta} - \frac{d\boldsymbol{u}}{d\eta} + \frac{d\boldsymbol{v}}{d\eta}\right], \frac{d\hat{\boldsymbol{u}}}{d\eta} - \frac{d\hat{\boldsymbol{v}}}{d\eta} - \frac{d\boldsymbol{u}}{d\eta} + \frac{d\boldsymbol{v}}{d\eta}\right) d\eta \\
&= -\sum_{j=1}^N h_j \left(A(x_j)\left(u_{\bar{x},j} - v_{\bar{x},j}\right), u_{\bar{x},j} - v_{\bar{x},j}\right) \\
&\quad + \sum_{j=1}^N \int_{x_{j-1}}^{x_j} \left(K(\eta)\left[\frac{d\boldsymbol{u}}{d\eta} - \frac{d\boldsymbol{v}}{d\eta}\right], \frac{d\boldsymbol{u}}{d\eta} - \frac{d\boldsymbol{v}}{d\eta}\right) d\eta \\
&= -\left(A\left(\boldsymbol{u}_{\bar{x}} - \boldsymbol{v}_{\bar{x}}\right), \boldsymbol{u}_{\bar{x}} - \boldsymbol{v}_{\bar{x}}\right)_{\hat{\omega}_h^+} + \int_0^1 \left(K(\eta)\left[\frac{d\boldsymbol{u}}{d\eta} - \frac{d\boldsymbol{v}}{d\eta}\right], \frac{d\boldsymbol{u}}{d\eta} - \frac{d\boldsymbol{v}}{d\eta}\right) d\eta \geq 0,
\end{aligned}$$

we get

$$\left\|\frac{d\boldsymbol{u}}{dx} - \frac{d\boldsymbol{v}}{dx}\right\|^2_{0,2,(0,1)} \geq \frac{1}{c_2}\left(A\left(\boldsymbol{u}_{\bar{x}} - \boldsymbol{v}_{\bar{x}}\right), \boldsymbol{u}_{\bar{x}} - \boldsymbol{v}_{\bar{x}}\right)_{\hat{\omega}_h^+}.$$

Thus,

$$
\begin{aligned}
(A_h(x,\boldsymbol{u}) - A_h(x,\boldsymbol{v}), \boldsymbol{u}-\boldsymbol{v})_{\hat{\omega}_h} &\geq \frac{1+\pi^2}{\pi^2}\frac{c_4}{c_2}\left(A(\boldsymbol{u}_{\bar{x}} - \boldsymbol{v}_{\bar{x}}), \boldsymbol{u}_{\bar{x}} - \boldsymbol{v}_{\bar{x}}\right)_{\hat{\omega}_h^+} \\
&\geq \frac{1+\pi^2}{\pi^2}\frac{c_4 c_1}{c_2}\|\boldsymbol{u}_{\bar{x}} - \boldsymbol{v}_{\bar{x}}\|^2_{0,2,\hat{\omega}_h^+} \geq 8\frac{1+\pi^2}{\pi^2}\frac{c_4 c_1}{c_2}\|\boldsymbol{u}-\boldsymbol{v}\|^2_{0,2,\hat{\omega}_h},
\end{aligned}
\tag{4.11}
$$

i.e. the operator $A_h(x,\boldsymbol{u})$ is strongly monotone. This implies (see, e.g. [82], p.461) that the equation $A_h(x,\boldsymbol{u}) = 0$ possesses a unique solution. ■

The EDS represents a nonlinear operator equation which can be iteratively solved as stated in the following lemma.

Lemma 4.2

Let the assumptions of Theorem 4.1 be satisfied. Suppose that

$$\|\boldsymbol{f}(x,\boldsymbol{u},\boldsymbol{\xi}) - \boldsymbol{f}(x,\boldsymbol{v},\boldsymbol{\eta})\| \leq L\{\|\boldsymbol{u}-\boldsymbol{v}\| + \|\boldsymbol{\xi}-\boldsymbol{\eta}\|\} \text{ for all } x\in(0,1),\ \boldsymbol{u},\boldsymbol{v},\boldsymbol{\xi},\boldsymbol{\eta}\in\mathbb{R}^n.$$

Then the stationary iteration method

$$
\begin{aligned}
&B_h\frac{\boldsymbol{u}^{(n)} - \boldsymbol{u}^{(n-1)}}{\tau} + A_h(x,\boldsymbol{u}^{(n-1)}) = 0, \quad x\in\hat{\omega}_h, \\
&\boldsymbol{u}^{(n)}(0) = \boldsymbol{\mu}_1, \quad \boldsymbol{u}^{(n)}(1) = \boldsymbol{\mu}_2, \quad n = 1,2,\ldots, \\
&\boldsymbol{u}^{(0)}(x) = V_2(x)[V_1(1)]^{-1}\boldsymbol{\mu}_1 + V_1(x)[V_1(1)]^{-1}\boldsymbol{\mu}_2, \\
&B_h\boldsymbol{u} \stackrel{\text{def}}{=} -(A\boldsymbol{u}_{\bar{x}})_{\hat{x}}, \quad A_h(x,\boldsymbol{u}) \stackrel{\text{def}}{=} B_h\boldsymbol{u} - \hat{T}^x\left(\boldsymbol{f}\left(\xi,\boldsymbol{u}(\xi),\frac{d\boldsymbol{u}}{d\xi}\right)\right), \\
&\tau \stackrel{\text{def}}{=} \tau_0 = \frac{1+\pi^2}{\pi^2}\frac{c_4}{c_2}\left[1+\sqrt{\frac{3}{2}}L\frac{c_2^2}{c_1^3}\left(1+\frac{\sqrt{2}L}{c_4}\right)\right]^{-2},
\end{aligned}
\tag{4.12}
$$

converges in the space H_{B_h}. Moreover, the following error estimation holds:

$$\left\|\boldsymbol{u}^{(n)} - \boldsymbol{u}\right\|_{B_h} \leq q^n\left\|\boldsymbol{u}^{(0)} - \boldsymbol{u}\right\|_{B_h}, \tag{4.13}$$

with

$$q \stackrel{\text{def}}{=} \sqrt{1 - \frac{1+\pi^2}{\pi^2}\frac{c_4}{c_2}\tau_0}, \quad c_4 \stackrel{\text{def}}{=} \frac{\pi^2}{1+\pi^2}c_1 - c_3 > 0, \quad \|\boldsymbol{u}\|_{B_h} \stackrel{\text{def}}{=} (B_h\boldsymbol{u},\boldsymbol{u})^{1/2}_{\hat{\omega}_h}.$$

Proof. The inequality (4.11) implies

$$\left(A_h(x,\boldsymbol{u}) - A_h(x,\boldsymbol{v}), \boldsymbol{u}-\boldsymbol{v}\right)_{\hat{\omega}_h} \geq \frac{1+\pi^2}{\pi^2}\frac{c_4}{c_2}\|\boldsymbol{u}-\boldsymbol{v}\|^2_{B_h}. \tag{4.14}$$

Using the Cauchy-Bunyakovsky-Schwarz inequality (see Theorem 1.8) we obtain

$$
\begin{aligned}
&(A_h(x,\boldsymbol{u}) - A_h(x,\boldsymbol{v}), \boldsymbol{z})_{\hat{\omega}_h} \\
&= (B_h\boldsymbol{u} - B_h\boldsymbol{v}, \boldsymbol{z})_{\hat{\omega}_h} \qquad (4.15)\\
&\quad - \sum_{\xi\in\hat{\omega}_h} \hbar(\xi)\left(\hat{T}^\xi\left(\boldsymbol{f}\left(\eta,\boldsymbol{u}(\eta),\frac{d\boldsymbol{u}}{d\eta}\right) - \boldsymbol{f}\left(\eta,\boldsymbol{v}(\eta),\frac{d\boldsymbol{v}}{d\eta}\right)\right),\boldsymbol{z}(\xi)\right)\\
&= (B_h\boldsymbol{u} - B_h\boldsymbol{v}, \boldsymbol{z})_{\hat{\omega}_h} - \int_0^1\left(\boldsymbol{f}\left(\eta,\boldsymbol{u}(\eta),\frac{d\boldsymbol{u}}{d\eta}\right) - \boldsymbol{f}\left(\eta,\boldsymbol{v}(\eta),\frac{d\boldsymbol{v}}{d\eta}\right),\hat{\boldsymbol{z}}(\eta)\right)d\eta\\
&\le \|\boldsymbol{u}-\boldsymbol{v}\|_{B_h}\|\boldsymbol{z}\|_{B_h} \qquad (4.16)\\
&\quad + \left(\int_0^1\left\|\boldsymbol{f}\left(\eta,\boldsymbol{u},\frac{d\boldsymbol{u}}{d\eta}\right) - \boldsymbol{f}\left(\eta,\boldsymbol{v},\frac{d\boldsymbol{v}}{d\eta}\right)\right\|^2 d\eta\right)^{\frac{1}{2}}\left(\int_0^1\|\hat{\boldsymbol{z}}(\eta)\|^2 d\eta\right)^{\frac{1}{2}}\\
&\le \|\boldsymbol{u}-\boldsymbol{v}\|_{B_h}\|\boldsymbol{z}\|_{B_h} + \sqrt{2}L\|\boldsymbol{u}-\boldsymbol{v}\|_{1,2,(0,1)}\|\hat{\boldsymbol{z}}\|_{0,2,(0,1)}. \qquad (4.17)
\end{aligned}
$$

Formula (4.2) implies

$$
\left[V_1^j(x_j)\right]^* = V_1^j(x_j), \quad \left(V_1^j(x_j)\boldsymbol{u},\boldsymbol{u}\right) = \int_{x_{j-1}}^{x_j}\left(K^{-1}(t)\boldsymbol{u},\boldsymbol{u}\right)dt \ge \frac{h_j}{c_2}\|\boldsymbol{u}\|^2,
$$

and further

$$
\left\|\left[V_1^j(x_j)\right]^{-1}\right\| \le \frac{c_2}{h_j}, \quad \left\|K^{-1}(t)\right\| \le \frac{1}{c_1}.
$$

Using the last estimate as well as the inequality

$$
\left\|V_\alpha^j(x)\right\| \le (-1)^{\alpha+1}\int_{x_{j+(-1)^\alpha}}^{x}\left\|K^{-1}(t)\right\|dt \le \frac{h_{j-1+\alpha}}{c_1}, \; x\in\bar{e}_\alpha^j, \quad \alpha=1,2,
$$

we obtain

$$
\begin{aligned}
\|\hat{\boldsymbol{z}}\|_{0,2,(0,1)}^2 &= \int_0^1\left\|V_1^j(x)\left[V_1^j(x_j)\right]^{-1}\boldsymbol{z}_j + V_2^{j-1}(x)\left[V_1^j(x_j)\right]^{-1}\boldsymbol{z}_{j-1}\right\|^2 dx\\
&\le 2\sum_{j=1}^N\int_{x_{j-1}}^{x_j}\left[\left(\left\|V_1^j(x)\right\|\left\|\left[V_1^j(x_j)\right]^{-1}\right\|\|\boldsymbol{z}_j\|\right)^2\right.\\
&\quad\left.+\left(\left\|V_2^{j-1}(x)\right\|\left\|\left[V_1^j(x_j)\right]^{-1}\right\|\|\boldsymbol{z}_{j-1}\|\right)^2\right]dx = \frac{4c_2^2}{c_1^2}\|\boldsymbol{z}\|_{0,2,\hat{\omega}_h}^2, \qquad (4.18)
\end{aligned}
$$

and

$$\begin{aligned}
\left\|\frac{d\hat{\boldsymbol{z}}}{dx}\right\|^2_{0,2,(0,1)} &= \int\limits_0^1 \left\| K^{-1}(x)\left[V_1^j(x_j)\right]^{-1}(\boldsymbol{z}_j - \boldsymbol{z}_{j-1})\right\|^2 dx \\
&\leq \sum_{j=1}^{N} \int\limits_{x_{j-1}}^{x_j} \left[\left\|K^{-1}(x)\right\| \left\|\left[V_1^j(x_j)\right]^{-1}\right\| h_j \left\|\boldsymbol{z}_{\bar{x},j}\right\|\right]^2 \\
&\leq \frac{c_2^2}{c_1^2}\left\|\boldsymbol{z}_{\bar{x}}\right\|^2_{0,2,\hat{\omega}_h^+}.
\end{aligned} \tag{4.19}$$

Let us prove the estimate

$$\|\boldsymbol{u} - \boldsymbol{v}\|_{1,2,(0,1)} \leq \sqrt{\frac{3}{2}\frac{c_2}{c_1}}\left(1 + \frac{\sqrt{2}L}{c_4}\right)\|\boldsymbol{u}_{\bar{x}} - \boldsymbol{v}_{\bar{x}}\|_{0,2,\hat{\omega}_h^+}. \tag{4.20}$$

To achieve this we write $\boldsymbol{u}(x) = \tilde{\boldsymbol{u}}(x) + \hat{\boldsymbol{u}}(x)$, $x_{j-1} \leq x \leq x_{j+1}$, and reduce the problem

$$\begin{gathered}
\frac{d}{dx}\left[K(x)\frac{d\boldsymbol{u}}{dx}\right] = -\boldsymbol{f}\left(x, \boldsymbol{u}, \frac{d\boldsymbol{u}}{dx}\right), \quad x \in e_\alpha^j, \\
\boldsymbol{u}(x_{j-2+\alpha}) = \boldsymbol{u}_{j-2+\alpha}, \quad \boldsymbol{u}(x_{j-1+\alpha}) = \boldsymbol{u}_{j-1+\alpha}, \quad \alpha = 1, 2.
\end{gathered}$$

to the problem

$$\begin{gathered}
\frac{d}{dx}\left[K(x)\frac{d\tilde{\boldsymbol{u}}}{dx}\right] = -\boldsymbol{f}\left(x, \tilde{\boldsymbol{u}}(x) + \hat{\boldsymbol{u}}(x), \frac{d\tilde{\boldsymbol{u}}}{dx} + \frac{d\hat{\boldsymbol{u}}}{dx}\right), \quad x \in e_\alpha^j, \\
\tilde{\boldsymbol{u}}(x_{j-2+\alpha}) = 0, \quad \tilde{\boldsymbol{u}}(x_{j-1+\alpha}) = 0, \quad \alpha = 1, 2.
\end{gathered}$$

Using (4.3), (4.5) and a Lipschitz condition for all vectors $\tilde{\boldsymbol{u}}(x)$, $\tilde{\boldsymbol{v}}(x)$ with $\tilde{u}_r(x), \tilde{v}_r(x) \in \overset{\circ}{W}{}_2^1(e)$, $r = 1, \ldots, n$, we have

$$\begin{aligned}
&c_1\frac{\pi^2}{1+\pi^2}\|\tilde{\boldsymbol{u}} - \tilde{\boldsymbol{v}}\|^2_{1,2,e_\alpha} \\
&\leq c_1\left\|\frac{d\tilde{\boldsymbol{u}}}{dx} - \frac{d\tilde{\boldsymbol{v}}}{dx}\right\|^2_{0,2,e_\alpha} \\
&\leq \int\limits_{x_{j-2+\alpha}}^{x_{j-1+\alpha}} \left(K(x)\left(\frac{d\tilde{\boldsymbol{u}}}{dx} - \frac{d\tilde{\boldsymbol{v}}}{dx}\right), \frac{d\tilde{\boldsymbol{u}}}{dx} - \frac{d\tilde{\boldsymbol{v}}}{dx}\right) dx \\
&= \int\limits_{x_{j-2+\alpha}}^{x_{j-1+\alpha}} \left(\boldsymbol{f}\left(x, \tilde{\boldsymbol{u}} + \hat{\boldsymbol{u}}, \frac{d\tilde{\boldsymbol{u}}}{dx} + \frac{d\hat{\boldsymbol{u}}}{dx}\right) - \boldsymbol{f}\left(x, \tilde{\boldsymbol{v}} + \hat{\boldsymbol{v}}, \frac{d\tilde{\boldsymbol{v}}}{dx} + \frac{d\hat{\boldsymbol{v}}}{dx}\right), \tilde{\boldsymbol{u}}(x) - \tilde{\boldsymbol{v}}(x)\right) dx
\end{aligned}$$

$$= \int_{x_{j-2+\alpha}}^{x_{j-1+\alpha}} \left(\boldsymbol{f}\left(x, \tilde{\boldsymbol{u}}+\hat{\boldsymbol{u}}, \frac{d\tilde{\boldsymbol{u}}}{dx}+\frac{d\hat{\boldsymbol{u}}}{dx}\right) - \boldsymbol{f}\left(x, \tilde{\boldsymbol{v}}+\hat{\boldsymbol{u}}, \frac{d\tilde{\boldsymbol{v}}}{dx}+\frac{d\hat{\boldsymbol{u}}}{dx}\right), \tilde{\boldsymbol{u}}(x) - \tilde{\boldsymbol{v}}(x) \right) dx$$

$$+ \int_{x_{j-2+\alpha}}^{x_{j-1+\alpha}} \left(\boldsymbol{f}\left(x, \tilde{\boldsymbol{v}}+\hat{\boldsymbol{u}}, \frac{d\tilde{\boldsymbol{v}}}{dx}+\frac{d\hat{\boldsymbol{u}}}{dx}\right) - \boldsymbol{f}\left(x, \tilde{\boldsymbol{v}}+\hat{\boldsymbol{v}}, \frac{d\tilde{\boldsymbol{v}}}{dx}+\frac{d\hat{\boldsymbol{v}}}{dx}\right), \tilde{\boldsymbol{u}}(x) - \tilde{\boldsymbol{v}}(x) \right) dx$$

$$\leq \left(\int_{x_{j-2+\alpha}}^{x_{j-1+\alpha}} \left\| \boldsymbol{f}\left(x, \tilde{\boldsymbol{v}}+\hat{\boldsymbol{u}}, \frac{d\tilde{\boldsymbol{v}}}{dx}+\frac{d\hat{\boldsymbol{u}}}{dx}\right) - \boldsymbol{f}\left(x, \tilde{\boldsymbol{v}}+\hat{\boldsymbol{v}}, \frac{d\tilde{\boldsymbol{v}}}{dx}+\frac{d\hat{\boldsymbol{v}}}{dx}\right) \right\|^2 dx \right)^{1/2}$$

$$\times \left(\int_{x_{j-2+\alpha}}^{x_{j-1+\alpha}} \|\tilde{\boldsymbol{u}}(x) - \tilde{\boldsymbol{v}}(x)\|^2 dx \right)^{1/2} + c_3 \|\tilde{\boldsymbol{u}} - \tilde{\boldsymbol{v}}\|_{1,2,e_\alpha}^2$$

$$\leq \sqrt{2}L \|\hat{\boldsymbol{u}} - \hat{\boldsymbol{v}}\|_{1,2,e_\alpha} \|\tilde{\boldsymbol{u}} - \tilde{\boldsymbol{v}}\|_{1,2,e_\alpha} + c_3 \|\tilde{\boldsymbol{u}} - \tilde{\boldsymbol{v}}\|_{1,2,e_\alpha}^2,$$

which yields

$$\left(c_1 \frac{\pi^2}{1+\pi^2} - c_3 \right) \|\tilde{\boldsymbol{u}} - \tilde{\boldsymbol{v}}\|_{1,2,e_\alpha} \leq \sqrt{2}L \|\hat{\boldsymbol{u}} - \hat{\boldsymbol{v}}\|_{1,2,e_\alpha},$$

$$\|\tilde{\boldsymbol{u}} - \tilde{\boldsymbol{v}}\|_{1,2,e_\alpha} \leq \frac{\sqrt{2}L}{c_4} \|\hat{\boldsymbol{u}} - \hat{\boldsymbol{v}}\|_{1,2,e_\alpha}.$$

The formulas (4.18), (4.19) and the inequality $\|\boldsymbol{u}\|_{0,2,\hat{\omega}_h}^2 \leq \frac{1}{8}\|\boldsymbol{u}_{\bar{x}}\|_{0,2,\hat{\omega}_h^+}^2$ imply

$$\|\boldsymbol{u} - \boldsymbol{v}\|_{1,2,(0,1)}$$

$$\leq \|\tilde{\boldsymbol{u}} - \tilde{\boldsymbol{v}}\|_{1,2,(0,1)} + \|\hat{\boldsymbol{u}} - \hat{\boldsymbol{v}}\|_{1,2,(0,1)}$$

$$\leq \left(1 + \frac{\sqrt{2}L}{c_4} \right) \|\hat{\boldsymbol{u}} - \hat{\boldsymbol{v}}\|_{1,2,(0,1)} \leq \sqrt{\frac{3}{2}\frac{c_2}{c_1}} \left(1 + \frac{\sqrt{2}L}{c_4} \right) \|\boldsymbol{u}_x - \boldsymbol{v}_{\bar{x}}\|_{0,2,\hat{\omega}_h^+}.$$

Taking into account (4.18), (4.20) we obtain, with an arbitrary vector $\boldsymbol{z}$,

$$(A_h(x, \boldsymbol{u}) - A_h(x, \boldsymbol{v}), \boldsymbol{z})_{\hat{\omega}_h}$$

$$\leq \|\boldsymbol{u} - \boldsymbol{v}\|_{B_h} \|\boldsymbol{z}\|_{B_h} + 2\sqrt{3}L\frac{c_2^2}{c_1^2}\left(1 + \frac{\sqrt{2}L}{c_4} \right) \|\boldsymbol{u}_{\bar{x}} - \boldsymbol{v}_{\bar{x}}\|_{0,2,\hat{\omega}_h^+} \|\boldsymbol{z}\|_{0,2,\hat{\omega}_h}$$

$$\leq \|\boldsymbol{u} - \boldsymbol{v}\|_{B_h} \|\boldsymbol{z}\|_{B_h} + \sqrt{\frac{3}{2}}L\frac{c_2^2}{c_1^2}\left(1 + \frac{\sqrt{2}L}{c_4} \right) \|\boldsymbol{u}_{\bar{x}} - \boldsymbol{v}_{\bar{x}}\|_{0,2,\hat{\omega}_h^+} \|\boldsymbol{z}_{\bar{x}}\|_{0,2,\hat{\omega}_h^+}$$

$$\leq \left[1 + \sqrt{\frac{3}{2}}L\frac{c_2^2}{c_1^3}\left(1 + \frac{\sqrt{2}L}{c_4} \right) \right] \|\boldsymbol{u} - \boldsymbol{v}\|_{B_h} \|\boldsymbol{z}\|_{B_h}.$$

Setting here $\boldsymbol{z} \stackrel{\text{def}}{=} B_h^{-1}(A_h(x,\boldsymbol{u}) - A_h(x,\boldsymbol{v}))$ we get

$$\|B_h^{-1}(A_h(x,\boldsymbol{u}) - A_h(x,\boldsymbol{v}))\|_{B_h} \leq \left[1 + \sqrt{\frac{3}{2}} L \frac{c_2^2}{c_1^3}\left(1 + \frac{\sqrt{2}L}{c_4}\right)\right] \|\boldsymbol{u} - \boldsymbol{v}\|_{B_h}. \quad (4.21)$$

Now the estimates (4.21), (4.14) imply

$$\begin{aligned}
&\left(A_h(x,\boldsymbol{u}) - A_h(x,\boldsymbol{v}), B_h^{-1}(A_h(x,\boldsymbol{u}) - A_h(x,\boldsymbol{v}))\right)_{\hat{\omega}_h} \\
&\leq \left[1 + \sqrt{\frac{3}{2}} L \frac{c_2^2}{c_1^3}\left(1 + \frac{\sqrt{2}L}{c_4}\right)\right]^2 \|\boldsymbol{u} - \boldsymbol{v}\|_{B_h}^2 \\
&\leq \frac{\pi^2}{1+\pi^2}\frac{c_2}{c_4}\left[1 + \sqrt{\frac{3}{2}} L \frac{c_2^2}{c_1^3}\left(1 + \frac{\sqrt{2}L}{c_4}\right)\right]^2 (A_h(x,\boldsymbol{u}) - A_h(x,\boldsymbol{v}), \boldsymbol{u} - \boldsymbol{v})_{\hat{\omega}_h}
\end{aligned}$$

and due to the results from [74], p.502, the iteration method (4.12) converges in H_{B_h} and the error estimate (4.13) holds. ■

Let us remember that the space H_{B_h} is equivalent to the space $\overset{\circ}{W}{}_2^1(\hat{\omega}_h)$, i.e. the corresponding norms satisfy

$$\gamma_1\|\boldsymbol{u}\|_{1,2,\hat{\omega}_h} \leq \|\boldsymbol{u}\|_{B_h} \leq \gamma_2\|\boldsymbol{u}\|_{1,2,\hat{\omega}_h}.$$

We can now formulate the following lemma.

Lemma 4.3

Let the assumptions of Lemma 4.2 be satisfied. Then, the iteration method (4.12) converges and together with the estimate (4.13) the following relation holds:

$$\left\|K\frac{d\boldsymbol{u}^{(n)}}{dx} - K\frac{d\boldsymbol{u}}{dx}\right\|_{0,2,\hat{\omega}_h} \leq M\left\|\boldsymbol{u}^{(n)} - \boldsymbol{u}\right\|_{1,2,\hat{\omega}_h} \leq Mq^n,$$

where

$$\begin{aligned}
\|\boldsymbol{u}\|_{0,2,\hat{\omega}_h} &= \left\{\sum_{j=1}^{N-1} \hbar_j\|\boldsymbol{u}_j\|^2 + \frac{1}{2}h_1\|\boldsymbol{u}_0\|^2 + \frac{1}{2}h_N\|\boldsymbol{u}_N\|^2\right\}^{1/2} \\
&= \left\{\frac{1}{2}\sum_{j=1}^{N} h_j\left[\|\boldsymbol{u}_j\|^2 + \|\boldsymbol{u}_{j-1}\|^2\right]\right\}^{1/2}.
\end{aligned}$$

Proof. Using the relations

$$K\frac{d\boldsymbol{u}}{dx}\bigg|_{x=x_j} = A_j\boldsymbol{u}_{\bar{x},j} + \left[V_1^j(x_j)\right]^{-1}\int\limits_{x_{j-1}}^{x_j} V_1^j(\xi)\frac{d}{d\xi}\left[K(\xi)\frac{d\boldsymbol{u}}{d\xi}\right]d\xi$$

$$= A_j\boldsymbol{u}_{\bar{x},j} - \left[V_1^j(x_j)\right]^{-1}\int\limits_{x_{j-1}}^{x_j} V_1^j(\xi)\boldsymbol{f}\left(\xi,\boldsymbol{u}(\xi),\frac{d\boldsymbol{u}}{d\xi}\right)d\xi,$$

$$K\frac{d\boldsymbol{u}}{dx}\bigg|_{x=x_{j-1}} = A_j\boldsymbol{u}_{\bar{x},j} - \left[V_2^{j-1}(x_{j-1})\right]^{-1}\int\limits_{x_{j-1}}^{x_j} V_2^{j-1}(\xi)\frac{d}{d\xi}\left[K(\xi)\frac{d\boldsymbol{u}}{d\xi}\right]d\xi$$

$$= A_j\boldsymbol{u}_{\bar{x},j} + \left[V_1^j(x_j)\right]^{-1}\int\limits_{x_{j-1}}^{x_j} V_2^{j-1}(\xi)\boldsymbol{f}\left(\xi,\boldsymbol{u}(\xi),\frac{d\boldsymbol{u}}{d\xi}\right)d\xi,$$

together with $(a+b)^2 \le 2(a^2+b^2)$, the Cauchy-Bunyakovsky-Schwarz inequality (see Theorem 1.8) as well as a Lipschitz condition, we deduce

$$\left\|K\frac{d\boldsymbol{u}^{(n)}}{dx} - K\frac{d\boldsymbol{u}}{dx}\right\|_{0,2,\hat{\omega}_h}$$

$$= \Bigg\{\frac{1}{2}\sum_{j=1}^{N} h_j\Bigg\| A_j\boldsymbol{u}_{\bar{x},j}^{(n)} - A_j\boldsymbol{u}_{\bar{x},j}$$

$$- \left[V_1^j(x_j)\right]^{-1}\int\limits_{x_{j-1}}^{x_j} V_1^j(\xi)\left[\boldsymbol{f}\left(\xi,\boldsymbol{u}^{(n)}(\xi),\frac{d\boldsymbol{u}^{(n)}}{d\xi}\right) - \boldsymbol{f}\left(\xi,\boldsymbol{u}(\xi),\frac{d\boldsymbol{u}}{d\xi}\right)\right]d\xi\Bigg\|^2$$

$$+ \frac{1}{2}\sum_{j=1}^{N} h_j\Bigg\| A_j\boldsymbol{u}_{\bar{x},j}^{(n)} - A_j\boldsymbol{u}_{\bar{x},j}$$

$$+ \left[V_1^j(x_j)\right]^{-1}\int\limits_{x_{j-1}}^{x_j} V_2^{j-1}(\xi)\left[\boldsymbol{f}\left(\xi,\boldsymbol{u}^{(n)}(\xi),\frac{d\boldsymbol{u}}{d\xi}\right) - \boldsymbol{f}\left(\xi,\boldsymbol{u}(\xi),\frac{d\boldsymbol{u}}{d\xi}\right)\right]d\xi\Bigg\|^2\Bigg\}^{1/2}$$

$$\le \Bigg\{2\sum_{j=1}^{N} h_j\left\|A_j\boldsymbol{u}_{\bar{x},j}^{(n)} - A_j\boldsymbol{u}_{\bar{x},j}\right\|^2 + \sum_{j=1}^{N} h_j\left\|\left[V_1^j(x_j)\right]^{-1}\right\|^2$$

$$\times \left[\int\limits_{x_{j-1}}^{x_j}\left\|V_1^j(\xi)\right\|\left\|\boldsymbol{f}\left(\xi,\boldsymbol{u}^{(n)}(\xi),\frac{d\boldsymbol{u}^{(n)}}{d\xi}\right) - \boldsymbol{f}\left(\xi,\boldsymbol{u}(\xi),\frac{d\boldsymbol{u}}{d\xi}\right)\right\|d\xi\right]^2$$

$$+\sum_{j=1}^{N} h_j \left\| \left[V_1^j(x_j)\right]^{-1} \right\|^2$$

$$\times \left[\int_{x_{j-1}}^{x_j} \left\|V_2^{j-1}(\xi)\right\| \left\| \boldsymbol{f}\left(\xi, \boldsymbol{u}^{(n)}(\xi), \frac{d\boldsymbol{u}^{(n)}}{d\xi}\right) - \boldsymbol{f}\left(\xi, \boldsymbol{u}(\xi), \frac{d\boldsymbol{u}}{d\xi}\right) \right\| d\xi \right]^2 \Bigg\}^{1/2}$$

$$\leq \Bigg\{ 2\sum_{j=1}^{N} h_j \left\| A_j \boldsymbol{u}_{\bar{x},j}^{(n)} - A_j \boldsymbol{u}_{\bar{x},j} \right\|^2 + \sum_{j=1}^{N} h_j \left\| \left[V_1^j(x_j)\right]^{-1} \right\|^2$$

$$\times \int_{x_{j-1}}^{x_j} \left[\left\|V_1^j(\xi)\right\|^2 + \left\|V_2^{j-1}(\xi)\right\|^2 \right] d\xi$$

$$\times \int_{x_{j-1}}^{x_j} \left\| \boldsymbol{f}\left(\xi, \boldsymbol{u}^{(n)}(\xi), \frac{d\boldsymbol{u}^{(n)}}{d\xi}\right) - \boldsymbol{f}\left(\xi, \boldsymbol{u}(\xi), \frac{d\boldsymbol{u}}{d\xi}\right) \right\|^2 d\xi \Bigg\}^{1/2}$$

$$\leq \Bigg\{ 2\sum_{j=1}^{N} h_j \left\| A_j \boldsymbol{u}_{\bar{x},j}^{(n)} - A_j \boldsymbol{u}_{\bar{x},j} \right\|^2$$

$$+ \frac{2c_2^2}{c_1^2} L^2 \sum_{j=1}^{N} h_j \int_{x_{j-1}}^{x_j} \left[\left\| \boldsymbol{u}^{(n)}(\xi) - \boldsymbol{u}(\xi) \right\| + \left\| \frac{d\boldsymbol{u}^{(n)}}{d\xi} - \frac{d\boldsymbol{u}}{d\xi} \right\| \right]^2 d\xi \Bigg\}^{1/2}$$

$$\leq \sqrt{2} c_2 \left\| \boldsymbol{u}_{\bar{x}}^{(n)} - \boldsymbol{u}_{\bar{x}} \right\|_{0,2,\hat{\omega}_h^+} + 2\frac{c_2}{c_1} L \left\| \boldsymbol{u}^{(n)} - \boldsymbol{u} \right\|_{1,2,(0,1)}.$$

Now, inequality (4.20) and Lemma 4.2 imply

$$\left\| K\frac{d\boldsymbol{u}^{(n)}}{dx} - K\frac{d\boldsymbol{u}}{dx} \right\|_{0,2,\hat{\omega}_h} \leq \sqrt{2}\left[c_2 + \sqrt{3}\, L \frac{c_2^2}{c_1^2}\left(1 + \frac{\sqrt{2}L}{c_4}\right) \right] \left\| \boldsymbol{u}^{(n)} - \boldsymbol{u} \right\|_{1,2,\hat{\omega}_h}$$

$$= M_1 \left\| \boldsymbol{u}^{(n)} - \boldsymbol{u} \right\|_{1,2,\hat{\omega}_h} \leq M\, q^n,$$

which proves the claim. ■

4.3 Implementation of the three-point EDS

To simplify the structure of the formulas let us again use the abbreviations

$$\beta \stackrel{\text{def}}{=} j + (-1)^\alpha, \quad \gamma \stackrel{\text{def}}{=} j - 1 + \alpha,$$

as we have already done in previous sections.

We begin our discussion with the following formula:

$$(-1)^{\alpha+1} \int\limits_{x_{j+(-1)^\alpha}}^{x_j} V_\alpha^j(\xi) \boldsymbol{f}\left(\xi, \boldsymbol{u}(\xi), \frac{d\boldsymbol{u}}{d\xi}\right) d\xi$$
$$= (-1)^\alpha V_\alpha^j(x_j) \boldsymbol{Z}_\alpha^j(x_j, \boldsymbol{u}) + \boldsymbol{Y}_\alpha^j(x_j, \boldsymbol{u}) - \boldsymbol{u}_\beta, \tag{4.22}$$

where $\boldsymbol{Y}_\alpha^j(x, \boldsymbol{u})$, $\boldsymbol{Z}_\alpha^j(x, \boldsymbol{u})$, $\alpha = 1, 2$, are the solutions of the IVPs

$$\begin{aligned}
&\frac{d\boldsymbol{Y}_\alpha^j(x, \boldsymbol{u})}{dx} = K^{-1}(x) \boldsymbol{Z}_\alpha^j(x, \boldsymbol{u}), \\
&\frac{d\boldsymbol{Z}_\alpha^j(x, \boldsymbol{u})}{dx} = -\boldsymbol{f}\left(x, \boldsymbol{Y}_\alpha^j(x, \boldsymbol{u}), K^{-1}(x) \boldsymbol{Z}_\alpha^j(x, \boldsymbol{u})\right), \quad x \in e_\alpha^j, \\
&\boldsymbol{Y}_\alpha^j(x_\beta, \boldsymbol{u}) = \boldsymbol{u}_\beta, \quad \boldsymbol{Z}_\alpha^j(x_\beta, \boldsymbol{u}) = K(x) \frac{d\boldsymbol{u}}{dx}\bigg|_{x=x_\beta}, \\
&j = 2-\alpha,\ 3-\alpha, \ldots, N+1-\alpha, \quad \alpha = 1, 2,
\end{aligned} \tag{4.23}$$

and the matrix functions $\bar{V}_\alpha^j(x) \stackrel{\text{def}}{=} (-1)^{\alpha+1} V_\alpha^j(x)$, $\alpha = 1, 2$, are the solutions of the IVPs

$$\begin{aligned}
&\frac{d\bar{V}_\alpha^j(x)}{dx} = K^{-1}(x), \quad x \in e_\alpha^j, \\
&\bar{V}_\alpha^j(x_\beta) = 0, \quad j = 2-\alpha,\ 3-\alpha, \ldots, N+1-\alpha, \quad \alpha = 1, 2.
\end{aligned} \tag{4.24}$$

Using (4.22) the right-hand side of the EDS (4.9) can be represented in the form

$$\begin{aligned}
\boldsymbol{\varphi}(x_j, \boldsymbol{u}) &\stackrel{\text{def}}{=} \hat{T}^{x_j}\left(\boldsymbol{f}\left(\xi, \boldsymbol{u}(\xi), \frac{d\boldsymbol{u}}{d\xi}\right)\right) \\
&= \hbar_j^{-1} \sum_{\alpha=1}^{2} (-1)^\alpha \left\{\boldsymbol{Z}_\alpha^j(x_j, \boldsymbol{u}) + (-1)^\alpha \left[V_\alpha^j(x_j)\right]^{-1} \left[\boldsymbol{Y}_\alpha^j(x_j, \boldsymbol{u}) - \boldsymbol{u}_\beta\right]\right\}.
\end{aligned} \tag{4.25}$$

Thus, in order to compute the input data of the EDS (4.9), (4.25) for all $x_j \in \hat{\bar{\omega}}_h$ the four IVPs (4.23), (4.24) with smooth right-hand sides should be solved on intervals whose lengths are proportional to the maximum step size.

The IVPs (4.23), (4.24) can be solved by a one-step method:

$$\begin{aligned}
&\boldsymbol{Y}_\alpha^{(\bar{m})j}(x_j, \boldsymbol{u}) = \boldsymbol{u}_\beta + (-1)^{\alpha+1} h_\gamma \boldsymbol{\Phi}_1\left(x_\beta, \boldsymbol{u}_\beta, (K\boldsymbol{u}')_\beta, (-1)^{\alpha+1} h_\gamma\right), \\
&\boldsymbol{Z}_\alpha^{(m)j}(x_j, \boldsymbol{u}) = (K\boldsymbol{u}')_\beta + (-1)^{\alpha+1} h_\gamma \boldsymbol{\Phi}_2\left(x_\beta, \boldsymbol{u}_\beta, (K\boldsymbol{u}')_\beta, (-1)^{\alpha+1} h_\gamma\right), \\
&\bar{V}_\alpha^{(\bar{m})j}(x_j) = (-1)^{\alpha+1} h_\gamma \Phi_3\left(x_\beta, 0, (-1)^{\alpha+1} h_\gamma\right),
\end{aligned} \tag{4.26}$$

where $\boldsymbol{\Phi}_1(x,\boldsymbol{u},\boldsymbol{\xi},h)$, $\boldsymbol{\Phi}_2(x,\boldsymbol{u},\boldsymbol{\xi},h)$, $\Phi_3(x,\boldsymbol{u},h)$ are the corresponding increment functions and

$$(K\boldsymbol{u}')_\beta = K(x)\frac{d\boldsymbol{u}}{dx}\bigg|_{x=x_\beta}.$$

Assume that $\boldsymbol{Z}_\alpha^{(m)j}(x_j,\boldsymbol{u})$ approximates $\boldsymbol{Z}_\alpha^j(x_j,\boldsymbol{u})$ with order m, and $\boldsymbol{Y}_\alpha^{(\bar m)j}(x_j,\boldsymbol{u})$, $V_\alpha^{(\bar m)j}(x_j)$ approximate $\boldsymbol{Y}_\alpha^j(x_j,\boldsymbol{u})$, $V_\alpha^j(x_j)$, respectively, with order $\bar m$.

If $K(x)$ and the right-hand side $f(x,\boldsymbol{u},\boldsymbol{\xi})$ are sufficiently smooth then there exist the expansions

$$\begin{aligned}
\boldsymbol{Y}_\alpha^j(x_j,\boldsymbol{u}) &= \boldsymbol{Y}_\alpha^{(\bar m)j}(x_j,\boldsymbol{u}) + [(-1)^{\alpha+1}h_\gamma]^{\bar m+1}\boldsymbol{\psi}_\alpha^j(x_\beta,\boldsymbol{u}) + O(h_\beta^{\bar m+2}),\\
\boldsymbol{Z}_\alpha^j(x_j,\boldsymbol{u}) &= \boldsymbol{Z}_\alpha^{(m)j}(x_j,\boldsymbol{u}) + [(-1)^{\alpha+1}h_\gamma]^{m+1}\tilde{\boldsymbol{\psi}}_\alpha^j(x_\beta,\boldsymbol{u}) + O(h_\gamma^{m+2}),\\
\bar V_\alpha^j(x_j) &= \bar V_\alpha^{(\bar m)j}(x_j) + [(-1)^{\alpha+1}h_\gamma]^{\bar m+1}\bar\psi_\alpha^j(x_\beta) + O(h_\gamma^{\bar m+2}).
\end{aligned}\tag{4.27}$$

In the case where the Taylor series method is used we have

$$\begin{aligned}
&\boldsymbol{\Phi}_1\left(x_\beta,\boldsymbol{u}_\beta,(K\boldsymbol{u}')_\beta,(-1)^{\alpha+1}h_\gamma\right)\\
&= \boldsymbol{u}'_\beta - \frac{(-1)^{\alpha+1}h_\gamma}{2}K_\beta^{-1}f(x_\beta,\boldsymbol{u}_\beta,\boldsymbol{u}'_\beta) + \sum_{p=3}^{\bar m}\frac{[(-1)^{\alpha+1}h_\gamma]^{p-1}}{p!}\frac{d^p\boldsymbol{Y}_\alpha^j(x_\beta,\boldsymbol{u})}{dx^p},\\
&\boldsymbol{\Phi}_2\left(x_\beta,\boldsymbol{u}_\beta,(K\boldsymbol{u}')_\beta,(-1)^{\alpha+1}h_\gamma\right)\\
&= -\boldsymbol{f}\left(x_\beta,\boldsymbol{u}_\beta,\boldsymbol{u}'_\beta\right) + \sum_{p=2}^{m}\frac{[(-1)^{\alpha+1}h_\gamma]^{p-1}}{p!}\frac{d^p\boldsymbol{Z}_\alpha^j(x_\beta,\boldsymbol{u})}{dx^p},\\
&\Phi_3\left(x_\beta,0,(-1)^{\alpha+1}h_\gamma\right)\\
&= K^{-1}(x_\beta) + \sum_{p=2}^{\bar m}\frac{[(-1)^{\alpha+1}h_\gamma]^{p-1}}{p!}\left[\frac{d^{p-1}}{dx^{p-1}}K^{-1}(x)\right]_{x=x_{j+(-1)^\alpha}}.
\end{aligned}$$

If an explicit Runge-Kutta method is used the increment functions are

$$\begin{aligned}
\boldsymbol{\Phi}_1\left(x_\beta,\boldsymbol{u}_\beta,(K\boldsymbol{u}')_\beta^\alpha,(-1)^{\alpha+1}h_\gamma\right) &= b_1\,\boldsymbol{g}_1 + b_2\,\boldsymbol{g}_2 + \cdots + b_s\,\boldsymbol{g}_s,\\
\boldsymbol{\Phi}_2\left(x_\beta,\boldsymbol{u}_\beta,(K\boldsymbol{u}')_\beta,(-1)^{\alpha+1}h_\gamma\right) &= b_1\,\bar{\boldsymbol{g}}_1 + b_2\,\bar{\boldsymbol{g}}_2 + \cdots + b_s\,\bar{\boldsymbol{g}}_s,\\
\Phi_3\left(x_\beta,0,(-1)^{\alpha+1}h_\gamma\right) &= b_1\,\tilde g_1 + b_2\,\tilde g_2 + \cdots + b_s\,\tilde g_s,
\end{aligned}$$

where

$$\begin{aligned}
\tilde g_i &= K^{-1}\left(x_\beta + c_i(-1)^{\alpha+1}h_\gamma\right),\\
\boldsymbol{g}_i &= \tilde g_i\left[(K\boldsymbol{u}')_\beta + (-1)^{\alpha+1}h_\gamma\sum_{p=1}^{i-1}a_{ip}\,\bar{\boldsymbol{g}}_p\right],
\end{aligned}$$

$$\bar{\boldsymbol{g}}_i = -\boldsymbol{f}\left(x_\beta + c_i(-1)^{\alpha+1}h_\gamma, \boldsymbol{u}_\beta + (-1)^{\alpha+1}h_\gamma \sum_{p=1}^{i-1} a_{ip}\,\boldsymbol{g}_p, \boldsymbol{g}_i\right), \quad i = 1, 2, \ldots, s.$$

In the next lemma the differences between $\boldsymbol{Y}_\alpha^j(x_j, \boldsymbol{u})$, $\boldsymbol{Z}_\alpha^j(x_j, \boldsymbol{u})$ and $\boldsymbol{Y}_\alpha^{(\bar{m})j}(x_j, \boldsymbol{u})$, $\boldsymbol{Z}_\alpha^{(m)j}(x_j, \boldsymbol{u})$, respectively, are studied.

Lemma 4.4

Suppose that

$$c_1 \|\boldsymbol{u}\|^2 \le (K(x)\boldsymbol{u}, \boldsymbol{u}) \le c_2 \|\boldsymbol{u}\|^2 \quad \textit{for all } \boldsymbol{u} \in \mathbb{R}^n, \quad k_{rs}(x) \in Q^{m+1}[0,1],$$

$$f_r(x, \boldsymbol{u}, \boldsymbol{\xi}) \in \bigcup_{j=1}^{N} \mathbb{C}^{m+1}\left([x_{j-1}, x_j] \times \mathbb{R}^{2n}\right).$$

If for the one-step method (4.26) *the expansions* (4.27) *exist, then*

$$\begin{aligned}
&\boldsymbol{Y}_\alpha^j(x_j, \boldsymbol{u}) = \boldsymbol{Y}_\alpha^{(\bar{m})j}(x_j, \boldsymbol{u}) + (-1)^{\alpha+1}h_\gamma^{\bar{m}+1}\boldsymbol{\psi}_1^\gamma(x_\beta, \boldsymbol{u}) + O(h_\gamma^{\bar{m}+2}),\\
&\boldsymbol{Z}_\alpha^j(x_j, \boldsymbol{u}) = \boldsymbol{Z}_\alpha^{(m)j}(x_j, \boldsymbol{u}) + [(-1)^{\alpha+1}h_\gamma]^{m+1}\tilde{\boldsymbol{\psi}}_1^\gamma(x_\beta, \boldsymbol{u}) + O(h_\gamma^{m+2}),\\
&V_\alpha^j(x_j) = V_\alpha^{(\bar{m})j}(x_j) + h_\gamma^{\bar{m}+1}\bar{\psi}_1^\gamma(x_\beta) + O(h_\gamma^{\bar{m}+2}),\\
&j = 2-\alpha,\ 3-\alpha, \ldots, N+1-\alpha, \quad \alpha = 1, 2.
\end{aligned} \tag{4.28}$$

Proof. Since the method (4.26) has order of accuracy $\bar{m}$ and the expansions (4.27) exist we have

$$\begin{aligned}
&\boldsymbol{Y}_1^j(x_j, \boldsymbol{u}) - \boldsymbol{Y}_1^{(\bar{m})j}(x_j, \boldsymbol{u})\\
&= \boldsymbol{Y}_1^j(x_j, \boldsymbol{u}) - \boldsymbol{u}(x_j - h_j) - h_j\boldsymbol{\Phi}_1\left(x_j - h_j, \boldsymbol{u}(x_j - h_j), K\boldsymbol{u}'|_{x=x_j-h_j}, h_j\right)\\
&= h_j^{\bar{m}+1}\boldsymbol{\psi}_1^j(x_j - h_j, \boldsymbol{u}) + O(h_j^{\bar{m}+2}),
\end{aligned} \tag{4.29}$$

and

$$\begin{aligned}
&\boldsymbol{Z}_1^j(x_j, \boldsymbol{u}) - \boldsymbol{Z}_1^{(m)j}(x_j, \boldsymbol{u})\\
&= \boldsymbol{Z}_1^j(x_j, \boldsymbol{u}) - K\boldsymbol{u}'|_{x=x_j-h_j} - h_j\boldsymbol{\Phi}_2\left(x_j - h_j, \boldsymbol{u}(x_j - h_j), K\boldsymbol{u}'|_{x=x_j-h_j}, h_j\right)\\
&= h_j^{m+1}\tilde{\boldsymbol{\psi}}_1^j(x_j - h_j, \boldsymbol{u}) + O(h_j^{m+2}).
\end{aligned} \tag{4.30}$$

Note that

$$\begin{aligned}
&\boldsymbol{Y}_2^j(x_j, \boldsymbol{u}) - \boldsymbol{Y}_2^{(\bar{m})j}(x_j, \boldsymbol{u})\\
&= \boldsymbol{Y}_2^j(x_j, \boldsymbol{u}) - \boldsymbol{u}_{j+1} + h_{j+1}\boldsymbol{\Phi}_1\left(x_{j+1}, \boldsymbol{u}_{j+1}, (K\boldsymbol{u}')_{j+1}, -h_{j+1}\right),
\end{aligned}$$

$$\begin{aligned}
&\boldsymbol{Z}_2^j(x_j,\boldsymbol{u}) - \boldsymbol{Z}_2^{(m)j}(x_j,\boldsymbol{u}) \\
&= \boldsymbol{Z}_2^j(x_j,\boldsymbol{u}) - (K\boldsymbol{u}')_{j+1} + h_{j+1}\boldsymbol{\Phi}_2\left(x_{j+1},\boldsymbol{u}_{j+1},(K\boldsymbol{u}')_{j+1},-h_{j+1}\right).
\end{aligned}$$

Substituting $-h_{j+1}$ instead of h_j into (4.29), (4.30) and taking into account that

$$\boldsymbol{Y}_1^j(x_j,\boldsymbol{u}) = \boldsymbol{Y}_2^j(x_j,\boldsymbol{u}), \quad \boldsymbol{Z}_1^j(x_j,\boldsymbol{u}) = \boldsymbol{Z}_2^j(x_j,\boldsymbol{u}),$$

we obtain

$$\begin{aligned}
&\boldsymbol{Y}_2^j(x_j,\boldsymbol{u}) - \boldsymbol{u}_{j+1} + h_{j+1}\boldsymbol{\Phi}_1\left(x_{j+1},\boldsymbol{u}_{j+1},(K\boldsymbol{u}')_{j+1},-h_{j+1}\right) \\
&= -h_{j+1}^{\bar{m}+1}\boldsymbol{\psi}_1^{j+1}(x_{j+1},\boldsymbol{u}) + O(h_{j+1}^{\bar{m}+2}),
\end{aligned}$$

$$\begin{aligned}
&\boldsymbol{Z}_2^j(x_j,\boldsymbol{u}) - (K\boldsymbol{u}')_{j+1} + h_{j+1}\boldsymbol{\Phi}_2\left(x_{j+1},\boldsymbol{u}_{j+1},(K\boldsymbol{u}')_{j+1},-h_{j+1}\right) \\
&= (-1)^{m+1}h_{j+1}^{m+1}\tilde{\boldsymbol{\psi}}_1^{j+1}(x_{j+1},\boldsymbol{u}) + O(h_{j+1}^{m+2}).
\end{aligned}$$

We have thus shown the first two relations in (4.28).

Analogously one can prove the equality

$$V_\alpha^j(x_j) = \bar{V}_\alpha^{(\bar{m})j}(x_j) + (-1)^{\alpha+1}h_\gamma^{\bar{m}+1}\bar{\psi}_1^{j-1+\alpha}(x_\beta) + O(h_\gamma^{\bar{m}+2}),$$

which together with $\bar{V}_\alpha^j(x) = (-1)^{\alpha+1}V_\alpha^j(x)$ gives the third relation in (4.28). ■

Based on the EDS (4.9), (4.25), we can now define the following TDS of the rank $\bar{m}$ (abbreviated to $\bar{m}$-TDS hereafter):

$$\begin{aligned}
&(A^{(\bar{m})}\boldsymbol{y}_{\bar{x}}^{(\bar{m})})_{\hat{x}} = -\boldsymbol{\varphi}^{(\bar{m})}(x,\boldsymbol{y}^{(\bar{m})}), \quad x\in\hat{\omega}_h, \\
&\boldsymbol{y}^{(\bar{m})}(0) = \boldsymbol{\mu}_1, \quad \boldsymbol{y}^{(\bar{m})}(1) = \boldsymbol{\mu}_2,
\end{aligned} \tag{4.31}$$

where

$$A^{(\bar{m})}(x_j) \stackrel{\text{def}}{=} h_j\left[V_1^{(\bar{m})j}(x_j)\right]^{-1},$$

$$\begin{aligned}
&\boldsymbol{\varphi}^{(\bar{m})}(x_j,\boldsymbol{u}) \\
&\stackrel{\text{def}}{=} \hbar_j^{-1}\sum_{\alpha=1}^{2}(-1)^\alpha\left\{\boldsymbol{Z}_\alpha^{(m)j}(x_j,\boldsymbol{u}) + (-1)^\alpha\left[V_\alpha^{(\bar{m})j}(x_j)\right]^{-1}\left[\boldsymbol{Y}_\alpha^{(\bar{m})j}(x_j,\boldsymbol{u}) - \boldsymbol{u}_\beta\right]\right\}.
\end{aligned}$$

In order to prove the order of accuracy of the $\bar{m}$-TDS (4.31) we need the following auxiliary statement.

Lemma 4.5

Let the assumptions of Lemma 4.4 be fulfilled. Then the following estimates hold:

$$\left\|A^{(\bar{m})}(x_j) - A(x_j)\right\| \leq M\, h^{\bar{m}}, \tag{4.32}$$

and

$$\begin{aligned} &\boldsymbol{\varphi}^{(\bar{m})}(x_j, \boldsymbol{u}) - \boldsymbol{\varphi}(x_j, \boldsymbol{u}) \\ &= \left\{ h_j^{m+1} \left[K(x) \left(\boldsymbol{\psi}_1^j(x, \boldsymbol{u}) - \bar{\psi}_1^j(x) K(x) \frac{d\boldsymbol{u}}{dx} \right) - \tilde{\boldsymbol{\psi}}_1^j(x, \boldsymbol{u}) \right]_{x=x_j+0} \right\}_{\hat{x}} \\ &\quad + O\left(\frac{h_j^{m+2} + h_{j+1}^{m+2}}{\hbar_j} \right), \end{aligned} \tag{4.33}$$

provided that m is odd, and

$$\begin{aligned} &\boldsymbol{\varphi}^{(\bar{m})}(x_j, \boldsymbol{u}) - \boldsymbol{\varphi}(x_j, \boldsymbol{u}) \\ &= \left\{ h_j^{m} \left[K(x) \left(\boldsymbol{\psi}_1^j(x, \boldsymbol{u}) - \bar{\psi}_1^j(x) K(x) \frac{d\boldsymbol{u}}{dx} \right) \right]_{x=x_j+0} \right\}_{\hat{x}} \\ &\quad + O\left(\frac{h_j^{m+1} + h_{j+1}^{m+1}}{\hbar_j} \right), \end{aligned} \tag{4.34}$$

provided that m is even.

Proof. The inequality (4.32) follows from (4.27) due to

$$A^{(\bar{m})}(x_j) - A(x_j) = h_j \left[V_1^j(x_j) \right]^{-1} \left[V_1^j(x_j) - V_1^{(\bar{m})j}(x_j) \right] \left[V_1^{(\bar{m})j}(x_j) \right]^{-1} = O(h_j^{\bar{m}}).$$

In order to prove (4.33), (4.34) we first want to note that

$$\begin{aligned} &\boldsymbol{\varphi}^{(\bar{m})}(x_j, \boldsymbol{u}) - \boldsymbol{\varphi}(x_j, \boldsymbol{u}) \\ &= \hbar_j^{-1} \sum_{\alpha=1}^{2} (-1)^\alpha \Bigg\{ \boldsymbol{Z}_\alpha^{(m)j}(x_j, \boldsymbol{u}) - \boldsymbol{Z}_\alpha^j(x_j, \boldsymbol{u}) + (-1)^\alpha \Bigg[\left(V_\alpha^{(\bar{m})j}(x_j) \right)^{-1} \\ &\quad \times \left(\boldsymbol{Y}_\alpha^{(\bar{m})j}(x_j, \boldsymbol{u}) - \boldsymbol{u}_\beta \right) - \left(V_\alpha^j(x_j) \right)^{-1} \left(\boldsymbol{Y}_\alpha^j(x_j, \boldsymbol{u}) - \boldsymbol{u}_\beta \right) \Bigg] \Bigg\}. \end{aligned} \tag{4.35}$$

Lemma 3.4 and the equalities

$$\boldsymbol{Y}_\alpha^j(x_j, \boldsymbol{u}) - \boldsymbol{u}_\beta = (-1)^{\alpha+1} h_\gamma \left. \frac{d\boldsymbol{u}}{dx} \right|_{x=x_\beta} + O(h_\gamma^2),$$

$$\left[V_\alpha^{(\bar{m})j}(x_j) \right]^{-1} \left[\boldsymbol{Y}_\alpha^{(\bar{m})j}(x_j, \boldsymbol{u}) - \boldsymbol{u}_\beta \right] - \left[V_\alpha^j(x_j) \right]^{-1} \left[\boldsymbol{Y}_\alpha^j(x_j, \boldsymbol{u}) - \boldsymbol{u}_\beta \right]$$

$$= \left[V_\alpha^j(x_j)\right]^{-1}\left[\boldsymbol{Y}_\alpha^{(\bar m)j}(x_j,\boldsymbol{u}) - \boldsymbol{Y}_\alpha^j(x_j,\boldsymbol{u})\right]$$

$$+ \left[V_\alpha^j(x_j)\right]^{-1}\left[V_\alpha^j(x_j) - V_\alpha^{(\bar m)j}(x_j)\right]\left[V_\alpha^{(\bar m)j}(x_j)\right]^{-1}\left[\boldsymbol{Y}_\alpha^{(\bar m)j}(x_j,\boldsymbol{u}) - \boldsymbol{u}_\beta\right],$$

$$\left[V_\alpha^j(x_j)\right]^{-1} = h_\gamma^{-1}K_\beta + O(1), \quad \left[V_\alpha^{(\bar m)j}(x_j)\right]^{-1} = h_\gamma^{-1}K_\beta + O(1)$$

imply

$$\boldsymbol{Z}_\alpha^{(m)j}(x_j,\boldsymbol{u}) - \boldsymbol{Z}_\alpha^j(x_j,\boldsymbol{u}) = -\left[(-1)^{\alpha+1}h_\gamma\right]^{m+1}\tilde{\boldsymbol{\psi}}_1^\gamma(x_\beta,\boldsymbol{u}) + O(h_\gamma^{m+2}),$$

and

$$\left[V_\alpha^{(\bar m)j}(x_j)\right]^{-1}\left[\boldsymbol{Y}_\alpha^{(\bar m)j}(x_j,\boldsymbol{u}) - \boldsymbol{u}_\beta\right] - \left[V_\alpha^j(x_j)\right]^{-1}\left[\boldsymbol{Y}_\alpha^j(x_j,\boldsymbol{u}) - \boldsymbol{u}_\beta\right]$$
$$= -(-1)^{\alpha+1}h_\gamma^{\bar m}\left[K(x)\left(\boldsymbol{\psi}_1^\gamma(x,\boldsymbol{u}) - \bar\psi_1^\gamma(x)K(x)\frac{d\boldsymbol{u}}{dx}\right)\right]_{x=x_\beta} + O(h_\gamma^{\bar m+1}). \tag{4.36}$$

Taking into account (4.36) we obtain from (4.35)

$$\boldsymbol{\varphi}^{(\bar m)}(x_j,\boldsymbol{u}) - \boldsymbol{\varphi}(x_j,\boldsymbol{u})$$
$$= \frac{1}{\hbar_j}\left\{h_{j+1}^{m+1}\left[K(x)\left(\boldsymbol{\psi}_1^{j+1}(x,\boldsymbol{u}) - \bar\psi_1^{j+1}(x)K(x)\frac{d\boldsymbol{u}}{dx}\right) - \tilde{\boldsymbol{\psi}}_1^{j+1}(x,\boldsymbol{u})\right]_{x=x_{j+1}}\right.$$
$$\left. - h_j^{m+1}\left[K(x)\left(\boldsymbol{\psi}_1^j(x,\boldsymbol{u}) - \bar\psi_1^j(x)K(x)\frac{d\boldsymbol{u}}{dx}\right) - \tilde{\boldsymbol{\psi}}_1^j(x,\boldsymbol{u})\right]_{x=x_{j-1}}\right\}$$
$$+ O\left(\frac{h_j^{m+2} + h_{j+1}^{m+2}}{\hbar_j}\right) \tag{4.37}$$

provided that m is odd, and

$$\boldsymbol{\varphi}^{(\bar m)}(x_j,\boldsymbol{u}) - \boldsymbol{\varphi}(x_j,\boldsymbol{u})$$
$$= \frac{1}{\hbar_j}\left\{h_{j+1}^{m}\left[K(x)\left(\boldsymbol{\psi}_1^{j+1}(x,\boldsymbol{u}) - \bar\psi_1^{j+1}(x)K(x)\frac{d\boldsymbol{u}}{dx}\right)\right]_{x=x_{j+1}}\right.$$
$$\left. - h_j^{m}\left[K(x)\left(\boldsymbol{\psi}_1^j(x,\boldsymbol{u}) - \bar\psi_1^j(x)K(x)\frac{d\boldsymbol{u}}{dx}\right)\right]_{x=x_{j-1}}\right\} + O\left(\frac{h_j^{m+1} + h_{j+1}^{m+1}}{\hbar_j}\right) \tag{4.38}$$

provided that m is even. Due to

$$\left[K(x)\left(\boldsymbol{\psi}_1^j(x,\boldsymbol{u})-\bar{\psi}_1^j(x)K(x)\frac{d\boldsymbol{u}}{dx}\right)\right]_{x=x_{j-1}}$$

$$=\left[K(x)\left(\boldsymbol{\psi}_1^j(x,\boldsymbol{u})-\bar{\psi}_1^j(x)K(x)\frac{d\boldsymbol{u}}{dx}\right)\right]_{x=x_j}+O(h_j),$$

$$\tilde{\boldsymbol{\psi}}_1^j(x_{j-1},\boldsymbol{u})=\tilde{\boldsymbol{\psi}}_1^j(x_j,\boldsymbol{u})+O(h_j),$$

the estimates (4.37), (4.38) yield (4.33), (4.34). ∎

We are now in a position to prove the following claim.

Theorem 4.3

Let the assumptions of Theorem 4.1 and Lemma 4.4 be fulfilled. Then there exists an $h_0>0$ such that for all $\{h_j\}_{j=1}^N$, $h \overset{\text{def}}{=} \max_{1\le j\le N} h_j \le h_0$, the $\bar{m}$-TDS (4.31) has a unique solution for which the following error estimate holds:

$$\left\|\boldsymbol{y}^{(\bar{m})}-\boldsymbol{u}\right\|_{1,2,\hat{\omega}_h}^{*}=\left[\left\|\boldsymbol{y}^{(\bar{m})}-\boldsymbol{u}\right\|_{0,2,\hat{\omega}_h}^2+\left\|K\frac{d\boldsymbol{y}^{(\bar{m})}}{dx}-K\frac{d\boldsymbol{u}}{dx}\right\|_{0,2,\hat{\omega}_h}^2\right]^{1/2}\le M\,h^{\bar{m}},$$

where

$$\left[K\frac{d\boldsymbol{y}^{(\bar{m})}}{dx}\right]_{x=x_0}=\boldsymbol{Z}_2^{(m)0}(x_0,\boldsymbol{y}^{(\bar{m})})+\left[V_2^{(\bar{m})0}(x_0)\right]^{-1}\left[\boldsymbol{Y}_2^{(\bar{m})0}(x_0,\boldsymbol{y}^{(\bar{m})})-\boldsymbol{y}_0^{(\bar{m})}\right],$$

$$\left[K\frac{d\boldsymbol{y}^{(\bar{m})}}{dx}\right]_{x=x_j}=\boldsymbol{Z}_1^{(m)j}(x_j,\boldsymbol{y}^{(\bar{m})})+\left[V_1^{(\bar{m})j}(x_j)\right]^{-1}\left[\boldsymbol{y}_j^{(\bar{m})}-\boldsymbol{Y}_1^{(\bar{m})j}(x_j,\boldsymbol{y}^{(\bar{m})})\right],$$

$j=1,2,\ldots,N,$

and the constant M is independent of h.

Proof. Let us consider the operator

$$A_h^{(\bar{m})}(x,\boldsymbol{u}) \overset{\text{def}}{=} B_h^{(\bar{m})}\boldsymbol{u}-\boldsymbol{\varphi}^{(\bar{m})}(x,\boldsymbol{u}),\quad \text{with}\quad B_h^{(\bar{m})}\boldsymbol{u}\overset{\text{def}}{=}-(A^{\bar{m}}\boldsymbol{u}_{\bar{x}})_{\hat{x}}.$$

From (4.32) – (4.34) we have

$$\left(A_h^{(\bar{m})}(x,\boldsymbol{u})-A_h^{(\bar{m})}(x,\boldsymbol{v}),\boldsymbol{u}-\boldsymbol{v}\right)_{\hat{\omega}_h}$$

$$=\left(A^{(\bar{m})}\left(\boldsymbol{u}_{\bar{x}}-\boldsymbol{v}_{\bar{x}}\right),\boldsymbol{u}_{\bar{x}}-\boldsymbol{v}_{\bar{x}}\right)_{\hat{\omega}_h}-\left(\boldsymbol{\varphi}^{(\bar{m})}(x,\boldsymbol{u})-\boldsymbol{\varphi}^{(\bar{m})}(x,\boldsymbol{v}),\boldsymbol{u}-\boldsymbol{v}\right)_{\hat{\omega}_h}$$

$$=\left(A_h(x,\boldsymbol{u})-A_h(x,\boldsymbol{v}),\boldsymbol{u}-\boldsymbol{v}\right)_{\hat{\omega}_h}+O\left(h^{\bar{m}}\right)$$

and formula (4.11) shows that there exists a $h_0 > 0$ such that for all $\{h_j\}_{j=1}^N$, with $h \leq h_0$, it holds that

$$\begin{aligned}&0 < \tilde{c}_1 \|\boldsymbol{u}\|_{0,2,\hat{\omega}_h}^2 \leq \left(A^{(\bar{m})}\boldsymbol{u}, \boldsymbol{u}\right)_{\hat{\omega}_h},\\&\left(A_h^{(\bar{m})}(x,\boldsymbol{u}) - A_h^{(\bar{m})}(x,\boldsymbol{v}), \boldsymbol{u}-\boldsymbol{v}\right)_{\hat{\omega}_h} \geq c\|\boldsymbol{u}-\boldsymbol{v}\|_{B_h^{(\bar{m})}}^2 \geq 8c\tilde{c}_1\|\boldsymbol{u}-\boldsymbol{v}\|_{0,2,\hat{\omega}_h}^2,\end{aligned} \tag{4.39}$$

where $0 < c < 1$. Therefore, under the assumption $h \leq h_0$ the operator $A_h^{(\bar{m})}(x, \boldsymbol{u})$ is strongly monotone, i.e. for $h \leq h_0$ the $\bar{m}$-TDS (4.31) has a unique solution $y^{(\bar{m})}(x)$, $x \in \hat{\omega}_h$ (see, e.g. [82], p.461).

The error function $\boldsymbol{z}(x) \stackrel{\text{def}}{=} \boldsymbol{y}^{(\bar{m})}(x) - \boldsymbol{u}(x)$, $x \in \hat{\omega}_h$, is the solution of the problem

$$\begin{aligned}&\left[A^{(\bar{m})}(x)\boldsymbol{z}_{\bar{x}}(x)\right]_{\hat{x}} + \boldsymbol{\varphi}^{(\bar{m})}(x,\boldsymbol{y}^{(\bar{m})}) - \boldsymbol{\varphi}^{(\bar{m})}(x,\boldsymbol{u})\\&= \boldsymbol{\varphi}(x,\boldsymbol{u}) - \boldsymbol{\varphi}^{(\bar{m})}(x,\boldsymbol{u}) + \left[\left(A(x) - A^{(\bar{m})}(x)\right)\boldsymbol{u}_{\bar{x}}(x)\right]_{\hat{x}},\\&\boldsymbol{z}(0) = \boldsymbol{z}(1) = 0.\end{aligned} \tag{4.40}$$

From (4.40) we obtain

$$\begin{aligned}&\left(A_h^{(\bar{m})}(x,\boldsymbol{u}) - A_h^{(\bar{m})}(x,\boldsymbol{y}^{(\bar{m})}), \boldsymbol{z}\right)_{\hat{\omega}_h}\\&= \left(\boldsymbol{\varphi}(x,\boldsymbol{u}) - \boldsymbol{\varphi}^{(\bar{m})}(x,\boldsymbol{u}), \boldsymbol{z}\right)_{\hat{\omega}_h} + \left((A^{(\bar{m})} - A)\boldsymbol{u}_{\bar{x}}, \boldsymbol{z}_{\bar{x}}\right)_{\hat{\omega}_h^+}.\end{aligned} \tag{4.41}$$

Due to (4.39) we have

$$\left(A_h^{(\bar{m})}(x,\boldsymbol{u}) - A_h^{(\bar{m})}(x,\boldsymbol{y}^{(\bar{m})}), \boldsymbol{z}\right)_{\hat{\omega}_h} \geq c\|\boldsymbol{z}\|_{B_h^{(\bar{m})}}^2. \tag{4.42}$$

Using the Cauchy-Bunyakovsky-Schwarz inequality (see Theorem 1.8) and (4.32) – (4.34), we deduce for the right-hand side of (4.41)

$$\begin{aligned}\left((A^{(\bar{m})} - A)\boldsymbol{u}_{\bar{x}}, \boldsymbol{z}_{\bar{x}}\right)_{\hat{\omega}_h} &\leq \left\|A^{(\bar{m})} - A\right\|_{0,2,\hat{\omega}_h} \|\boldsymbol{u}_{\bar{x}}\|_{0,2,\hat{\omega}_h^+} \|\boldsymbol{z}_{\bar{x}}\|_{0,2,\hat{\omega}_h^+}\\&\leq M\,h^{\bar{m}}\|\boldsymbol{z}_{\bar{x}}\|_{0,2,\hat{\omega}_h^+} \leq \frac{M\,h^{\bar{m}}}{\tilde{c}_1}\|\boldsymbol{z}\|_{B_h^{(\bar{m})}},\end{aligned} \tag{4.43}$$

$$\left(\boldsymbol{\varphi}(x,\boldsymbol{u}) - \boldsymbol{\varphi}^{(\bar{m})}(x,\boldsymbol{u}), \boldsymbol{z}\right)_{\hat{\omega}_h} \leq M\,h^{m+1}\|\boldsymbol{z}_{\bar{x}}\|_{0,2,\hat{\omega}_h^+} \leq \frac{M\,h^{m+1}}{\tilde{c}_1}\|\boldsymbol{z}\|_{B_h^{(\bar{m})}}, \tag{4.44}$$

provided that m is odd, and

$$\left(\boldsymbol{\varphi}(x,\boldsymbol{u}) - \boldsymbol{\varphi}^{(\bar{m})}(x,\boldsymbol{u}), \boldsymbol{z}\right)_{\hat{\omega}_h} \leq M\,h^{m}\|\boldsymbol{z}_{\bar{x}}\|_{0,2,\hat{\omega}_h^+} \leq \frac{M\,h^{m}}{\tilde{c}_1}\|\boldsymbol{z}\|_{B_h^{(\bar{m})}}, \tag{4.45}$$

provided that m is even. The estimates (4.42) – (4.45) yield $\|\boldsymbol{z}\|_{B_h^{(\bar m)}} \le M\,h^{\bar m}$.

Taking into account the equivalence of the norms $\|\cdot\|_{1,2,\hat\omega_h}$ and $\|\cdot\|_{B_h^{(\bar m)}}$, we obtain $\|\boldsymbol{z}\|_{1,2,\hat\omega_h} \le M\,h^{\bar m}$.

Since $\boldsymbol{y}_0^{(\bar m)} = \boldsymbol{Y}_2^0(x_0, \boldsymbol{y}^{(\bar m)})$ and $\boldsymbol{y}_j^{(\bar m)} = \boldsymbol{Y}_1^j(x_j, \boldsymbol{y}^{(\bar m)})$, the formulas (4.28) imply

$$\begin{aligned}
\left\|\left[K\frac{d\boldsymbol{z}}{dx}\right]_{x=x_0}\right\| &\le \left\|\boldsymbol{Z}_2^{(m)0}(x_0,\boldsymbol{y}^{(\bar m)}) - \boldsymbol{Z}_2^0(x_0,\boldsymbol{y}^{(\bar m)})\right\| + \left\|\boldsymbol{Z}_2^0(x_0,\boldsymbol{y}^{(\bar m)}) - \boldsymbol{Z}_2^0(x_0,\boldsymbol{u})\right\| \\
&\quad + \left\|\left[V_2^{(\bar m)0}(x_0)\right]^{-1}\right\| \left\|\boldsymbol{Y}_2^{(\bar m)0}(x_0,\boldsymbol{y}^{(\bar m)}) - \boldsymbol{Y}_2^0(x_0,\boldsymbol{y}^{(\bar m)})\right\| \\
&\le M_1\,h^{\bar m} + \left\|\frac{\partial}{\partial\boldsymbol{u}}\boldsymbol{Z}_2^0(x_0,\boldsymbol{u})\Big|_{\boldsymbol{u}=\tilde{\boldsymbol{u}}}\right\| \|\boldsymbol{z}\|_{0,2,\hat\omega_h} \le M_2\,h^{\bar m},
\end{aligned}$$

$$\begin{aligned}
\left\|\left[K\frac{d\boldsymbol{z}}{dx}\right]_{x=x_j}\right\| &\le \left\|\boldsymbol{Z}_1^{(m)j}(x_j,\boldsymbol{y}^{(\bar m)}) - \boldsymbol{Z}_1^j(x_j,\boldsymbol{y}^{(\bar m)})\right\| + \left\|\boldsymbol{Z}_1^j(x_j,\boldsymbol{y}^{(\bar m)}) - \boldsymbol{Z}_1^j(x_j,\boldsymbol{u})\right\| \\
&\quad + \left\|\left[V_1^{(\bar m)j}(x_j)\right]^{-1}\right\| \left\|\boldsymbol{Y}_1^j(x_j,\boldsymbol{y}^{(\bar m)}) - \boldsymbol{Y}_1^{(\bar m)j}(x_j,\boldsymbol{y}^{(\bar m)})\right\| \\
&\le M_3\,h^{\bar m} + \left\|\frac{\partial}{\partial\boldsymbol{u}}\boldsymbol{Z}_1^j(x_j,\boldsymbol{u})\Big|_{\boldsymbol{u}=\tilde{\boldsymbol{u}}}\right\| \|\boldsymbol{z}\|_{0,2,\hat\omega_h} \le M_4\,h^{\bar m}, \\
&\quad j = 1,2,\ldots,N,
\end{aligned}$$

$$\left\|K\frac{d\boldsymbol{z}}{dx}\right\|_{0,2,\hat\omega_h} \le \max_{j=0,1,\ldots,N}\left\|\left[K\frac{d\boldsymbol{z}}{dx}\right]_{x=x_j}\right\| \le M\,h^{\bar m},$$

and we obtain $\|\boldsymbol{z}\|_{1,2,\hat\omega_h}^* \le Mh^{\bar m}$. This completes the proof. ∎

The nonlinear $\bar m$-TDS (4.31) can be solved by an iteration method which is given in the following theorem.

Theorem 4.4

Let the assumptions of Theorem 4.3 be fulfilled. Then:

- *the function $\boldsymbol{\varphi}^{(\bar m)}$ satisfies the Lipschitz condition*

$$\left\|\boldsymbol{\varphi}^{(\bar m)}(x,\boldsymbol{u}) - \boldsymbol{\varphi}^{(\bar m)}(x,\boldsymbol{v})\right\|_{0,2,\hat\omega_h} \le \tilde L\,\|\boldsymbol{u}-\boldsymbol{v}\|_{0,2,\hat\omega_h},$$

- *there exists a $h_0 > 0$ such that for all $\{h_j\}_{j=1}^N$, with $h \le h_0$, it holds that*

$$0 < \tilde c_1\|\boldsymbol{u}\|_{0,2,\hat\omega_h}^2 \le \left(A^{(\bar m)}\boldsymbol{u},\boldsymbol{u}\right)_{0,2,\hat\omega_h},$$

$$\left(A_h^{(\bar m)}(x,\boldsymbol{u}) - A_h^{(\bar m)}(x,\boldsymbol{v}),\boldsymbol{u}-\boldsymbol{v}\right)_{\hat\omega_h} \ge c\,\|\boldsymbol{u}-\boldsymbol{v}\|_{B_h^{(\bar m)}}^2, \quad 0 < c < 1,$$

- *the following iteration method converges:*

$$\boldsymbol{y}^{(\bar{m},0)}(x) = V_2(x)\,[V_1(1)]^{-1}\,\boldsymbol{\mu}_1 + V_1(x)\,[V_1(1)]^{-1}\,\boldsymbol{\mu}_2,$$

$$B_h^{(\bar{m})}\,\frac{\boldsymbol{y}^{(\bar{m},n)} - \boldsymbol{y}^{(\bar{m},n-1)}}{\tau} + A_h^{(\bar{m})}(x, \boldsymbol{y}^{(\bar{m},n-1)}) = 0, \quad x \in \hat{\omega}_h, \tag{4.46}$$

$$\boldsymbol{y}^{(\bar{m},n)}(0) = \boldsymbol{\mu}_1, \quad \boldsymbol{y}^{(\bar{m},n)}(1) = \boldsymbol{\mu}_2, \quad n = 1, 2, \ldots,$$

where

$$B_h^{(\bar{m})}\boldsymbol{y} \stackrel{\text{def}}{=} -(A^{(\bar{m})}\boldsymbol{y}_{\bar{x}})_{\hat{x}}, \quad A_h^{(\bar{m})}(x, \boldsymbol{y}) \stackrel{\text{def}}{=} B_h^{(\bar{m})}\boldsymbol{y} - \boldsymbol{\varphi}^{(\bar{m})}(x, \boldsymbol{y}),$$

$$\tau \stackrel{\text{def}}{=} \tau_0 = c\left(1 + \frac{\tilde{L}}{8\tilde{c}_1}\right)^{-2}.$$

- *the corresponding error can be estimated by*

$$\left\|\boldsymbol{y}^{(\bar{m},n)} - \boldsymbol{u}\right\|_{1,2,\hat{\omega}_h}^{*} \le M(h^m + q^n), \quad q \stackrel{\text{def}}{=} \sqrt{1 - c\tau_0}, \tag{4.47}$$

where

$$\left[K\frac{d\boldsymbol{y}^{(\bar{m},n)}}{dx}\right]_{x=x_0}$$
$$= \boldsymbol{Z}_2^{(m)0}(x_0, \boldsymbol{y}^{(\bar{m},n)}) + \left[V_2^{(\bar{m})0}(x_0)\right]^{-1}\left[\boldsymbol{Y}_2^{(\bar{m})0}(x_0, \boldsymbol{y}^{(\bar{m},n)}) - \boldsymbol{y}_0^{(\bar{m},n)}\right],$$

$$\left[K\frac{d\boldsymbol{y}^{(\bar{m},n)}}{dx}\right]_{x=x_j}$$
$$= \boldsymbol{Z}_1^{(m)j}(x_j, \boldsymbol{y}^{(\bar{m},n)}) + \left[V_1^{(\bar{m})j}(x_j)\right]^{-1}\left[\boldsymbol{y}_j^{(\bar{m},n)} - \boldsymbol{Y}_1^{(\bar{m})j}(x_j, \boldsymbol{y}^{(\bar{m},n)})\right],$$

$$j = 1, 2, \ldots, N, \tag{4.48}$$

and the constant M does not depend on h, m and n.

Proof. Due to Theorem 4.3 we have

$$\begin{aligned}\left\|\boldsymbol{y}^{(\bar{m},n)} - \boldsymbol{u}\right\|_{1,2,\hat{\omega}_h}^{*} &\le \left\|\boldsymbol{y}^{(\bar{m})} - \boldsymbol{u}\right\|_{1,2,\hat{\omega}_h}^{*} + \left\|\boldsymbol{y}^{(\bar{m},n)} - \boldsymbol{y}^{(\bar{m})}\right\|_{1,2,\hat{\omega}_h}^{*} \\ &\le Mh^{\bar{m}} + \left\|\boldsymbol{y}^{(\bar{m},n)} - \boldsymbol{y}^{(\bar{m})}\right\|_{1,2,\hat{\omega}_h}^{*}.\end{aligned} \tag{4.49}$$

The assumption $f_r(x, \boldsymbol{u}, \boldsymbol{\xi}) \in \bigcup_{j=1}^{N} \mathbb{C}^m\left([x_{j-1}, x_j] \times \mathbb{R}^{2n}\right)$ yields

$$\left|\boldsymbol{\varphi}^{(\bar{m})}(x, \boldsymbol{u}) - \boldsymbol{\varphi}^{(\bar{m})}(x, \boldsymbol{v})\right| \le \tilde{L}\,|\boldsymbol{u} - \boldsymbol{v}|\,.$$

Using the Cauchy-Bunyakovsky-Schwarz inequality (see Theorem 1.8) we obtain

$$\begin{aligned}
&\left(A_h^{(\bar{m})}(x, \boldsymbol{u}) - A_h^{(\bar{m})}(x, \boldsymbol{v}), \boldsymbol{w}\right)_{\hat{\omega}_h} \\
&\le \|\boldsymbol{u} - \boldsymbol{v}\|_{B_h^{(\bar{m})}} \|\boldsymbol{w}\|_{B_h^{(\bar{m})}} + \left\|\boldsymbol{\varphi}^{(\bar{m})}(x, \boldsymbol{u}) - \boldsymbol{\varphi}^{(\bar{m})}(x, \boldsymbol{v})\right\|_{0,2,\hat{\omega}_h} \|\boldsymbol{w}\|_{0,2,\hat{\omega}_h} \\
&\le \|\boldsymbol{u} - \boldsymbol{v}\|_{B_h^{(\bar{m})}} \|\boldsymbol{w}\|_{B_h^{(\bar{m})}} + \tilde{L}\,\|\boldsymbol{u} - \boldsymbol{v}\|_{0,2,\hat{\omega}_h} \|\boldsymbol{w}\|_{0,2,\hat{\omega}_h} \\
&\le \|\boldsymbol{u} - \boldsymbol{v}\|_{B_h^{(\bar{m})}} \|\boldsymbol{w}\|_{B_h^{(\bar{m})}} + \frac{\tilde{L}}{8} \|\boldsymbol{u}_{\bar{x}} - \boldsymbol{v}_{\bar{x}}\|_{0,2,\hat{\omega}_h^+} \|\boldsymbol{w}_{\bar{x}}\|_{0,2,\hat{\omega}_h^+} \\
&\le \left(1 + \frac{\tilde{L}}{8\tilde{c}_1}\right) \|\boldsymbol{u} - \boldsymbol{v}\|_{B_h^{(\bar{m})}} \|\boldsymbol{w}\|_{B_h^{(\bar{m})}}\,.
\end{aligned}$$

With $\boldsymbol{w} \stackrel{\text{def}}{=} \left(B_h^{(\bar{m})}\right)^{-1}\left(A_h^{(\bar{m})}(x, \boldsymbol{u}) - A_h^{(\bar{m})}(x, \boldsymbol{v})\right)$, this yields

$$\left\|\left(B_h^{(\bar{m})}\right)^{-1}\left(A_h^{(\bar{m})}(x, \boldsymbol{u}) - A_h^{(\bar{m})}(x, \boldsymbol{v})\right)\right\|_{B_h^{(\bar{m})}} \le \left(1 + \frac{\tilde{L}}{8\tilde{c}_1}\right) \|\boldsymbol{u} - \boldsymbol{v}\|_{B_h^{(\bar{m})}}\,. \tag{4.50}$$

From the relations (4.40) and (4.50) we have

$$\begin{aligned}
&\left(A_h^{(\bar{m})}(x, \boldsymbol{u}) - A_h^{(\bar{m})}(x, \boldsymbol{v}), \left(B_h^{(\bar{m})}\right)^{-1}\left(A_h^{(\bar{m})}(x, \boldsymbol{u}) - A_h^{(\bar{m})}(x, \boldsymbol{v})\right)\right)_{\hat{\omega}_h} \\
&\le \left(1 + \frac{\tilde{L}}{8\tilde{c}_1}\right)^2 \|\boldsymbol{u} - \boldsymbol{v}\|^2_{B_h^{(\bar{m})}} \\
&\le \frac{1}{c}\left(1 + \frac{\tilde{L}}{8\tilde{c}_1}\right)^2 \left(A_h^{(\bar{m})}(x, \boldsymbol{u}) - A_h^{(\bar{m})}(x, \boldsymbol{v}), \boldsymbol{u} - \boldsymbol{v}\right)_{\hat{\omega}_h}\,.
\end{aligned}$$

Therefore, (see, e.g. [74], p.502) the iteration method (4.46) converges in the space $H_{B_h^{(\bar{m})}}$, which is equivalent to the space $\overset{\circ}{W}{}_2^1(\hat{\omega}_h)$, and the following estimate holds:

$$\left\|\boldsymbol{y}^{(\bar{m},n)} - \boldsymbol{y}^{(\bar{m})}\right\|_{1,2,\hat{\omega}_h} \le M_1 q^n.$$

Moreover, we have

$$\left\|\left[K\frac{d\boldsymbol{y}^{(\bar{m},n)}}{dx}\right]_{x=x_0} - \left[K\frac{d\boldsymbol{y}^{(\bar{m})}}{dx}\right]_{x=x_0}\right\|$$

$$\leq \left\| \boldsymbol{Z}_2^{(m)0}(x_0, \boldsymbol{y}^{(\bar{m},n)}) - \boldsymbol{Z}_2^{(m)0}(x_0, \boldsymbol{y}^{(\bar{m})}) \right\|$$

$$+ \left\| \left[V_2^{(\bar{m})0}(x_0) \right]^{-1} \right\| \left\| \boldsymbol{Y}_2^{(\bar{m})0}(x_0, \boldsymbol{y}^{(\bar{m},n)}) - \boldsymbol{Y}_2^{(\bar{m})0}(x_0, \boldsymbol{y}^{(\bar{m})}) \right\|$$

$$\leq \left[\left\| \frac{\partial}{\partial \boldsymbol{u}} \boldsymbol{Z}_2^{(m)0}(x_0, \boldsymbol{u}) \Big|_{\boldsymbol{u}=\tilde{\boldsymbol{y}}} \right\| + \left\| \left[V_2^{(\bar{m})0}(x_0) \right]^{-1} \right\| \left\| \frac{\partial}{\partial \boldsymbol{u}} \boldsymbol{Y}_2^{(\bar{m})0}(x_0, \boldsymbol{u}) \Big|_{\boldsymbol{u}=\bar{\boldsymbol{y}}} \right\| \right]$$

$$\times \left\| \boldsymbol{y}^{(\bar{m},n)} - \boldsymbol{y}^{(\bar{m})} \right\|_{0,2,\hat{\omega}_h}$$

$$\leq M_1 \left\| \boldsymbol{y}^{(\bar{m},n)} - \boldsymbol{y}^{(\bar{m})} \right\|_{1,2,\hat{\omega}_h},$$

$$\left\| \left[K \frac{d\boldsymbol{y}^{(\bar{m},n)}}{dx} \right]_{x=x_j} - \left[K \frac{d\boldsymbol{y}^{(\bar{m})}}{dx} \right]_{x=x_j} \right\|$$

$$\leq \left\| \boldsymbol{Z}_1^{(m)j}(x_j, \boldsymbol{y}^{(\bar{m},n)}) - \boldsymbol{Z}_1^{(m)j}(x_j, \boldsymbol{y}^{(\bar{m})}) \right\|$$

$$+ \left\| \left[V_1^{(\bar{m})j}(x_j) \right]^{-1} \right\| \left\| \boldsymbol{Y}_1^{(\bar{m})j}(x_j, \boldsymbol{y}^{(\bar{m},n)}) - \boldsymbol{Y}_1^{(\bar{m})j}(x_j, \boldsymbol{y}^{(\bar{m})}) \right\|$$

$$\leq \left[\left\| \frac{\partial}{\partial \boldsymbol{u}} \boldsymbol{Z}_1^{(m)j}(x_j, \boldsymbol{u}) \Big|_{\boldsymbol{u}=\tilde{\boldsymbol{y}}} \right\| + \left\| \left[V_1^{(\bar{m})j}(x_j) \right]^{-1} \right\| \left\| \frac{\partial}{\partial \boldsymbol{u}} \boldsymbol{Y}_1^{(\bar{m})j}(x_j, \boldsymbol{u}) \Big|_{\boldsymbol{u}=\bar{\boldsymbol{y}}} \right\| \right]$$

$$\times \left\| \boldsymbol{y}^{(\bar{m},n)} - \boldsymbol{y}^{(\bar{m})} \right\|_{0,2,\hat{\omega}_h}$$

$$\leq M_2 \left\| \boldsymbol{y}^{(\bar{m},n)} - \boldsymbol{y}^{(\bar{m})} \right\|_{0,2,\hat{\omega}_h}, \quad j = 1, 2, \ldots, N,$$

$$\left\| K \frac{d\boldsymbol{y}^{(\bar{m},n)}}{dx} - K \frac{d\boldsymbol{y}^{(\bar{m})}}{dx} \right\|_{0,2,\hat{\bar{\omega}}_h} \leq \max_{j=0,1,\ldots,N} \left\| \left[K \frac{d\boldsymbol{y}^{(\bar{m},n)}}{dx} \right]_{x=x_j} - \left[K \frac{d\boldsymbol{y}^{(\bar{m})}}{dx} \right]_{x=x_j} \right\|$$

$$\leq M \left\| \boldsymbol{y}^{(\bar{m},n)} - \boldsymbol{y}^{(\bar{m})} \right\|_{1,2,\hat{\omega}_h},$$

which yields

$$\left\| \boldsymbol{y}^{(\bar{m},n)} - \boldsymbol{y}^{(\bar{m})} \right\|^*_{1,2,\hat{\omega}_h} \leq M q^n. \tag{4.51}$$

The estimates (4.49) and (4.51) imply (4.47). ■

A derivative-free variant of Newton's method is usually used to numerically solve the $\bar{m}$-TDS (4.31). It can be realized as follows. Substituting the approximate solutions (4.26) into the right-hand side of the difference scheme (4.31), we obtain

$$\boldsymbol{\varphi}^{(\bar{m})}(x_j, \boldsymbol{u})$$

$$= \hbar_j^{-1} \sum_{\alpha=1}^{2} (-1)^\alpha \{(K\boldsymbol{u}')_\beta + (-1)^{\alpha+1} h_\gamma \boldsymbol{\Phi}_2(x_\beta, \boldsymbol{u}_\beta, (K\boldsymbol{u}')_\beta, (-1)^{\alpha+1} h_\gamma)$$

$$- [\Phi_3(x_\beta, 0, (-1)^{\alpha+1} h_\gamma)]^{-1} \boldsymbol{\Phi}_1(x_\beta, \boldsymbol{u}_\beta, (K\boldsymbol{u}')_\beta, (-1)^{\alpha+1} h_\gamma)\}.$$

From (4.27) we get

$$\boldsymbol{\Phi}_1(x, \boldsymbol{u}, \boldsymbol{\xi}, 0) = K^{-1}(x)\boldsymbol{\xi}, \quad \frac{\partial \boldsymbol{\Phi}_1(x, \boldsymbol{u}, \boldsymbol{\xi}, 0)}{\partial h} = -\frac{1}{2} K^{-1}(x) \boldsymbol{f}(x, \boldsymbol{u}, \boldsymbol{\xi}),$$

$$\boldsymbol{\Phi}_2(x, \boldsymbol{u}, \boldsymbol{\xi}, 0) = -\boldsymbol{f}(x, \boldsymbol{u}, \boldsymbol{\xi}), \quad \Phi_3(x, 0, 0) = K^{-1}(x),$$

and therefore

$$\boldsymbol{\varphi}^{(\bar{m})}(x_j, \boldsymbol{y}^{(\bar{m})}) = \boldsymbol{f}\left(x_j, \boldsymbol{y}_j^{(\bar{m})}, \left.\frac{d\boldsymbol{y}^{(\bar{m})}}{dx}\right|_{x=x_j} \right) + O\left(\frac{h_\gamma^2}{\hbar_j}\right),$$

$$\left.\frac{d\boldsymbol{y}^{(\bar{m})}}{dx}\right|_{x=x_j} = \boldsymbol{y}_{\bar{x},j}^{(\bar{m})} + O\left(\frac{h_\gamma^2}{\hbar_j}\right).$$

The discrete Newton's method now reads

$$\begin{aligned}
&\nabla \boldsymbol{y}_0^{(\bar{m},n)} = 0, \quad \nabla \boldsymbol{y}_N^{(\bar{m},n)} = 0, \\
&\left(A^{(\bar{m})} \nabla \boldsymbol{y}_{\bar{x}}^{(\bar{m},n)}\right)_{\hat{x},j} + \frac{\partial \boldsymbol{f}\left(x_j, \boldsymbol{y}_j^{(\bar{m},n-1)}, \dot{\boldsymbol{y}}_j^{(\bar{m},n-1)}\right)}{\partial \boldsymbol{u}} \nabla \boldsymbol{y}_j^{(\bar{m},n)} \\
&\quad + \frac{\partial \boldsymbol{f}\left(x_j, \boldsymbol{y}_j^{(\bar{m},n-1)}, \dot{\boldsymbol{y}}_j^{(\bar{m},n-1)}\right)}{\partial \boldsymbol{\xi}} \nabla \boldsymbol{y}_{\bar{x},j}^{(\bar{m},n)} \\
&= -\boldsymbol{\varphi}^{(\bar{m})}(x_j, \boldsymbol{y}^{(\bar{m},n-1)}) - \left(A^{(\bar{m})} \boldsymbol{y}_{\bar{x}}^{(\bar{m},n-1)}\right)_{\hat{x},j}, \quad j = 1, 2, \ldots, N-1, \\
&\boldsymbol{y}_j^{(\bar{m},n)} = \boldsymbol{y}_j^{(\bar{m},n-1)} + \nabla \boldsymbol{y}_j^{(\bar{m},n)}, \quad j = 0, 1, \ldots, N, \quad n = 1, 2, \ldots,
\end{aligned} \tag{4.52}$$

where the derivatives

$$\dot{\boldsymbol{y}}_j^{(\bar{m},n-1)} \stackrel{\text{def}}{=} \left.\frac{d\boldsymbol{y}^{(\bar{m},n-1)}}{dx}\right|_{x=x_j}$$

can be computed by formula (4.48).

Note that the convergence of Newton's method for systems of ODEs with a monotone operator has been investigated in [48].

4.4 Numerical examples

Example 4.1. Let us consider the BVP [43]

$$\begin{gathered}\frac{d^2u_1}{dx^2} = u_1 + 3\exp(u_2), \quad \frac{d^2u_2}{dx^2} = u_1 - \exp(u_2) + \exp(-x), \\ u_1(0) = 0, \quad u_1(1) = \exp(-2) - \exp(-1), \quad u_2(0) = 0, \quad u_2(1) = -2\end{gathered} \tag{4.53}$$

with the exact solution

$$u_1(x) = \exp(-2x) - \exp(-x), \quad u_2(x) = -2x.$$

In order to solve problem (4.53) numerically on the equidistant grid ω_h we used the following 6-TDS:

$$\boldsymbol{y}^{(6)}_{\bar{x}x} = -\boldsymbol{\varphi}^{(6)}(x, \boldsymbol{y}^{(6)}), \quad x \in \omega_h, \quad \boldsymbol{y}^{(6)}(0) = \boldsymbol{\mu}_1, \quad \boldsymbol{y}^{(6)}(1) = \boldsymbol{\mu}_2, \tag{4.54}$$

where

$$\boldsymbol{\varphi}^{(6)}(x_j, \boldsymbol{u}) = h^{-1} \sum_{\alpha=1}^{2} (-1)^\alpha \left\{ \boldsymbol{Z}_\alpha^{(6)j}(x_j, \boldsymbol{u}) + \frac{(-1)^\alpha}{h} \left[\boldsymbol{Y}_\alpha^{(6)j}(x_j, \boldsymbol{u}) - \boldsymbol{u}_\beta \right] \right\},$$

and $\boldsymbol{Y}_\alpha^{(6)j}(x_j, \boldsymbol{u})$, $\boldsymbol{Z}_\alpha^{(6)j}(x_j, \boldsymbol{u})$ are the numerical solutions of the IVPs

$$\begin{aligned}&\frac{d\boldsymbol{Y}_\alpha^j(x, \boldsymbol{u})}{dx} = \boldsymbol{Z}_\alpha^j(x, \boldsymbol{u}), \quad \frac{d\boldsymbol{Z}_\alpha^j(x, \boldsymbol{u})}{dx} = -\boldsymbol{f}(x, \boldsymbol{Y}_\alpha^j(x, \boldsymbol{u}), \boldsymbol{Z}_\alpha^j(x, \boldsymbol{u})), \quad x \in e_\alpha^j, \\ &\boldsymbol{Y}_\alpha^j(x_\beta, \boldsymbol{u}) = \boldsymbol{u}_\beta, \quad \boldsymbol{Z}_\alpha^j(x_\beta, \boldsymbol{u}) = \boldsymbol{u}'_\beta, \\ &j = 2 - \alpha, 3 - \alpha, \ldots, N + 1 - \alpha, \quad \alpha = 1, 2,\end{aligned} \tag{4.55}$$

with

$$\boldsymbol{f}(x, \boldsymbol{Y}_\alpha^j(x, \boldsymbol{u}), \boldsymbol{Z}_\alpha^j(x, \boldsymbol{u})) \stackrel{\text{def}}{=} -\begin{pmatrix} Y_{\alpha,1}^j + 3\exp(Y_{\alpha,2}^j) \\ Y_{\alpha,1}^j - \exp(Y_{\alpha,2}^j) + \exp(-x) \end{pmatrix}.$$

We have solved the IVP (4.55) with the explicit 7-stage Runge-Kutta method of order 6 which is characterized by the Butcher matrix given in Table 2.4. For the numerical solution of the TDS (4.54) we used the Newton method (4.52). Table 4.1 contains the numerical results obtained by this TDS. Here, we have used the formulas

$$\mathtt{er} \stackrel{\text{def}}{=} \left\| z^{(6)} \right\|^*_{1,2,\omega_h} = \left\| y^{(6)} - u \right\|^*_{1,2,\omega_h} \quad \text{and} \quad \mathtt{p} \stackrel{\text{def}}{=} \log_2 \frac{\left\| z^{(6)} \right\|^*_{1,2,\omega_h}}{\left\| z^{(6)} \right\|^*_{1,2,\omega_{h/2}}}$$

to measure the error and the order of convergence, respectively. One can see that the numerical results are in a good agreement with our theory.

N	er	p
4	$0.2848\,E-5$	
8	$0.4400\,E-7$	6.0
16	$0.6907\,E-9$	6.0
32	$0.1081\,E-10$	6.0

Table 4.1: Numerical results for problem (4.53)

Example 4.2. Let us consider the problem

$$\frac{d^2u_1}{dx^2} = u_2, \quad \frac{d^2u_2}{dx^2} = 6\exp(-4u_1) - \frac{12}{(1+x)^4}, \quad 0 < x < 1, \tag{4.56}$$

$$u_1(0) = 0, \quad u(1) = \ln(2), \quad u_2(0) = -1, \quad u_2(1) = -0.25,$$

with the exact solution

$$u_1(x) = \ln(1+x), \quad u_2(x) = -\frac{1}{(1+x)^2}.$$

This BVP was solved by the above mentioned 6-TDS on the equidistant grid ω_h using the same IVP-solver as in the previous examples. In order to gain the prescribed order of accuracy EPS, Runge's $h - h/2$-strategy was used. Table 4.2 contains numerical results which are in complete agreement with our theory.

EPS	N	Error
$1.0\,E-4$	4	$0.1365\,E-5$
$1.0\,E-6$	4	$0.2634\,E-6$
$1.0\,E-8$	16	$0.2341\,E-9$
$1.0\,E-10$	32	$0.2398\,E-11$

Table 4.2: Numerical results for problem (4.56)

Example 4.3. The next example is the BVP (see [29])

$$\begin{aligned} \frac{d^2u_1}{dx^2} &= \lambda^2 u_1 + u_2 + x + (1-\lambda^2)\exp(-x), \\ \frac{d^2u_2}{dx^2} &= -u_1 + \exp(u_2) + \exp(-\lambda x), \\ u_1(0) &= 2, \quad u_1(1) = \exp(-\lambda) + \exp(-1), \\ u_2(0) &= 0, \quad u_2(1) = -1, \end{aligned} \tag{4.57}$$

with the exact solution

$$u_1(x) = \exp(-\lambda x) + \exp(-x), \quad u_2(x) = -x.$$

The BVP (4.57) is monotone since

$$\begin{aligned}(\boldsymbol{f}(x,\boldsymbol{u},\boldsymbol{\xi}) - \boldsymbol{f}(x,\boldsymbol{v},\boldsymbol{\eta}), \boldsymbol{u}-\boldsymbol{v})\\ = -\lambda^2(u_1-v_1)^2 - \exp(\theta u_2 + (1-\theta)v_2)(u_2-v_2)^2 \le 0.\end{aligned}$$

Note that the Lipschitz constant of the right-hand side of the ODE (4.57) in a neighborhood of the exact solution is $L = \sqrt{\lambda^4+3} > 1$.

Table 4.3 and Table 4.4 contain the numerical results obtained by the 6-TDS (4.54) for the BVP (4.57) with $\lambda = 500$ and $\lambda = 1000$, respectively.

EPS	N	Error
$1.0\,E-4$	1000	$0.4406\,E-5$
$1.0\,E-6$	2000	$0.1315\,E-6$
$1.0\,E-8$	4000	$0.1380\,E-9$

Table 4.3: Numerical results for problem (4.57) with $\lambda = 500$

EPS	N	Error
$1.0\,E-4$	2000	$0.4411\,E-5$
$1.0\,E-6$	4000	$0.1316\,E-7$
$1.0\,E-8$	8000	$0.2239\,E-9$

Table 4.4: Numerical results for problem (4.57) with $\lambda = 1000$

Example 4.4. Finally, let us consider the following singularly perturbed BVP (see [85]):

$$\begin{aligned}\varepsilon\frac{d^2u_1}{dx^2} &= \frac{du_1}{dx} + u_1 + 0.5\exp(-u_2) + f_1(x),\\ \varepsilon\frac{d^2u_2}{dx^2} &= 2\frac{du_2}{dx} + 0.5u_1 + \exp(u_2) + f_2(x),\\ u_1(0) &= \exp(-1/\varepsilon) + 1, \quad u_1(1) = 1,\\ u_2(0) &= \exp(-2/\varepsilon + 1) + 1, \quad u_2(1) = 0,\end{aligned} \tag{4.58}$$

with the exact solution

$$u_1(x) = \exp\left((x-1)/\varepsilon\right) + \cos(0.5\pi x), \quad u_2(x) = -\exp\left(2(x-1)/\varepsilon + 1\right) + \exp(x).$$

The numerical results obtained by the 6-TDS (4.54) are given in Table 4.5.

ε	EPS	N	Error
$1.0\,E1$	$1.0\,E-3$	100	$0.3731\,E-3$
$1.0\,E1$	$1.0\,E-5$	400	$0.3701\,E-5$
$1.0\,E-1$	$1.0\,E-3$	6400	$0.2684\,E-4$
$1.0\,E-1$	$1.0\,E-5$	25600	$0.5148\,E-5$

Table 4.5: Numerical results for problem (4.58)

Chapter 5

Difference schemes for nonlinear BVPs on the half-axis

One thing I have learned in a long life:
that all our science, measured against reality, is
primitive and childlike - and yet it is the most
precious thing we have.

Albert Einstein (1879–1955)

In this chapter we generalize the idea of the exact difference schemes to BVPs which are defined on the half axis. Let us consider the following scalar nonlinear BVP on the infinite interval $[0, \infty)$,

$$\begin{aligned} &\frac{d^2u}{dx^2} - m^2 u = -f(x,u), \quad x \in (0,\infty), \quad u(x) \in \mathbb{R}, \\ &u(0) = \mu_1, \quad \lim_{x\to\infty} u(x) = 0, \end{aligned} \tag{5.1}$$

where $m \neq 0$ is a real constant. We will develop three-point EDS which are defined on *non-uniform* grids under the assumption that the function $f(x,u)$ in (5.1) is sufficiently smooth between a finite number of discontinuity points with respect to the first variable. Moreover, the practical implementation of the EDS by n-TDS of order of accuracy $\bar{n} = 2[(n+1)/2]$ is proposed. As before the freely selectable natural number n is called the rank of the TDS.

5.1 Existence and uniqueness of the solution

In [1, p. 83] sufficient conditions for the existence of a solution of the (vector-) problem

$$\frac{d^2\boldsymbol{u}}{dx^2} - m^2\boldsymbol{u} = -\boldsymbol{f}(x,\boldsymbol{u}), \quad x \in (0,\infty), \quad \boldsymbol{u}(x) \in \mathbb{R}^n,$$
$$\boldsymbol{u}(0) = \boldsymbol{\mu}_1, \quad \lim_{x\to\infty} \boldsymbol{u}(x) = \boldsymbol{0},$$

are given. Below, we use Banach's Fixed Point Theorem (see Theorem 1.1) to find more constructive conditions guaranteeing not only the existence but also the uniqueness of solutions of the BVP (5.1).

Let us introduce the function

$$u^{(0)}(x) \stackrel{\text{def}}{=} \mu_1 \exp(-m\,x), \tag{5.2}$$

the set

$$\Omega(D,\beta) \stackrel{\text{def}}{=} \left\{u(x) : u \in \mathbb{C}^1[0,\infty),\ \left\|u - u^{(0)}\right\|_{1,D} \leq \beta,\ D \subseteq [0,\infty)\right\},$$

$$\|u\|_{1,D} \stackrel{\text{def}}{=} \max\left\{\|u\|_{0,D}, \left\|\frac{du}{dx}\right\|_{0,D}\right\}, \qquad \|u\|_{0,D} \stackrel{\text{def}}{=} \max_{x\in D}|u(x)|,$$

and the class $Q^{(0)}[0,\infty)$ of piecewise continuous functions with a finite number of discontinuity points of first kind.

The next theorem gives sufficient conditions under which the problem (5.1) has a unique solution in the sphere $\Omega(D,r)$.

Theorem 5.1

Suppose the following hypotheses are satisfied:

- *for all $x \in [0,\infty)$ and for all $u \in \Omega([0,\infty),r)$, $r \stackrel{\text{def}}{=} K_1 \max\{1/m^2, 1/m\}$, it holds that*

$$f_u(x) \stackrel{\text{def}}{=} f(x,u) \in \mathcal{Q}^0[0,\infty), \quad |f(x,u)| \leq K(x) \leq K_1, \tag{5.3}$$

- *it is*

$$\lim_{x\to\infty} e^{-mx} \int_0^x e^{m\xi} K(\xi)d\xi = \lim_{x\to\infty} e^{mx} \int_x^\infty e^{-m\xi} K(\xi)d\xi = 0, \tag{5.4}$$

- *for all $x \in [0,\infty)$ and for all $u, v \in \Omega([0,\infty),r)$ it holds that*

$$|f(x,u) - f(x,v)| \leq L_1|u - v|, \tag{5.5}$$

- *and*

$$q \stackrel{\text{def}}{=} L_1 \max\{1/m^2, 1/m\} < 1. \tag{5.6}$$

Under these hypotheses the BVP (5.1) has a unique solution $u(x) \in \Omega([0,\infty), r)$ which is the limit of the sequence $\{u^{(k)}(x)\}_{k=0}^{\infty}$ defined by the starting function (5.2) and the fixed point iteration

$$\begin{aligned} &\frac{d^2 u^{(k)}}{dx^2} - m^2 u^{(k)} = -f\left(x, u^{(k-1)}\right), \quad x \in (0,\infty), \\ &u^{(k)}(0) = \mu_1, \quad \lim_{x\to\infty} u^{(k)}(x) = 0, \quad k = 1, 2, \dots . \end{aligned} \tag{5.7}$$

Moreover, the error estimate

$$\left\| u^{(k)} - u \right\|_{1,[0,\infty)} \le \frac{q^k}{1-q} r \tag{5.8}$$

holds.

Proof. We transform problem (5.1) into the equivalent integral form

$$u(x) = \Re(x, u(\cdot)) = \int_0^\infty G(x,\xi) f(\xi, u(\xi)) d\xi + u^{(0)}(x), \quad x \ge 0, \tag{5.9}$$

where

$$G(x,\xi) = \begin{cases} \dfrac{\sinh(mx)\exp(-m\xi)}{m}, & 0 \le x \le \xi, \\ \dfrac{\exp(-mx)\sinh(m\xi)}{m}, & x \ge \xi \end{cases} \tag{5.10}$$

is the Green's function of the linear, homogeneous part of problem (5.1).

Let us show that the operator (5.9) transforms the set $\Omega([0,\infty), r)$ into itself. Taking into account the equalities

$$\int_0^\infty G(x,\xi) d\xi = \frac{1 - \exp(-mx)}{m^2}, \qquad \int_0^\infty \frac{\partial G(x,\xi)}{\partial x} d\xi = \frac{\exp(-mx)}{m},$$

we get, for all $v \in \Omega([0,\infty), r)$,

$$\left\| \Re(x, v(\cdot)) - u^{(0)} \right\|_{1,[0,\infty)} \le K_1 \left\| \int_0^\infty G(x,\xi) d\xi \right\|_{1,[0,\infty)} \le r.$$

Moreover, the operator $\Re(x, u(\cdot))$ is contractive on the set $\Omega([0,\infty), r)$, since for all $u, v \in \Omega([0,\infty), r)$ we have

$$\|\Re(x, u(\cdot)) - \Re(x, v(\cdot))\|_{1,[0,\infty)} \le q \, \|u - v\|_{1,[0,\infty)},$$

provided that $q \stackrel{\text{def}}{=} L_1 \max\{1/m^2, 1/m\} < 1$.

Thus, $\Re(x, u(\cdot))$ satisfies all the conditions of Banach's Fixed Point Theorem (see Theorem 1.1) with $q < 1$. This implies that the equation (5.9) possesses a unique solution which is the fixed point of the sequence (5.7) with the error estimate (5.8). By standard arguments it is easy to show that the solution of (5.9) satisfies the ODE and the boundary condition $u(0) = \mu_1$, and as a result of (5.3) it also satisfies the boundary condition at infinity. ■

Example 5.1. Let us consider the BVP

$$\begin{gathered}\frac{d^2u}{dx^2} - m^2u = -f(x, u), \\ u(0) = \mu_1, \quad \lim_{x\to\infty} u(x) = 0,\end{gathered} \tag{5.11}$$

with

$$f(x, u) \stackrel{\text{def}}{=} \frac{u^2}{1 + x^2} \tag{5.12}$$

and

$$m > 4\mu_1, \quad \mu_1 > 0, \quad m > 1. \tag{5.13}$$

In that case the set $\Omega([0, \infty), r)$ is the ball of all $u(x)$ satisfying

$$|u(x) - \mu_1 e^{-mx}| \leq r,$$

i.e.,

$$|u(x)| \leq \mu_1 e^{-mx} + r \leq \mu_1 + r. \tag{5.14}$$

This implies the inequality

$$|f(x, u)| \leq \frac{(\mu_1 + r)^2}{1 + x^2} \stackrel{\text{def}}{=} K(x) \leq (\mu_1 + r)^2 \stackrel{\text{def}}{=} K_1 \tag{5.15}$$

which together with the condition

$$r = K_1 \max\left\{\frac{1}{m^2}, \frac{1}{m}\right\} = \frac{K_1}{m}$$

determines the value of r in dependence of the input data μ_1 and m. Thus we have

$$r = \frac{(\mu_1 + r)^2}{m}$$

which implies

$$r_{1,2} = -\mu_1 + \frac{m}{2} \pm \sqrt{m\left(\frac{m}{4} - \mu_1\right)}.$$

We choose the root

$$r = -\mu_1 + \frac{m}{2} - \sqrt{m\left(\frac{m}{4} - \mu_1\right)} = \frac{\mu_1^2}{-\mu_1 + \frac{m}{2} + \sqrt{m\left(\frac{m}{4} - \mu_1\right)}} > 0.$$

From the inequality

$$|f(x,u) - f(x,v)| \le \frac{|(u-v)(u+v)|}{1+x^2} \le 2(\mu_1 + r)\,|u-v|$$

we obtain the Lipschitz constant $L_1 \stackrel{\text{def}}{=} 2(\mu_1 + r)$. The inequality (5.6) now reads

$$q \stackrel{\text{def}}{=} \frac{L_1}{m} = \frac{2(\mu_1+r)}{m} = \frac{2}{m}\left(\frac{m}{2} - \sqrt{m\left(\frac{m}{4} - \mu_1\right)}\right) = 1 - \sqrt{1 - \frac{4\mu_1}{m}}$$

and due to (5.13) we have $q < 1$, i.e. the condition (5.6) is fulfilled.

It remains to check the conditions (5.4). Using (5.4) and L'Hospital's rule we obtain

$$\lim_{x\to\infty} e^{-mx} \int_0^x e^{m\xi} \frac{(\mu_1+r)^2}{1+\xi^2}\, d\xi = \lim_{x\to\infty} \frac{e^{mx}\dfrac{(\mu_1+r)^2}{1+x^2}}{me^{mx}} = 0,$$

$$\lim_{x\to\infty} e^{mx} \int_x^\infty e^{-m\xi} \frac{(\mu_1+r)^2}{1+\xi^2}\, d\xi = \lim_{x\to\infty} \frac{e^{-mx}\dfrac{(\mu_1+r)^2}{1+x^2}}{me^{-mx}} = 0.$$

Thus, the assumptions of Theorem 5.1 are satisfied and problem (5.11), under the assumptions (5.13), possesses a unique solution in $\Omega([0,\infty), r)$ which can be determined by the fixed point iteration. □

In many cases the determination of variable bounds on the solution is of great interest. In order to develop such bounds we introduce the tubular set

$$\Omega([0,\infty), p(x), r(x)) \stackrel{\text{def}}{=} \{u(x) \in \mathbb{C}[0,\infty) : p(x) \le u(x) \le r(x)\}$$

and postulate the following conditions on the function $f(x,u)$ in (5.1):

- Let for all $u(x), v(x) \in \Omega([0,\infty), p(x), r(x))$ and $x \in [0,\infty)$ exist functions $p(x)$, $r(x)$ and $L(x)$ such that

$$p(x) \le \int_0^\infty G(x,\xi) f(\xi, u(\xi)) d\xi + u^{(0)}(x) \le r(x), \tag{5.16}$$

  and

$$|f(x,u) - f(x,v)| \le L(x)\,|u-v|, \tag{5.17}$$

  where $G(x,\xi)$ is Green's function given in (5.10).

- Assume that the function $L(x)$ satisfies

$$\int_0^\infty G(x,\xi)\, L(\xi) d\xi \le L_1 < 1, \text{ for all } x \in [0,\infty). \tag{5.18}$$

In the following theorem sufficient conditions for the unique solvability of problem (5.1) in $\Omega([0,\infty),p(x),r(x))$ are given.

Theorem 5.2

Suppose that for all $x \in [0,\infty)$ and for all $u \in \Omega([0,\infty),p(x),r(x))$ it holds that

$$f_u(x) \stackrel{\text{def}}{=} f(x,u) \in \mathcal{Q}^0[0,\infty), \quad |f(x,u)| \leq K(x).$$

Moreover, let the conditions (5.4), (5.16)–(5.18) be satisfied. Then problem (5.1) has a unique solution $u(x) \in \Omega([0,\infty),p(x),r(x))$ which can be determined by the fixed point iteration.

Proof. Under the assumptions formulated above the operator

$$\Re(x,u(\cdot)) = \int\limits_0^\infty G(x,\xi)f(\xi,u(\xi))d\xi + u^{(0)}(x), \quad x \geq 0,$$

transforms the set $\Omega([0,\infty),p(x),r(x))$ into itself and is contractive. Banach's Fixed Point Theorem (see Theorem 1.1) yields the claim of the theorem. ∎

The next example illustrates Theorem 5.2.

Example 5.2. We consider again the BVP (5.11) for $\mu_1 > 0$ and $2\mu_1/m^2 < 1$, with the objective of finding functions $p(x)$ and $r(x)$ which serve as estimates for the exact solution at each local point x.

Since $G(x,\xi)$ and the function $f(x,u)$ are positive we can choose $p(x) \equiv 0$ in (5.16). The second inequality in (5.16) implies the following condition on the function $r(x)$:

$$\int_0^x G(x,\xi)\frac{u^2(\xi)}{1+\xi^2}d\xi + \mu_1 e^{-mx} \leq \int_0^x G(x,\xi)r^2(\xi)d\xi + \mu_1 e^{-mx} \leq r(x).$$

We look for a solution of the second inequality and use the ansatz $r(x) = c\,e^{-mx}$, where c is an unknown constant. Substituting this ansatz into the inequality we obtain

$$c^2 \int_0^x G(x,\xi)e^{-2m\xi}d\xi + \mu_1 e^{-mx} \leq ce^{-mx}.$$

Obviously, to calculate the integral it is sufficient to solve the BVP

$$\frac{d^2u}{dx^2} - m^2u = -e^{-2mx}, \quad u(0) = 0,\ u(\infty) = 0.$$

Thus,

$$\frac{c^2}{2m^2}\left(e^{-mx} - e^{-2mx}\right) + \mu_1 e^{-mx} \leq ce^{-mx}. \tag{5.19}$$

It follows that

$$\frac{c^2}{2m^2}\left(1-e^{-mx}\right)+\mu_1-c\le 0,$$

which in particular holds true for c satisfying

$$\frac{c^2}{2m^2}-c+\mu_1=0.$$

We choose

$$c=\frac{1-\sqrt{1-\frac{2\mu_1}{m^2}}}{\frac{1}{m^2}}=\frac{2\mu_1}{1+\sqrt{1-\frac{2\mu_1}{m^2}}}$$

and obtain

$$r(x)=\frac{2\mu_1}{1+\sqrt{1-\frac{2\mu_1}{m^2}}}e^{-mx}.$$

The estimate

$$|f(x,u)-f(x,v)|\le\frac{|u-v|\,|u+v|}{1+x^2}\le\frac{2r(x)}{1+x^2}|u-v|$$

and formula (5.17) imply

$$L(x)=\frac{2r(x)}{1+x^2}=2c\frac{e^{\;mx}}{1+x^2}.$$

It can easily be seen that

$$|f(x,u)|\le K(x)$$

with

$$K(x)=\frac{r^2(x)}{1+x^2}.$$

The conditions (5.4) can be checked in the same way as in Example 5.1. Condition (5.18) is fulfilled since

$$\int_0^\infty G(x,\xi)L(\xi)d\xi=2c\int_0^\infty G(x,\xi)\frac{e^{-m\xi}}{1+\xi^2}d\xi\le 2c\int_0^\infty G(x,\xi)e^{-m\xi}d\xi$$

$$=\frac{2c}{2m}xe^{-mx}\le\frac{c}{m}\frac{1}{m}e^{-1}=\frac{1}{m^2e}\frac{2\mu_1}{1+\sqrt{1-\frac{2\mu_1}{m^2}}}$$

$$=\frac{1}{e}\frac{\alpha}{1+\sqrt{1-\alpha}}<1\quad\text{for } \alpha=\frac{2\mu_1}{m^2}<1.$$

Now, Theorem 5.2 states that

$$0\le u(x)\le\frac{2\mu_1}{1+\sqrt{1-\frac{2\mu_1}{m^2}}}e^{-mx}\quad\text{for all } x\in[0,\infty).$$

□

5.2 Existence of a three-point EDS

On the interval $[0,\infty)$ we use the non-uniform closed grid $\hat{\bar{\omega}}_h$ and the corresponding open grid $\hat{\omega}_h$ such that the discontinuity points (with respect to the first argument) of the function $f(x,u)$ coincide with grid points. This means that N has to be chosen such that $\rho \subseteq \hat{\omega}_h$, where ρ denotes the set of all discontinuity points.

Let $h \stackrel{\text{def}}{=} h_{max}$ and h_{min} denote the maximum and minimum step size, respectively. We assume that the step sizes h_j and the corresponding grid $\hat{\bar{\omega}}_h$ satisfy

$$c_1 \le \frac{h_{max}}{h_{min}} \le c_2, \tag{5.20}$$

where c_1 and c_2 are real constants. In order to obtain maximum order of convergence of our EDS (see Theorem 5.4), we postulate

$$\frac{1}{h_{max}} \le x_N \le \frac{1}{h_{min}}. \tag{5.21}$$

The inequality $h_{min}\, N \le x_N = h_1 + h_2 + \cdots + h_N \le h_{max} N$ together with (5.21) imply

$$h_{min} \le \frac{1}{x_N} \le \frac{1}{N\, h_{min}}, \qquad \frac{1}{N\, h_{max}} \le \frac{1}{x_N} \le h_{max}.$$

Due to (5.20) we further obtain

$$\frac{h_{max}}{c_2} \le h_{min} \le \frac{1}{\sqrt{N}}, \qquad c_2 h_{min} \ge h_{max} \ge \frac{1}{\sqrt{N}},$$

which yields

$$\begin{gathered} h_{max} \le \frac{c_2}{\sqrt{N}}, \qquad h_{min} \ge \frac{1}{c_2\sqrt{N}}, \\ \frac{\sqrt{N}}{c_2} \le h_{min}\, N \le x_N \le h_{max}\, N \le c_2\sqrt{N}. \end{gathered} \tag{5.22}$$

Thus, we have $h_{max} \to 0$ and $x_N \to \infty$ as $N \to \infty$.

We now introduce the set of grid functions

$$\Omega(\hat{\bar{\omega}}_h, \beta) \stackrel{\text{def}}{=} \left\{ v(x),\ x \in \hat{\bar{\omega}}_h : \ \|v - u^{(0)}\|_{1,\hat{\omega}_h^+} \le \beta \right\},$$

where

$$\|y\|_{1,\hat{\omega}_h^+} \stackrel{\text{def}}{=} \max\left\{ \|y\|_{0,\hat{\omega}_h^+}, \|y_{\bar{x}}\|_{0,\hat{\omega}_h^+} \right\}, \qquad \|y\|_{0,\hat{\omega}_h^+} \stackrel{\text{def}}{=} \max_{1\le j\le N} |y_j|,$$

$$\|y\|_{0,\hat{\bar{\omega}}_h} \stackrel{\text{def}}{=} \max_{0\le j\le N} |y_j|, \qquad y_{\bar{x},j} \stackrel{\text{def}}{=} (y_j - y_{j-1})/h_j, \qquad \hat{\omega}_h^+ \stackrel{\text{def}}{=} \hat{\omega}_h \cup x_N.$$

Moreover, in the following we use again the abbreviations

$$\begin{gathered} \bar{e}_\alpha^j \stackrel{\text{def}}{=} [x_{j-2+\alpha}, x_{j-1+\alpha}], \qquad \bar{e}^j \stackrel{\text{def}}{=} [x_{j-1}, x_j], \qquad \bar{e}_2^N \stackrel{\text{def}}{=} [x_N, \infty); \\ e_\alpha^j,\ e^j, \text{ and } e_2^N \text{ are the corresponding open intervals, resp.} \end{gathered} \tag{5.23}$$

Let the BVPs

$$\frac{d^2Y_\alpha^j(x,v)}{dx^2} - m^2Y_\alpha^j(x,v) = -f\left(x, Y_\alpha^j(x,v)\right), \quad x \in e_\alpha^j,$$
$$Y_\alpha^j(x_{j-2+\alpha}, v) = v(x_{j-2+\alpha}), \quad Y_\alpha^j(x_{j-1+\alpha}, v) = v(x_{j-1+\alpha}), \tag{5.24}$$
$$j = 2-\alpha, \ldots, N+1-\alpha, \quad \alpha = 1, 2,$$

on subintervals of the length $O(h)$ and the problem

$$\frac{d^2Y_2^N(x,v)}{dx^2} - m^2\, Y_2^N(x,v) = -f\left(x, Y_2^N(x,v)\right), \quad x \in e_2^N,$$
$$Y_2^N(x_N, v) = v(x_N), \quad \lim_{x\to\infty} Y_2^N(x,v) = 0, \tag{5.25}$$

on the interval $[x_N, \infty)$ be given. The following lemma shows that the exact solution of problem (5.1) can be expressed on each subinterval by the solutions of (5.24) and (5.25).

Lemma 5.1

Suppose that the hypotheses of Theorem 5.1 are satisfied. Let $v(x) \in \Omega(\hat{\bar{\omega}}_h, r)$ be an arbitrary grid function. Then problems (5.24) and (5.25) have unique solutions $Y_\alpha^j(x,v) \in \Omega(\bar{e}_\alpha^j, r)$, $j = 2-\alpha, \ldots, N+1-\alpha$, $\alpha = 1, 2$ and $Y_2^N(x,v) \in \Omega(\bar{e}_2^N, r)$. Furthermore, the solution of problem (5.1) can be represented in the form

$$u(x) = \begin{cases} Y_\alpha^j(x,u), & x \in \bar{e}_\alpha^j, \quad j = 2-\alpha(1)N+1-\alpha, \quad \alpha = 1, 2, \\ Y_2^N(x,u), & x \in \bar{e}_2^N, \end{cases} \tag{5.26}$$

provided that condition (5.6) is satisfied.

Proof. Problems (5.24) and (5.25) are equivalent to the operator equations

$$U_\alpha^j(x) = \Re_\alpha^j(x, v, U_\alpha^j) \stackrel{\text{def}}{=} \int_{x_{j-2+\alpha}}^{x_{j-1+\alpha}} G^{j-1+\alpha}(x,\xi) f\left(\xi, U_\alpha^j(\xi)\right) d\xi + \hat{v}(x), \quad x \in e_\alpha^j,$$

$$j = 2-\alpha, \ldots, N+1-\alpha, \quad \alpha = 1, 2,$$

$$\hat{v}(x) \stackrel{\text{def}}{=} \frac{v(x_j)\sinh(m(x - x_{j-1})) + v(x_{j-1})\sinh(m(x_j - x))}{\sinh(mh_j)}, \quad x \in [x_{j-1}, x_j],$$

$$j = 1, \ldots, N,$$

$$U_2^N(x) = \Re_2^N(x, v, U_2^N) \stackrel{\text{def}}{=} \int_{x_N}^{\infty} G^\infty(x,\xi) f\left(\xi, U_2^N(\xi)\right) d\xi + v(x_N)e^{-m(x-x_N)},$$

$$x \in e_2^N,$$

where

$$G^{j-1+\alpha}(x,\xi)$$
$$\stackrel{\text{def}}{=} \begin{cases} \dfrac{\sinh(m(x-x_{j-2+\alpha}))\sinh(m(x_{j-1+\alpha}-\xi))}{m\sinh(mh_{j-1+\alpha})}, & x_{j-2+\alpha} \le x \le \xi, \\ \dfrac{\sinh(m(x_{j-1+\alpha}-x))\sinh(m(\xi-x_{j-2+\alpha}))}{m\sinh(mh_{j-1+\alpha})}, & \xi \le x \le x_{j-1+\alpha}, \end{cases}$$

$$G^{\infty}(x,\xi)$$
$$\stackrel{\text{def}}{=} \begin{cases} \dfrac{\sinh(m(x-x_N))\exp(-m(\xi-x_N))}{m}, & x_N \le x \le \xi, \\ \dfrac{\exp(-m(x-x_N))\sinh(m(\xi-x_N))}{m}, & x \ge \xi. \end{cases}$$

Note that

$$\begin{aligned} u^{(0)}(x) &= \mu_1 \exp(-mx) \\ &= \frac{u^{(0)}(x_j)\sinh(m(x-x_{j-1})) + u^{(0)}(x_{j-1})\sinh(m(x_j-x))}{\sinh(mh_j)} \\ &= \hat{u}^{(0)}(x). \end{aligned}$$

We now study the properties of the operators

$$\Re_\alpha^j(x,v,U_\alpha^j),\ \ \alpha=1,2,\ \ j=2-\alpha,\dots,N+1-\alpha, \quad \text{and} \quad \Re_2^N(x,v,U_2^N).$$

Let $U_\alpha^j(x) \in \Omega(e_\alpha^j, r)$ and $U_2^N(x) \in \Omega(e_2^N, r)$, then the equalities

$$\begin{aligned} &\int\limits_{x_{j-2+\alpha}}^{x_{j-1+\alpha}} G^{j-1+\alpha}(x,\xi)d\xi \\ &\quad = \frac{1}{m^2}\left[1 - \frac{\sinh(m(x-x_{j-2+\alpha})) + \sinh(m(x_{j-1+\alpha}-x))}{\sinh(mh_{j-1+\alpha})}\right], \\ &\int\limits_{x_N}^{\infty} G^{\infty}(x,\xi)d\xi = \frac{1-\exp(-m(x-x_N))}{m^2}, \end{aligned}$$

imply

$$\left\|\Re_\alpha^j(x,v,U_\alpha^j) - u^{(0)}\right\|_{1,e_\alpha^j} \le$$

$$\left\|\frac{\left|v_{j-1+\alpha} - u_{j-1+\alpha}^{(0)}\right| \sinh(m(x-x_{j-2+\alpha}))}{\sinh(mh_{j-1+\alpha})}\right.$$

$$+\frac{\left|v_{j-2+\alpha}-u^{(0)}_{j-2+\alpha}\right|\sinh(m(x_{j-1+\alpha}-x))}{\sinh(mh_{j-1+\alpha})}+K_1\int\limits_{x_{j-2+\alpha}}^{x_{j-1+\alpha}}G^{j-1+\alpha}(x,\xi)\,d\xi\Bigg\|_{1,e^j_\alpha}\leq r$$

for all $v\in\Omega(\hat{\hat{\omega}}_h,r)$,

$$\left\|\Re^N_2(x,v,U^N_2)-u^{(0)}\right\|_{1,e^N_2}$$

$$\leq\left\|\left|v_N-u^{(0)}(x_N)\right|\exp(-m(x-x_N))+K_1\int\limits_{x_N}^{\infty}G^\infty(x,\xi)d\xi\right\|_{1,e^N_2}$$

$$\leq r\quad\text{for all } v\in\Omega(\hat{\hat{\omega}}_h,r).$$

Thus, for all $v\in\Omega(\hat{\hat{\omega}}_h,r)$ the operators $\Re^j_\alpha(x,v,U^j_\alpha)$, $j=2-\alpha,\ldots,N+1-\alpha$, $\alpha=1,2$, and $\Re^N_2(x,v,U^N_2)$ transform the sets $\Omega(e^j_\alpha,r)$ and $\Omega(e^N_2,r)$, respectively, into themselves.

Moreover, for all $U^j_\alpha(x),\tilde{U}^j_\alpha(x)\in\Omega(e^j_\alpha,r)$ we have the estimates

$$\|\Re^j_\alpha\left(x,v,U^j_\alpha\right)-\Re^j_\alpha\left(x,v,\tilde{U}^j_\alpha\right)\|_{1,e^j_\alpha}$$

$$\leq\left\|1-\frac{\sinh(m(x-x_{j-2+\alpha}))+\sinh(m(x_{j-1+\alpha}-x))}{\sinh(mh_{j-1+\alpha})}\right\|_{1,e^j_\alpha}\frac{L_1}{m^2}\|U^j_\alpha-\tilde{U}^j_\alpha\|_{1,e^j_\alpha}$$

$$\leq q\|U^j_\alpha-\tilde{U}^j_\alpha\|_{1,e^j_\alpha},$$

and for all $U^N_2(x),\tilde{U}^N_2(x)\in\Omega(e^N_2,r)$ it holds that

$$\|\Re^N_2\left(x,v,U^N_2\right)-\Re^N_2\left(x,v,\tilde{U}^N_2\right)\|_{1,e^N_2}$$

$$\leq\frac{L_1}{m^2}\|1-\exp(-m(x-x_N))\|_{1,e^N_2}\|U^N_2-\tilde{U}^N_2\|_{1,e^N_2}$$

$$\leq q\|U^N_2-\tilde{U}^N_2\|_{1,e^N_2}.$$

We see that the operators $\Re^j_\alpha\left(x,v,U^j_\alpha\right)$, $j=2-\alpha,\ldots,N+1-\alpha$, $\alpha=1,2$, are contractive on $\Omega(e^j_\alpha,r)$, and the operator $\Re^N_2\left(x,v,U^N_2\right)$ is contractive on $\Omega(e^N_2,r)$. This completes the proof. ■

We are now in a position to prove the first main result of this chapter.

Theorem 5.3

Suppose that the assumptions of Theorem 5.1 are satisfied. Then, for problem (5.1) there exists an EDS which is of the form

$$(au_{\bar{x}})_{\hat{x},j}-d(x_j)u_j=-\varphi(x_j,u),\quad j=1,\ldots,N-1,\tag{5.27}$$

$$u_0 = \mu_1, \quad -a(x_N)u_{\bar{x},N} = \beta_2 u_N - \mu_2(x_N, u), \tag{5.28}$$

where

$$a(x_j) \stackrel{\text{def}}{=} \frac{mh_j}{\sinh(mh_j)}, \quad j = 1, \dots, N, \quad \beta_2 \stackrel{\text{def}}{=} m\frac{\exp(mh_N) - 1}{\sinh(mh_N)},$$

$$d(x_j) \stackrel{\text{def}}{=} \frac{m}{\hbar_j}\left\{\frac{\cosh(mh_j) - 1}{\sinh(mh_j)} + \frac{\cosh(mh_{j+1}) - 1}{\sinh(mh_{j+1}}\right\}, \quad j = 1, \dots, N-1, \tag{5.29}$$

$$\varphi(x_j, u) = \hat{T}^{x_j}(f(\xi, u(\xi))), \quad j = 1, \dots, N-1,$$

$$\mu_2(x_N, u) = \hat{\hat{T}}^{x_N}(f(\xi, u(\xi))), \tag{5.30}$$

with

$$\hat{T}^{x_j}(f(\xi, u(\xi))) \stackrel{\text{def}}{=} \frac{1}{\hbar_j \sinh(mh_j)} \int_{x_{j-1}}^{x_j} \sinh(m(\xi - x_{j-1}))f(\xi, u(\xi))d\xi$$

$$+ \frac{1}{\hbar_j \sinh(mh_{j+1})} \int_{x_j}^{x_{j+1}} \sinh(m(x_{j+1} - \xi))f(\xi, u(\xi))d\xi,$$

$$j = 1, \dots, N-1,$$

$$\hat{\hat{T}}^{x_N}(f(\xi, u(\xi))) \stackrel{\text{def}}{=} \frac{1}{\sinh(mh_N)} \int_{x_{N-1}}^{x_N} \sinh(m(\xi - x_{N-1}))f(\xi, u(\xi))d\xi$$

$$+ \int_{x_N}^{\infty} \exp(-m(\xi - x_N))f(\xi, u(\xi))d\xi.$$

The function $u(\xi)$ on the right-hand side of (5.27) is given by (5.26) and depends only on $u_0, u_1, \dots, u_N$.

Proof. It can easily be seen that

$$\hat{T}^{x_j}(u'' - m^2 u) = (au_{\bar{x}})_{\hat{x},j} - d(x_j)u_j, \quad j = 1, \dots, N-1,$$

$$\hat{\hat{T}}^{x_N}(u'' - m^2 u) = -a(x_N)u_{\bar{x},N} - \beta_2 u_N.$$

Applying the operators $\hat{T}^{x_j}$, $j = 1, \dots, N-1$, and $\hat{\hat{T}}^{x_N}$ to the ODE $u'' - m^2 u = -f(x, u)$, we obtain the three-point EDS (5.27). ■

The following lemma claims the uniqueness of the solution of the EDS (5.27).

Lemma 5.2

Suppose that the assumptions of Theorem 5.3 are satisfied. Then there exists an $h_0 > 0$ such that for all $h \le h_0$ and all grid functions $u(x) \in \Omega(\hat{\bar{\omega}}_h, r)$ the EDS (5.27) has a unique solution which is the limit of the sequence $\{u^{(k)}(x)\}_{k=0}^{\infty}$ defined by

$$\begin{aligned}
&\left(a u_{\bar{x}}^{(k)}\right)_{\hat{x},j} - d(x_j)u_j^{(k)} = -\hat{T}^{x_j}\left(f\left(\xi, u^{(k-1)}(\xi)\right)\right), \quad j = 1,\dots,N-1,\\
&u_0^{(k)} = \mu_1, \quad -a(x_N)u_{\bar{x},N}^{(k)} = \beta_2 u_N^{(k)} - \hat{\hat{T}}^{x_N}\left(f\left(\xi, u^{(k-1)}(\xi)\right)\right),
\end{aligned} \tag{5.31}$$

$$u^{(k)}(x) = \begin{cases} Y_\alpha^j\left(x, u^{(k)}\right), & x \in \bar{e}_\alpha^j, \quad j = 2-\alpha,\dots,N+1-\alpha, \quad \alpha = 1,2,\\ Y_2^N\left(x, u^{(k)}\right), & x \in \bar{e}_2^N \end{cases}$$

$$u^{(0)}(x) = \mu_1 \exp(-mx).$$

Moreover, the following error estimate holds:

$$\left\|u^{(k)} - u\right\|_{1,\hat{\omega}_h^+}^* = \max\left\{\left\|u^{(k)} - u\right\|_{0,\hat{\omega}_h^+}, \left\|\frac{du^{(k)}}{dx} - \frac{du}{dx}\right\|_{0,\hat{\omega}_h^+}\right\} \le M q_1^k, \tag{5.32}$$

where $q_1 \stackrel{\text{def}}{=} q + M_1 h < 1$ and M, M_1 are some constants.

Proof. For the proof we use Banach's Fixed Point Theorem (see Theorem 1.1). We represent the solution of the auxiliary problem

$$\begin{aligned}
&(a\tilde{u}_{\bar{x}})_{\hat{x}} - d(x)\tilde{u} = -\hat{T}^x(\tilde{u}'' - m^2\tilde{u}), \quad x \in \hat{\omega}_h^+,\\
&\tilde{u}(0) = \mu_1, \quad \tilde{u}(x_{N+1}) = 0
\end{aligned} \tag{5.33}$$

as

$$\tilde{u}(x) = \sum_{i=1}^{N} \hbar_i \tilde{G}^h(x, x_i)\hat{T}^{x_i}(f(\eta, \tilde{u}(\eta))) + \frac{\sinh(m(x_{N+1} - x))}{\sinh(m x_{N+1})}\mu_1, \quad x \in \hat{\omega}_h^+, \tag{5.34}$$

where $\tilde{G}^h(x, \xi)$ is Green's function of problem (5.33), i.e.,

$$\tilde{G}^h(x,\xi) = \begin{cases} \dfrac{\sinh(m\,x)\sinh(m(x_{N+1} - \xi))}{m\sinh(m\,x_{N+1})}, & 0 \le x \le \xi,\\[2ex] \dfrac{\sinh(m(x_{N+1} - x))\sinh(m\,\xi)}{m\sinh(m\,x_{N+1})}, & \xi \le x \le x_{N+1}. \end{cases}$$

Considering the equation (5.34) for $x_{N+1} \to \infty$, we get the solution of problem (5.27)

$$u(x) = \Re_h(x,u) \stackrel{\text{def}}{=} \sum_{i=1}^{N} \hbar_i G(x,x_i)\hat{T}^{x_i}(f(\eta,u(\eta))) + G(x,x_N)\hat{\hat{T}}^{x_N}(f(\eta,u(\eta))) + u^{(0)}(x),$$

for all $x \in \hat{\omega}_h^+$. Here $G(x,\xi)$ is Green's function of the linear homogeneous part of problem (5.1).

Let us study the operator $\Re_h(x,u)$. We first wish to point out that the operators $\hat{T}^{x_i}$, $i = 1,\dots,N-1$, and $\hat{\hat{T}}^{x_N}$ have the following property. For all $Q(\xi) \in H_h$, $w(\eta) \in L_1(0,\infty)$ (H_h is the space of all grid functions $v(x)$, $x \in \hat{\omega}_h^+$, such that $v(0) = 0$) the following identity holds:

$$\begin{aligned}
&\sum_{i=1}^{N-1} \hbar_i \hat{T}^{x_i}(w(\eta))Q(x_i) + \hat{\hat{T}}^{x_N}(w(\eta))Q(x_N) \\
&= \sum_{i=1}^{N} Q(x_i)\frac{1}{\sinh(m\,h_i)} \int_{x_{i-1}}^{x_i} \sinh(m(\eta - x_{i-1}))w(\eta)d\eta \\
&\quad + \sum_{i=1}^{N-1} Q(x_i)\frac{1}{\sinh(m\,h_{i+1})} \int_{x_i}^{x_{i+1}} \sinh(m(x_{i+1} - \eta))w(\eta)d\eta \\
&\quad + Q(x_N)\int_{x_N}^{\infty} \exp(-m(\eta - x_N))w(\eta)d\eta \qquad (5.35) \\
&= \sum_{i=1}^{N} \int_{x_{i-1}}^{x_i} \left\{ Q(x_i)\frac{\sinh(m(\eta - x_{i-1}))}{\sinh(m\,h_i)} + Q(x_{i-1})\frac{\sinh(m(x_i - \eta))}{\sinh(m\,h_i)} \right\} w(\eta)d\eta \\
&\quad + Q(x_N)\int_{x_N}^{\infty} \exp(-m(\eta - x_N))w(\eta)d\eta.
\end{aligned}$$

Due to (5.35), we have

$$\begin{aligned}
&\Re_h(x,u) \\
&= \sum_{i=1}^{N} \int_{x_{i-1}}^{x_i} \left\{ G(x,x_i)\frac{\sinh(m(\eta - x_{i-1}))}{\sinh(m\,h_i)} + G(x,x_{i-1})\frac{\sinh(m(x_i - \eta))}{\sinh(m\,h_i)} \right\} \qquad (5.36) \\
&\quad \times f(\eta,u(\eta))d\eta \\
&\quad + \frac{\sinh(m\,x)\exp(-m\,x_N)}{m} \int_{x_N}^{\infty} \exp(-m(\eta - x_N))f(\eta,u(\eta))d\eta + u^{(0)}(x)
\end{aligned}$$

$$= \sum_{i=1}^{N} \int_{x_{i-1}}^{x_i} G(x,\eta) f(\eta, u(\eta)) d\eta + \int_{x_N}^{\infty} G(x,\eta) f(\eta, u(\eta)) d\eta + u^{(0)}(x) \tag{5.37}$$

$$= \int_{0}^{\infty} G(x,\eta) f(\eta, u(\eta)) d\eta + u^{(0)}(x), \quad x \in \hat{\omega}_h^+,$$

where

$$u(\eta) = Y_1^i(\eta, u), \quad \eta \in \bar{e}^i, \quad i = 1, \dots, N,$$
$$\hat{u}(\eta) = Y_2^N(\eta, u), \quad \eta \in \bar{e}_2^N.$$

The operator (5.37) transforms the set $\Omega(\hat{\bar{\omega}}_h, r)$ into itself. Let $v \in \Omega(\hat{\bar{\omega}}_h, r)$, then

$$v(x) = Y_1^i(x, v) \in \Omega(\bar{e}^j, r), \quad \hat{v}(x) = Y_2^N(x, v) \in \Omega(\bar{e}_2^N, r)$$

(see the proof of Lemma 5.1). Since

$$\int_0^{\infty} G(x,\eta) d\eta = \frac{1 - \exp(-m\,x)}{m^2},$$

$$\int_0^{\infty} G_{\bar{x}}(x,\eta) d\eta = \frac{1}{m\,h(x)} \int_{x-h(x)}^{x} \exp(-m\,\eta) d\eta \le \frac{\exp(-m(x - h(x)))}{m},$$

we have, for all $v \in \Omega(\hat{\bar{\omega}}_h, r)$,

$$\left\| \Re_h(x,v) - u^{(0)} \right\|_{1,\hat{\omega}_h^+} \le K_1 \left\| \int_0^{\infty} G(x,\eta) d\eta \right\|_{1,\hat{\omega}_h^+} \le r.$$

Furthermore we have, for all $u, v \in \Omega(\hat{\bar{\omega}}_h, r)$,

$$\| \Re_h(x,u) - \Re_h(x,v) \|_{1,\hat{\omega}_h^+} \le q \, \|u - v\|_{0,[0,\infty)}.$$

Let us now show that

$$\|u - v\|_{0,[0,\infty)} \le (1 + M\,h) \|u - v\|_{0,\hat{\omega}_h^+}. \tag{5.38}$$

Using the splitting

$$u(x) = \hat{u}(x) + \breve{u}(x),$$

where

$$\hat{u}(x) \stackrel{\text{def}}{=} \frac{u(x_j)\sinh(m(x - x_{j-1})) - u(x_{j-1})\sinh(m(x_j - x))}{\sinh(mh_j)},$$

$$\breve{u}(x) \stackrel{\text{def}}{=} Y_1^j(x,u) - \hat{u}(x) \text{ for } x \in \bar{e}^j, \quad j = 1, \ldots, N,$$

and

$$\hat{u}(x) \stackrel{\text{def}}{=} u(x_N)\exp(-m(x-x_N)), \quad \breve{u}(x) \stackrel{\text{def}}{=} Y_2^N(x,u) - \hat{u}(x) \text{ for } x \in \bar{e}_2^N,$$

we transform the problems

$$\frac{d^2u}{dx^2} - m^2u = -f(x,u), \quad x \in e^j, \quad u(x_{j-1}) = u_{j-1}, \quad u(x_j) = u_j, \quad j = 1, \ldots, N,$$

$$\frac{d^2u}{dx^2} - m^2u = -f(x,u), \quad x \in e_2^N, \quad u(x_N) = u_N, \quad \lim_{x\to\infty} u(x) = 0$$

into the form

$$\frac{d^2\breve{u}}{dx^2} - m^2\breve{u} = -f(x,\hat{u}(x)+\breve{u}(x)), \quad x \in e^j, \ \breve{u}(x_{j-1}) = \breve{u}(x_j) = 0, \ j = 1, \ldots, N,$$

$$\frac{d^2\breve{u}}{dx^2} - m^2\breve{u} = -f(x,\hat{u}(x)+\breve{u}(x)), \quad x \in e_2^N, \quad \breve{u}(x_N) = 0, \quad \lim_{x\to\infty} \breve{u}(x) = 0.$$

Then

$$\breve{u}(x) = \int_{x_{j-1}}^{x_j} G^j(x,\xi)f(\xi,\hat{u}(\xi)+\breve{u}(\xi))d\xi, \quad x \in e^j, \quad j = 1, \ldots, N,$$

$$\breve{u}(x) = \int_{x_N}^{\infty} G^{\infty}(x,\xi)f(\xi,\hat{u}(\xi)+\breve{u}(\xi))d\xi, \quad x \in e_2^N,$$

where

$$G^j(x,\xi) = \begin{cases} \dfrac{\sinh(m(x-x_{j-1}))\sinh(m(x_j-\xi))}{m\sinh(m\,h_j)}, & x_{j-1} \le x \le \xi, \\[2ex] \dfrac{\sinh(m(x_j-x))\sinh(m(\xi-x_{j-1}))}{m\sinh(m\,h_j)}, & \xi \le x \le x_j, \end{cases}$$

$$G^{\infty}(x,\xi) = \begin{cases} \dfrac{\sinh(m(x-x_N))\exp(-m(\xi-x_N))}{m}, & x_N \le x \le \xi, \\[2ex] \dfrac{\exp(-m(x-x_N))\sinh(m(\xi-x_N))}{m}, & x \ge \xi. \end{cases}$$

The Lipschitz condition yields

$$\|\breve{u}-\breve{v}\|_{1,\bar{e}^j}$$

$$\le L_1\left[\|\hat{u}-\hat{v}\|_{0,\bar{e}^j} + \|\breve{u}-\breve{v}\|_{0,\bar{e}^j}\right]\left\|\int_{x_{j-1}}^{x_j} G^j(x,\xi)d\xi\right\|_{1,\bar{e}^j}$$

$$= \frac{L_1}{m^2} \left\| 1 - \frac{\cosh(m(x_j + x_{j-1})/2 - x)}{\cosh(m\, h_j/2)} \right\|_{1,\bar{e}^j} \left[\|\hat{u} - \hat{v}\|_{0,\bar{e}^j} + \|\breve{u} - \breve{v}\|_{0,\bar{e}^j} \right]$$

$$\leq h\, q \left(\|u - v\|_{0,\hat{\omega}_h^+} + \|\breve{u} - \breve{v}\|_{1,\bar{e}^j} \right), \quad j - 1, \ldots, N,$$

$$\|\breve{u} - \breve{v}\|_{1,\bar{e}_2^N}$$

$$\leq L_1 \left[|u(x_N) - v(x_N)| + \|\breve{u} - \breve{v}\|_{0,\bar{e}_2^N} \right] \left\| \int_{x_N}^{\infty} G^\infty(x,\xi) d\xi \right\|_{1,\bar{e}_2^N}$$

$$\leq h\, q \left(\|u - v\|_{0,\hat{\omega}_h^+} + \|\breve{u} - \breve{v}\|_{1,\bar{e}_2^N} \right).$$

This implies

$$\begin{aligned} \|\breve{u} - \breve{v}\|_{1,\bar{e}^j} &\leq \frac{h\, q}{1 - h\, q} \|u - v\|_{0,\hat{\omega}_h^+} \leq h\, M_2 \|u - v\|_{0,\hat{\omega}_h^+}, \\ \|\breve{u} - \breve{v}\|_{1,\bar{e}_2^N} &< \frac{h\, q}{1 - h\, q} \|u - v\|_{0,\hat{\omega}_h^+} \leq h\, M_3 \|u - v\|_{0,\hat{\omega}_h^+}. \end{aligned} \tag{5.39}$$

Now, the inequalities

$$\begin{aligned} \|u - v\|_{0,\bar{e}^j} &\leq \|\hat{u} - \hat{v}\|_{0,\bar{e}^j} + \|\breve{u} - \breve{v}\|_{0,\bar{e}^j} \\ &\leq \|u - v\|_{0,\hat{\omega}_h^+} + h\, M_2 \|u - v\|_{0,\hat{\omega}_h^+}, \\ \|u - v\|_{0,\bar{e}_2^N} &\leq \|u - v\|_{0,\hat{\omega}_h^+} + h\, M_3 \|u - v\|_{0,\hat{\omega}_h^+} \end{aligned}$$

yield the estimate (5.38).

Taking into account the inequality (5.38), we get

$$\|\Re_h(x,u) - \Re_h(x,v)\|_{1,\hat{\omega}_h^+} \leq (q + M_1\, h)\, \|u - v\|_{1,\hat{\omega}_h^+} = q_1 \|u - v\|_{1,\hat{\omega}_h^+}.$$

Due to (5.6) it holds that $q < 1$. If h_0 is small enough we have $q_1 \stackrel{\text{def}}{=} q + M_1\, h < 1$, and the operator (5.37) is a contraction for all $u, v \in \Omega(\hat{\hat{\omega}}_h, r)$. Thus, due to Banach's Fixed Point Theorem (see Theorem 1.1), for h_0 small enough the EDS (5.27) has a unique solution which is the fixed point of the sequence (5.31) with the error estimate

$$\left\| u^{(k)} - u \right\|_{1,\hat{\omega}_h^+} \leq \frac{q_1^k}{1 - q_1} r. \tag{5.40}$$

Moreover, due to (5.39), (5.40) we have

$$\begin{aligned}
&\left\|\frac{du^{(k)}}{dx}-\frac{du}{dx}\right\|_{0,\hat\omega_h^+}\\
&\le \left\|\frac{d\hat u^{(k)}}{dx}-\frac{d\hat u}{dx}\right\|_{0,\hat\omega_h^+}+\left\|\frac{d\breve u^{(k)}}{dx}-\frac{d\breve u}{dx}\right\|_{0,\hat\omega_h^+}\\
&\le M_4\left\|u_{\bar x}^{(k)}-u_{\bar x}\right\|_{0,\hat\omega_h^+}+M_5\left\|u^{(k)}-u\right\|_{0,\hat\omega_h^+}+h\,\max\{M_2,M_3\}\left\|\breve u^{(k)}-\breve u\right\|_{1,\bar e_2^N}\\
&\le M_6\left\|u^{(k)}-u\right\|_{1,\hat\omega_h^+}\le M\,q_1^k,
\end{aligned} \tag{5.41}$$

which proves the estimate (5.32). ■

5.3 Implementation of the three-point EDS

First we want to remember that the right-hand side of the EDS is given by

$$\begin{aligned}
\varphi(x_j,u)&=\hat T^{x_j}(f(\xi,u(\xi)))\\
&=\frac{1}{\hbar_j}\Bigg[Z_2^j(x_j,u)-Z_1^j(x_j,u)+\frac{m(\cosh(m\,h_j)Y_1^j(x_j,u)-u_{j-1})}{\sinh(m\,h_j)}\\
&\qquad+\frac{m(\cosh(m\,h_{j+1})Y_2^j(x_j,u)-u_{j+1})}{\sinh(m\,h_{j+1})}\Bigg],\quad j=1,\dots,N-1,
\end{aligned}$$

where

$$Z_\alpha^j(x,u)\overset{\text{def}}{=}\frac{dY_\alpha^j(x,u)}{dx},\qquad \alpha=1,2.$$

The right-hand side of the difference equation at x_N is given by

$$\mu_2(x_N,u)=\hat{\hat T}^{x_N}(f(\xi,u(\xi)),$$

with

$$\hat{\hat{T}}^{x_N}(f(\xi,u(\xi)))$$

$$= Z_2^N(x_N,u) + m\,Y_2^N(x_N,u) - Z_1^N(x_N,u) + \frac{m(\cosh(m\,h_N)Y_1^N(x_N,u) - u_{N-1})}{\sinh(m\,h_N)}.$$

In order to evaluate the EDS (5.27) for all $x_j \in \hat{\omega}_h$ it is necessary to solve the problems (5.24) and (5.25) with $v_j = u_j$, $j = 0,\ldots,N$. It is algorithmically more preferable (because of the well-developed theory and software) to deal with IVPs instead of BVPs. Therefore, we transform (5.24) and (5.25) into the following IVPs on subintervals of the length $O(h)$:

$$\frac{dY_\alpha^j(x,u)}{dx} = Z_\alpha^j(x,u), \quad \frac{dZ_\alpha^j(x,u)}{dx} - m^2\,Y_\alpha^j(x,u) = -f\left(x, Y_\alpha^j(x,u)\right), \quad x \in e_\alpha^j,$$

$$Y_\alpha^j(x_\beta,u) = u(x_\beta), \; Z_\alpha^j(x_\beta,u) = \left.\frac{du}{dx}\right|_{x=x_\beta}, \; j = 2-\alpha,\ldots,N+1-\alpha, \; \alpha = 1,2, \tag{5.42}$$

and

$$\frac{dY_2^N(x,u)}{dx} = Z_2^N(x,u), \quad \frac{dZ_2^N(x,u)}{dx} - m^2\,Y_2^N(x,u) = -f(x, Y_2^N(x,u)), \quad x \in e_2^N,$$

$$Y_2^N(x_N,u) = u(x_N), \quad \lim_{x\to\infty} Y_2^N(x,u) = 0. \tag{5.43}$$

For the solution of problems (5.42) (which are equivalent to (5.24)) we use a one-step method (e.g., the Taylor series method or a Runge-Kutta method) of order n with the corresponding increment function $\mathbf{\Phi} = (\Phi_1, \Phi_2)^T$:

$$\begin{aligned} Y_\alpha^{(\bar{n})j}(x_j,u) &= u_\beta + (-1)^{\alpha+1}h_\gamma\Phi_1(x_\beta,u_\beta,u'_\beta,(-1)^{\alpha+1}h_\gamma), \\ Z_\alpha^{(n)j}(x_j,u) &= u'_\beta + (-1)^{\alpha+1}h_\gamma\Phi_2(x_\beta,u_\beta,u'_\beta,(-1)^{\alpha+1}h_\gamma). \end{aligned} \tag{5.44}$$

Thus, the value $Z_\alpha^{(n)j}(x_j,u)$ approximates $Z_\alpha^j(x_j,u)$ at least with order of accuracy n and $Y_\alpha^{(\bar{n})j}(x_j,u)$ approximates $Y_\alpha^j(x_j,u)$ with order of accuracy $\bar{n} \stackrel{\text{def}}{=} 2[(n+1)/2]$. If the function $f(x,u)$ is sufficiently smooth then there exist expansions

$$Y_\alpha^j(x_j,u) = Y_\alpha^{(\bar{n})j}(x_j,u) + \psi_\alpha^j(x_\beta,u)\left[(-1)^{\alpha+1}h_\gamma\right]^{\bar{n}+1} + O(h_\gamma^{\bar{n}+2}) \tag{5.45}$$

and

$$Z_\alpha^j(x_j,u) = Z_\alpha^{(n)j}(x_j,u) + \tilde{\psi}_\alpha^j(x_\beta,u)\left[(-1)^{\alpha+1}h_\gamma\right]^{n+1} + O(h_\gamma^{n+2}). \tag{5.46}$$

For the Taylor series method we have

$$\Phi_1\left(x_\beta,u_\beta,u'_\beta,(-1)^{\alpha+1}h_\gamma\right)$$

$$= u'_\beta + \frac{(-1)^{\alpha+1}h_\gamma}{2}\left(m^2u_\beta - f_\beta\right) + \sum_{p=3}^{\bar{n}} \frac{\left[(-1)^{\alpha+1}h_\gamma\right]^{p-1}}{p!}\frac{d^pY_\alpha^j(x_\beta,u)}{dx^p},$$

$$\Phi_2\left(x_\beta, u_\beta, u'_\beta, (-1)^{\alpha+1}h_\gamma\right) = m^2 u_\beta - f_\beta + \sum_{p=2}^{n} \frac{\left[(-1)^{\alpha+1}h_\gamma\right]^{p-1}}{p!} \frac{d^p Z^j_\alpha(x_\beta, u)}{dx^p},$$

and for an explicit Runge-Kutta-Nystrom method it holds that

$$\Phi_1\left(x_\beta, u_\beta, u'_\beta, (-1)^{\alpha+1}h_\gamma\right) = u'_\beta + (-1)^{\alpha+1}h_\gamma(\bar{b}_1 k_1 + \bar{b}_2 k_2 + \cdots + \bar{b}_s k_s),$$

$$\Phi_2\left(x_\beta, u_\beta, u'_\beta, (-1)^{\alpha+1}h_\gamma\right) = b_1 k_1 + b_2 k_2 + \cdots + b_s k_s,$$

$$k_1 = \tilde{f}(x_\beta, u_\beta),$$

$$k_2 = \tilde{f}(x_\beta + c_2(-1)^{\alpha+1}h_\gamma, u_\beta + c_2(-1)^{\alpha+1}h_\gamma u'_\beta + h_\gamma^2 a_{21} k_1),$$

$$\vdots$$

$$k_s = \tilde{f}(x_\beta + c_s(-1)^{\alpha+1}h_\gamma, u_\beta + c_s(-1)^{\alpha+1}h_\gamma u'_\beta + h_\gamma^2(a_{s1}k_1 + a_{s2}k_2 + \cdots + a_{s,s-1}k_{s-1})).$$

$$\tilde{f}(x,u) = m^2 u - f(x,u).$$

Let us now find an approximation of the solution of problem (5.43) which is equivalent to (5.25). The following assumptions on the function $f(x,u)$ will be significant for our subsequent considerations.

Assumption 5.1.

(a) $f(x,u)$ is analytic in a neighborhood of the point $(\infty, 0)$, and

(b) $\lim_{x\to\infty} f'_u(x,0) = 0.$ □

We represent the exact solution of problem (5.43) in the form

$$Y_2^N(x,u) = \sum_{i=1}^{\infty} \frac{A_i}{x^i} + r(x), \quad Z_2^N(x,u) = -\sum_{i=1}^{\infty} \frac{iA_i}{x^{i+1}} + r'(x), \tag{5.47}$$

where $r(x) \in \mathbb{C}^\infty[x_N, \infty)$ and $\lim_{x\to\infty} x^n r(x) = 0$ for all $n \in \mathbb{N}_0$.

Inserting $Y_2^N(x,u)$ into (5.43), we obtain

$$F(x, \{A\}) \stackrel{\text{def}}{=} \sum_{i=1}^{\infty} \frac{i(i+1)A_i}{x^{i+2}} - m^2 \sum_{i=1}^{\infty} \frac{A_i}{x^i} + f\left(x, \sum_{i=1}^{\infty} \frac{A_i}{x^i} + r(x)\right) + r''(x) - m^2 r(x) = 0,$$

where $\{A\} \stackrel{\text{def}}{=} A_1, A_2, \ldots$.

We change the variables by $t = 1/x$ and write $\tilde{F}(t, \{A\}) \stackrel{\text{def}}{=} F(1/t, \{A\})$. Taking into account Assumption 5.1, we get for the coefficients A_i, $i = 1, 2, \ldots$, the recurrent system of equations

$$\left.\frac{d^k \tilde{F}(t, \{A\})}{dt^k}\right|_{t=0} = 0, \quad k = 1, 2, \ldots . \tag{5.48}$$

Note, if the ODE in (5.1) is autonomous, then system (5.48) possesses only the trivial solution.

A look at formula (5.47) suggests the use of the following ansatz for the approximate solution of problem (5.43):

$$\begin{aligned} Y_2^{(\bar{n}-1)N}(x, u) &= \frac{A_1}{x} + \frac{A_2}{x^2} + \cdots + \frac{A_{\bar{n}-1}}{x^{\bar{n}-1}}, \\ Z_2^{(\bar{n}-1)N}(x, u) &= -\frac{A_1}{x^2} - \frac{2A_2}{x^3} - \cdots - \frac{(\bar{n}-2)A_{\bar{n}-2}}{x^{\bar{n}-1}}. \end{aligned} \tag{5.49}$$

The unknown coefficients $A_1, A_2, \ldots, A_{\bar{n}-1}$ can successively be determined from (5.48), whereby two cases are possible. In the first case all the coefficients $A_1, A_2, \ldots, A_{\bar{n}-1}$ are equal to zero. Then

$$Y_2^{(\bar{n}-1)N}(x, u) \equiv 0, \quad Z_2^{(n-1)N}(x, u) \equiv 0.$$

In the second case at least one of the coefficients $A_1, A_2, \ldots, A_{\bar{n}-1}$ is not equal to zero. This situation arises if

$$f(x, 0) = \sum_{i=1}^{\infty} \frac{f_i}{x^i}, \qquad \sum_{i=1}^{\infty} |f_i|^2 \neq 0$$

and the solution is represented by (5.49).

The method just described can be realized with a computer algebra system, like MAPLE or MATHEMATICA. However, the coefficients A_i are often known from the given problem. It is also possible to compute the coefficients numerically. For instance, in the paper [67] a numerical algorithm is proposed by which the coefficients can be computed automatically.

The following lemma compares the approximate values $Y_\alpha^{(\bar{n})j}$, $Z_\alpha^{(\bar{n})j}$ with the exact values Y_α^j, Z_α^j, respectively.

Lemma 5.3

Suppose that

$$f(x, u) \in \bigcup_{j=1}^{N} \mathbb{C}^{n+1}(\bar{e}^j \times \Omega([0, \infty), r + \Delta)) \cup \mathbb{C}^{n+1}(\bar{e}_2^N \times \Omega([0, \infty), r + \Delta)),$$

Assumption 5.1 *is satisfied, and there exist the expansions* (5.45), (5.46) *for the numerical method (5.44). Then*

$$Y_\alpha^j(x_j,u) = Y_\alpha^{(\bar{n})j}(x_j,u) + (-1)^{\alpha+1}\psi_1^\gamma(x_\beta,u)h_\gamma^{\bar{n}+1} + O(h_\gamma^{\bar{n}+2}), \tag{5.50}$$

$$Z_\alpha^j(x_j,u) = Z_\alpha^{(n)j}(x_j,u) + \tilde{\psi}_1^\gamma(x_\beta,u)\left[(-1)^{\alpha+1}h_\gamma\right]^{n+1} + O(h_\gamma^{n+2}), \tag{5.51}$$

$$Y_2^N(x_N,u) = Y_2^{(\bar{n}-1)N}(x_N,u) + O\left(\frac{1}{x_N}\right)^{\bar{n}}, \tag{5.52}$$

$$Z_2^N(x_N,u) = Z_2^{(\bar{n}-1)N}(x_N,u) + O\left(\frac{1}{x_N}\right)^{\bar{n}}. \tag{5.53}$$

Proof. Due to (5.45), (5.46) we have

$$\begin{aligned} &Y_1^j(x_j,u) - Y_1^{(\bar{n})j}(x_j,u) \\ &= Y_1^j(x_j,u) - u(x_j-h_j) - h_j\Phi_1(x_j-h_j,u(x_j-h_j),u'(x_j-h_j),h_j) \\ &= \psi_1^j(x_j-h_j,u)h_j^{\bar{n}+1} + O\left(h_j^{\bar{n}+2}\right), \end{aligned} \tag{5.54}$$

$$\begin{aligned} &Z_1^j(x_j,u) - Z_1^{(n)j}(x_j,u) \\ &= Z_1^j(x_j,u) - u'(x_j-h_j) - h_j\Phi_2(x_j-h_j,u(x_j-h_j),u'(x_j-h_j),h_j) \\ &= \tilde{\psi}_1^j(x_j-h_j,u)h_j^{n+1} + O\left(h_j^{n+2}\right). \end{aligned}$$

We note that

$$Y_2^j(x_j,u) - Y_2^{(\bar{n})j}(x_j,u) = Y_2^j(x_j,u) - u_{j+1} + h_{j+1}\Phi_1(x_{j+1},u_{j+1},u'_{j+1},-h_{j+1}),$$

$$Z_2^j(x_j,u) - Z_2^{(n)j}(x_j,u) = Z_2^j(x_j,u) - u'_{j+1} + h_{j+1}\Phi_2(x_{j+1},u_{j+1},u'_{j+1},-h_{j+1}).$$

Let us substitute $-h_{j+1}$ instead of h_j into (5.54) and take into account that $Y_1^j(x_j,u) = Y_2^j(x_j,u)$, $Z_1^j(x_j,u) = Z_2^j(x_j,u)$ holds. We obtain

$$\begin{aligned} &Y_2^j(x_j,u) - u_{j+1} + h_{j+1}\Phi_1(x_{j+1},u_{j+1},u'_{j+1},-h_{j+1}) \\ &= -\psi_1^{j+1}(x_{j+1},u)h_{j+1}^{\bar{n}+1} + O\left(h_j^{\bar{n}+2}\right), \end{aligned}$$

$$\begin{aligned} &Z_2^j(x_j,u) - u'_{j+1} + h_{j+1}\Phi_2(x_{j+1},u_{j+1},u'_{j+1},-h_{j+1}) \\ &= (-1)^{n+1}\tilde{\psi}_1^{j+1}(x_{j+1},u)h_{j+1}^{n+1} + O\left(h_j^{n+2}\right), \end{aligned}$$

which yields (5.50), (5.51).

The relations (5.52), (5.53) are true due to the expansions (5.47) and the equalities (5.49). Thus, the proof is complete. ∎

Now, instead of the EDS (5.27) we can use the n-TDS

$$\left(ay_{\bar{x}}^{(\bar{n})}\right)_{\hat{x},j} - d(x_j)y_j^{(\bar{n})} = -\varphi^{(\bar{n})}\left(x_j, y^{(\bar{n})}\right), \quad j = 1, \dots, N-1,$$
$$y_0^{(n)} = \mu_1, \quad -a(x_N)y_{\bar{x},N}^{(\bar{n})} = \beta_2\, y_N^{(\bar{n})} - \mu_2^{(\bar{n})}\left(x_N, y^{(\bar{n})}\right), \tag{5.55}$$

where

$$\begin{aligned}
&\varphi^{(\bar{n})}(x_j, u)\\
&= \frac{1}{\hbar_j}\left[Z_2^{(n)j}(x_j,u) - Z_1^{(n)j}(x_j,u) + \frac{m\left(\cosh(m\,h_j)Y_1^{(\bar{n})j}(x_j,u) - u_{j-1}\right)}{\sinh(m\,h_j)}\right.\\
&\left. + \frac{m\left(\cosh(m\,h_{j+1})Y_2^{(\bar{n})j}(x_j,u) - u_{j+1}\right)}{\sinh(m\,h_{j+1})}\right].
\end{aligned} \tag{5.56}$$

If at least one of the coefficients $A_1, A_2, \dots, A_{\bar{n}-1}$ is not equal to zero, we have

$$\begin{aligned}
&\mu_2^{(\bar{n})}(x_N, u)\\
&= Z_2^{(\bar{n}-1)N}(x_N,u) + mY_2^{(\bar{n}-1)N}(x_N,u)\\
&\quad - Z_1^{(n)N}(x_N,u) + \frac{m\left(\cosh(m\,h_N)Y_1^{(\bar{n})N}(x_N,u) - u_{N-1}\right)}{\sinh(m\,h_N)}\\
&= \frac{mA_1}{x_N} + \frac{mA_2 - A_1}{x_N^2} + \dots + \frac{mA_{\bar{n}-1} - (\bar{n}-2)A_{\bar{n}-2}}{x_N^{\bar{n}-1}}\\
&\quad - Z_1^{(n)N}(x_N,u) + \frac{m\left(\cosh(m\,h_N)Y_1^{(\bar{n})N}(x_N,u) - u_{N-1}\right)}{\sinh(m\,h_N)}.
\end{aligned} \tag{5.57}$$

Otherwise it holds that

$$\mu_2^{(\bar{n})}(x_N,u) = -Z_1^{(n)N}(x_N,u) + \frac{m\left(\cosh(m\,h_N)Y_1^{(\bar{n})N}(x_N,u) - u_{N-1}\right)}{\sinh(m\,h_N)}. \tag{5.58}$$

In order to obtain an error estimate for the n-TDS (5.55) we need the following lemma.

Lemma 5.4

Suppose that the assumptions of Lemma 5.3 are satisfied. Then

$$\begin{aligned}
&\varphi^{(\bar{n})}(x_j,u) - \varphi(x_j,u)\\
&= \left\{h_j^{n+1}\left[\frac{m\,h_j\cosh(m\,h_j)}{\sinh(m\,h_j)}\psi_1^j(x,u)\Big|_{x=x_j+0} - \tilde{\psi}_1^j(x,u)\Big|_{x=x_j+0}\right]\right\}_{\hat{x}}\\
&\quad + O\left(\frac{h_j^{n+2} + h_{j+1}^{n+2}}{\hbar_j}\right)
\end{aligned} \tag{5.59}$$

if n is odd, and

$$\varphi^{(\bar{n})}(x_j,u)-\varphi(x_j,u)$$
$$=\left\{h_j^n\frac{m\,h_j\cosh(m\,h_j)}{\sinh(m\,h_j)}\,\psi_1^j(x,u)\bigg|_{x=x_j+0}\right\}_{\hat{x}}+O\left(\frac{h_j^{n+1}+h_{j+1}^{n+1}}{\hbar_j}\right) \tag{5.60}$$

if n is even. Furthermore,

$$\left|\mu_2^{(\bar{n})}(x_N,u)-\mu_2(x_N,u)\right|\le M\left(\max\left\{h^{\bar{n}},\left(\frac{1}{x_N}\right)^{\bar{n}}\right\}\right),\quad N\to\infty, \tag{5.61}$$

$$\left|\varphi^{(\bar{n})}(x_j,u)\right|\le K_1+M\,h\quad \text{for all } u\in\Omega(\hat{\omega}_h,r+\Delta), \tag{5.62}$$

$$\left|\varphi^{(\bar{n})}(x_j,u)-\varphi^{(\bar{n})}(x_j,v)\right|\le(L_1+M\,h)\|u-v\|_{0,\hat{\omega}_h}$$
$$\text{for all } u,v\in\Omega(\hat{\bar{\omega}}_h,r+\Delta), \tag{5.63}$$

$$\left|\mu_2^{(\bar{n})}(x_N,u)\right|\le K_1+M\max\left\{h,\frac{1}{x_N}\right\}$$
$$\text{for all } u\in\Omega(\hat{\bar{\omega}}_h,r+\Delta),\quad N\to\infty, \tag{5.64}$$

$$\left|\mu_2^{(\bar{n})}(x_N,u)-\mu_2^{(\bar{n})}(x_N,v)\right|\le\left(L_1+M\max\left\{h,\frac{1}{x_N}\right\}\right)\|u-v\|_{0,\hat{\omega}_h^+},$$
$$\text{for all } u,v\in\Omega(\hat{\bar{\omega}}_h,r+\Delta),\quad N\to\infty, \tag{5.65}$$

where the constant M is independent of h and $1/x_N$.

Proof. We begin with the proof of (5.59)–(5.61). We note that

$$\varphi^{(\bar{n})}(x_j,u)-\varphi(x_j,u)$$
$$=\hbar_j^{-1}\sum_{\alpha=1}^{2}(-1)^\alpha\Bigg[Z_\alpha^{(n)j}(x_j,u)-Z_\alpha^j(x_j,u) \tag{5.66}$$
$$+(-1)^\alpha\frac{m\cosh(m\,h_\gamma)\left(Y_\alpha^{(\bar{n})j}(x_j,u)-Y_\alpha^j(x_j,u)\right)}{\sinh(m\,h_\gamma)}\Bigg],$$

$$\mu_2^{(\bar{n})}(x_N,u)-\mu_2(x_N,u)$$
$$=Z_2^{(\bar{n}-1)N}(x_N,u)-Z_2^N(x_N,u)+m\left(Y_2^{(\bar{n}-1)N}(x_N,u)-Y_2^N(x_N,u)\right) \tag{5.67}$$
$$-Z_1^{(n)N}(x_N,u)$$

$$+ Z_1^N(x_N,u) + \frac{m\cosh(m\,h_N)\left[Y_1^{(\bar{n})N}(x_N,u) - Y_1^N(x_N,u)\right]}{\sinh(m\,h_N)}. \tag{5.68}$$

The relations (5.66) for n odd imply

$$\varphi^{(\bar{n})}(x_j,u) - \varphi(x_j,u)$$

$$= \frac{1}{\hbar_j}\left[h_{j+1}^{n+2}\frac{m\cosh(m\,h_{j+1})}{\sinh(m\,h_{j+1})}\psi_1^{j+1}(x_{j+1},u) - h_j^{n+2}\frac{m\cosh(m\,h_j)}{\sinh(m\,h_j)}\psi_1^j(x_{j-1},u)\right.$$

$$\left. - h_{j+1}^{n+1}\tilde{\psi}_1^{j+1}(x_{j+1},u) + h_j^{n+1}\tilde{\psi}_1^j(x_{j-1},u)\right] + O\left(\frac{h_j^{n+2}+h_{j+1}^{n+2}}{\hbar_j}\right) \tag{5.69}$$

and for n even imply

$$\varphi^{(\bar{n})}(x_j,u) - \varphi(x_j,u)$$

$$= \frac{1}{\hbar_j}\left[h_{j+1}^{n+1}\frac{m\cosh(m\,h_{j+1})}{\sinh(m\,h_{j+1})}\psi_1^{j+1}(x_{j+1},u) - h_j^{n+1}\frac{m\cosh(m\,h_j)}{\sinh(m\,h_j)}\psi_1^j(x_{j-1},u)\right]$$

$$+ O\left(\frac{h_j^{n+1}+h_{j+1}^{n+1}}{\hbar_j}\right). \tag{5.70}$$

Taking into account

$$\psi_1^j(x_{j-1},u) = \psi_1^j(x_j,u) + O(h_j), \quad \tilde{\psi}_1^j(x_{j-1},u) = \tilde{\psi}_1^j(x_j,u) + O(h_j),$$

the equalities (5.69) and(5.70) yield (5.59), (5.60).

The inequality (5.61) is true due to (5.68) and because of the estimates (5.50) – (5.53).

Let us now prove the inequalities (5.62) – (5.65). The conditions (5.45), (5.46) imply

$$\Phi_1(x,u,u',0) = u', \quad \frac{\partial\Phi_1(x,u,u',0)}{\partial h} = \frac{1}{2}(m^2u - f(x,u)),$$

$$\Phi_2(x,u,u',0) = m^2u - f(x,u).$$

Thus, the following equations hold true:

$$Y_\alpha^{(\bar{n})j}(x_j,u)$$

$$= u_\beta + (-1)^{\alpha+1}h_\gamma\Phi_1(\boldsymbol{v}) + h_\gamma^2\frac{\partial\Phi_1(\boldsymbol{v})}{\partial h} + \frac{(-1)^{\alpha+1}h_\gamma^3}{2}\frac{\partial^2\Phi_1(\boldsymbol{w})}{\partial h^2}$$

$$= u_\beta + (-1)^{\alpha+1}h_\gamma u_\beta' + \frac{h_\gamma^2}{2}[m^2u_\beta - f(x_\beta,u_\beta)] + \frac{(-1)^{\alpha+1}h_\gamma^3}{2}\frac{\partial^2\Phi_1(\boldsymbol{w})}{\partial h^2},$$

$$
\begin{aligned}
&Z_\alpha^{(n)j}(x_j,u)\\
&= u'_\beta + (-1)^{\alpha+1} h_\gamma \Phi_2(\boldsymbol{v}) + h_\gamma^2 \frac{\partial \Phi_2(\boldsymbol{w})}{\partial h}\\
&= u'_\beta + (-1)^{\alpha+1} h_\gamma \left[m^2 u_\beta - f(x_\beta,u_\beta)\right] + h_\gamma^2 \frac{\partial \Phi_2(\boldsymbol{w})}{\partial h},
\end{aligned}
$$

where we have set

$$
\boldsymbol{v} \stackrel{\text{def}}{=} (x_\beta, u_\beta, u'_\beta, 0), \quad \boldsymbol{w} \stackrel{\text{def}}{=} (x_\beta, u_\beta, u'_\beta, \bar{h}).
$$

Since

$$
\begin{aligned}
&\varphi^{(\bar{n})}(x_j,u)\\
&= \hbar_j^{-1} \sum_{\alpha=1}^{2} \left\{ (-1)^\alpha \left[1 - \frac{m\,h_\gamma \cosh(m\,h_\gamma)}{\sinh(m\,h_\gamma)}\right] u'_\beta \right.\\
&\quad + \left[1 - \frac{m\,h_\gamma \cosh(m\,h_\gamma)}{2\sinh(m\,h_\gamma)}\right] h_\gamma f(x_\beta,u_\beta)\\
&\quad + \frac{m\left[\cosh(m\,h_\gamma)\left(1+\dfrac{m^2 h_\gamma^2}{2}\right) - 1 - m\,h_\gamma \sinh(m\,h_\gamma)\right]}{\sinh(m\,h_\gamma)} u_\beta\\
&\quad \left. + h_\gamma^2 \frac{\partial \Phi_2(\boldsymbol{w})}{\partial h} + \frac{m\,h_\gamma^3 \cosh(m\,h_\gamma)}{2\sinh(m\,h_\gamma)} \frac{\partial^2 \Phi_1(\boldsymbol{w})}{\partial h^2} \right\}\\
&= \frac{1}{\hbar_j} \sum_{\alpha=1}^{2} \left\{ \frac{h_\gamma}{2} f(x_\beta,u_\beta) \right\} + O\left(\frac{h_j^2 + h_{j+1}^2}{\hbar_j}\right),
\end{aligned}
$$

and

$$
\mu_2^{(\bar{n})}(x_N,u) = h_N \left[1 - \frac{m\,h_N \cosh(m\,h_N)}{2\sinh(m\,h_N)}\right] f_{N-1} + O\left(\max\left\{h_N, \frac{1}{x_N}\right\}\right),
$$

we have

$$
\left|\varphi^{(\bar{n})}(x_j,u)\right| \le K_1 + M\,h, \qquad \left|\mu_2^{(\bar{n})}(x_N,u)\right| \le K_1 + M \max\left\{h_N, \frac{1}{x_N}\right\}.
$$

Let us prove the estimate (5.63). Since

$$
\begin{aligned}
&\left|\varphi^{(\bar{n})}(x_j,u) - \varphi^{(\bar{n})}(x_j,v)\right|\\
&\le \frac{1}{\hbar_j} \sum_{\alpha=1}^{2} \left\{ \frac{h_\gamma}{2} |f(x_\beta,u_\beta) - f(x_\beta,v_\beta)| \left[1 + \frac{|\sinh(m\,h_\gamma) - m\,h_\gamma \cosh(m\,h_\gamma)|}{\sinh(m\,h_\gamma)}\right] \right.
\end{aligned}
$$

$$+\left|1-\frac{m\,h_{j\gamma}\cosh(m\,h_\gamma)}{\sinh(m\,h_\gamma)}\right|\left|u'_\beta-v'_\beta\right|$$

$$+\frac{m\left[\cosh(m\,h_\gamma)\left(1+\dfrac{m^2h_\gamma^2}{2}\right)-1-m\,h_\gamma\sinh(m\,h_\gamma)\right]}{\sinh(m\,h_\gamma)}|u_\beta-v_\beta|$$

$$+h_\gamma^2\left|\frac{\partial\Phi_2(x_\beta,u_\beta,u'_\beta,\tilde{h})}{\partial h}-\frac{\partial\Phi_2(x_\beta,v_\beta,v'_\beta,\tilde{h})}{\partial h}\right|$$

$$+\frac{m\,h_\gamma^3\cosh(m\,h_\gamma)}{2\sinh(m\,h_\gamma)}\left|\frac{\partial^2\Phi_1(x_\beta,u_\beta,u'_\beta,\bar{h})}{\partial h^2}-\frac{\partial^2\Phi_1(x_\beta,v_\beta,v'_\beta,\bar{h})}{\partial h^2}\right|\Bigg\},$$

the Mean Value Theorem guarantees that there exist $\bar{u}$, $\bar{u}'$, $\tilde{u}$ and $\tilde{u}'$ such that

$$\left|\varphi^{(\bar{n})}(x_j,u)-\varphi^{(\bar{n})}(x_j,v)\right|$$

$$\leq (L_1+M\,h)\,\|u-v\|_{0,\hat{\omega}_h^+}+M_2\,h\Bigg\{\left[1+\left|\frac{\partial^2\Phi_2(x_\beta,\bar{u},u'_\beta,\tilde{h})}{\partial h\partial u}\right|\right.$$

$$\left.+\left|\frac{\partial^3\Phi_1(x_\beta,\tilde{u},u'_\beta,\bar{h})}{\partial h^2\partial u}\right|\right]\|u-v\|_{0,\hat{\omega}_h^+}+\left[1+\left|\frac{\partial^2\Phi_2(x_\beta,u_\beta,\bar{u}',\tilde{h})}{\partial h\partial u'}\right|\right.$$

$$\left.+\left|\frac{\partial^3\Phi_1(x_\beta,u_\beta,\tilde{u}',\bar{h})}{\partial h^2\partial u}\right|\right]\|u-v\|_{0,\hat{\omega}_h^+}\Bigg\}.$$

This together with the inequality (5.41) yields the estimates (5.63).

Estimate (5.65) follows from

$$\left|\mu_2^{(\bar{n})}(x_N,u)-\mu_2^{(\bar{n})}(x_N,v)\right|$$

$$\leq\frac{h_N}{2}|f(x_{N-1},u_{N-1}-f(x_{N-1},v_{N-1})|+M_1\left(\max\left\{h_N^2,\frac{1}{x_N}\right\}\right)\|u-v\|_{0,\hat{\omega}_h^+}$$

$$\leq\left(L_1+M\max\left\{h_N,\frac{1}{x_N}\right\}\right)\|u-v\|_{0,\hat{\omega}_h^+}.$$

■

The auxiliary results above provide the following second main result of this chapter.

Theorem 5.4

Under the hypotheses of Theorem 5.1 and Lemma 5.3, there exists a constant $N_0 > 0$ such that for all $N \geq N_0$ the n-TDS (5.55) – (5.58), (5.44) possesses a unique solution. Moreover, if the grid satisfies the conditions (5.20) – (5.22), i.e., $\frac{1}{x_N} \leq h \leq c_2 N^{-1/2}$, then the following error estimate holds:

$$\begin{aligned} \left\| y^{(\bar{n})} - u \right\|^*_{1,\hat{\omega}_h^+} &= \max\left\{ \left\| y^{(\bar{n})} - u \right\|_{0,\hat{\omega}_h^+}, \left\| \frac{dy^{(\bar{n})}}{dx} - \frac{du}{dx} \right\|_{0,\hat{\omega}_h} \right\} \\ &\leq M \max\left\{ h^{\bar{n}}, \left(\frac{1}{x_N} \right)^{\bar{n}} \right\} \\ &\leq M\, h^{\bar{n}} \leq M\, N^{-\bar{n}/2}, \end{aligned}$$

where

$$\frac{dy^{(\bar{n})}(x_0)}{dx} = Z_2^{(n)0}\left(x_0, y^{(\bar{n})}\right) + \frac{m\cosh(m\,h_1)\left(Y_2^{(\bar{n})0}(x_0, y^{(\bar{n})}) - y_0^{(\bar{n})}\right)}{\sinh(m\,h_1)},$$

$$\frac{dy^{(\bar{n})}(x_j)}{dx} = Z_1^{(n)j}\left(x_j, y^{(\bar{n})}\right) + \frac{m\cosh(m\,h_j)\left(y_j^{(\bar{n})} - Y_1^{(\bar{n})j}(x_j, y^{(\bar{n})})\right)}{\sinh(m\,h_j)},$$

$j = 1, \ldots, N.$

and the constant M is independent of h and $1/x_N$.

Proof. To show that the n-TDS (5.55) has the unique solution $y_i^{(\bar{n})}$, $i = 1, \ldots, N$, we use Banach's Fixed Point Theorem (see Theorem 1.1). For this we consider the operator equation

$$y_i^{(\bar{n})} = \tilde{\Re}_h(x_i, y^{(\bar{n})}),$$

where

$$\tilde{\Re}_h(x_i, y^{(\bar{n})}) \stackrel{\text{def}}{=} \sum_{j=1}^{N-1} \hbar_j G(x_i, x_j)\varphi^{(\bar{n})}(x_j, y^{(\bar{n})}) + \mu^{(\bar{n})}(x_N, y^{(\bar{n})}) G(x_i, x_N) + u^{(0)}(x_i),$$

$i = 1, \ldots, N,$

and $G(x, \xi)$ is Green's function (5.10).

Majorizing Riemann's sums by integrals we get

$$\sum_{j=1}^{N-1} \hbar_j G(x_i, x_j) + G(x_i, x_N)$$

$$
\begin{aligned}
&= \frac{\exp(-mx_i)}{2m}\sum_{j=1}^{i} h_j \sinh(mx_j) + \frac{\exp(-mx_i)}{2m}\sum_{j=1}^{i} h_{j+1}\sinh(mx_j) \\
&\quad + \frac{\sinh(mx_i)}{2m}\sum_{j=i+1}^{N-1} h_j \exp(-mx_j) + \frac{\sinh(mx_i)}{2m}\sum_{j=i+1}^{N-1} h_{j+1}\exp(-mx_j) \\
&\quad + \frac{\sinh(mx_i)}{m}\exp(-mx_N) \\
&\le \frac{\exp(-mx_i)}{2m}\int_0^{x_i}\sinh(m\eta + h)d\eta + \frac{\exp(-mx_i)}{2m}\int_0^{x_i}\sinh(m\eta)d\eta \\
&\quad + \frac{\sinh(mx_i)}{2m}\int_{x_i}^{x_N}\exp(-m\eta)d\eta + \frac{\exp(-mx_i)}{2m}\int_{x_i}^{x_N}\exp(-m(\eta-h))d\eta \\
&\quad + \frac{\sinh(mx_i)}{2m}\int_{x_N}^{\infty}\exp(-m\eta)d\eta \\
&\le \int_0^{x_N} G(x_i,\xi)d\xi + M_1 h + \int_{x_N}^{\infty} G(x_i,\xi)d\xi \\
&\le \int_0^{\infty} G(x_i,\xi)d\xi + M_1 h = \frac{1-\exp(-mx_i)}{m^2} + M_1 h,
\end{aligned}
$$

and

$$
\sum_{j=1}^{N-1} \hbar_j G_{\bar{x}}(x_i,x_j) + G_{\bar{x}}(x_i,x_N) \le \int_0^{\infty} G_{\bar{x}}(x_i,\xi)d\xi + M_1 h \le \frac{\exp(-mx_i)}{m} + M_1 h.
$$

Now, due to (5.62), we get

$$
\begin{aligned}
&\left\|\Re_h(x,v) - u^{(0)}\right\|_{1,\hat{\omega}_h^+} \\
&\le \left(K_1 + M\max\left\{h, \frac{1}{x_N}\right\}\right)\left\|\sum_{j=1}^{N-1}\hbar_j G(x,x_j) + G(x,x_N)\right\|_{1,\hat{\omega}_h^+} \\
&\le r + M_2 \max\left\{h, \frac{1}{x_N}\right\} \\
&\le r + \Delta \quad \text{for all } v \in \Omega(\hat{\omega}_h, r+\Delta),
\end{aligned}
$$

i.e. the operator $\Re_h(x,v)$ transforms $\Omega(\hat{\hat{\omega}}_h, r+\Delta)$ into itself.

Moreover, taking into account the inequality (5.63) we have

$$\begin{aligned}
&\|\Re_h(x,u) - \Re_h(x,v)\|_{1,\hat{\omega}_h^+} \\
&\le \left(L_1 + M\max\left\{h, \frac{1}{x_N}\right\}\right)\|u-v\|_{1,\hat{\omega}_h^+}\left\|\sum_{j=1}^{N-1}\hbar_j G(x,x_j) + G(x,x_N)\right\|_{1,\hat{\omega}_h^+} \\
&\le q_2\|u-v\|_{1,\hat{\omega}_h^+} \quad \text{for all } u,v\in\Omega(\hat{\hat{\omega}}_h, r).
\end{aligned}$$

If we choose N_0 such that

$$q_2 \stackrel{\text{def}}{=} \max\left\{\frac{1}{m}, \frac{1}{m^2}\right\}\left(L_1 + M\max\left\{h, \frac{1}{x_N}\right\}\right) < 1,$$

then the mapping $\Re_h(x,u)$ is contractive.

The error $z_i \stackrel{\text{def}}{=} y_i^{(\bar{n})} - u(x_i)$, $i=0,\dots,N$, satisfies the problem

$$\begin{aligned}
&(az_{\bar{x}})_{\hat{x},i} - d(x_i)z_i = \varphi(x_i,u) - \varphi^{(\bar{n})}(x_i,y^{(\bar{n})}), \quad i=1,\dots,N-1, \\
&z_0 = 0, \quad -a(x_N)z_{\bar{x},N} = \beta_2 z_N + \mu_2(x_N,u) - \mu_2^{(\bar{n})}(x_N,y^{(\bar{n})}).
\end{aligned} \tag{5.71}$$

Using Green's function the solution of this problem and its first difference can be represented by

$$\begin{aligned}
z_i = &\sum_{j=1}^{N-1}\hbar_j G(x_i,x_j)\{\varphi(x_j,u) - \varphi^{(\bar{n})}(x_j,y^{(\bar{n})})\} \\
&+ G(x_i,x_N)\left[\mu_2(x_N,u) - \mu_2^{(\bar{n})}(x_N,y^{(\bar{n})})\right], \quad i=1,\dots,N,
\end{aligned} \tag{5.72}$$

$$\begin{aligned}
z_{\bar{x},i} = &\sum_{j=1}^{N-1}\hbar_j G_{\bar{x}}(x_i,x_j)\{\varphi(x_j,u) - \varphi^{(\bar{n})}(x_j,y^{(\bar{n})})\} \\
&+ G_{\bar{x}}(x_i,x_N)\left[\mu_2(x_N,u) - \mu_2^{(\bar{n})}(x_N,y^{(\bar{n})})\right], \quad i=1,\dots,N.
\end{aligned} \tag{5.73}$$

For n odd, due to (5.59), we get from (5.72), (5.73):

$$\begin{aligned}
z_i = &-\sum_{j=1}^{N} h_j G_{\bar{\xi}}(x_i,x_j)\left\{h_j^{n+1}\left[\frac{m\,h_j\cosh(m\,h_j)}{\sinh(m\,h_j)}\,\psi_1^j(x,u)\Big|_{x=x_{j0}}\right.\right. \\
&\left.\left. - \tilde{\psi}_1^j(x,u)\Big|_{x=x_{j0}}\right]\right\} \\
&+\sum_{j=1}^{N-1}\hbar_j G(x_i,x_j)\left[\varphi^{(\bar{n})}(x_j,u) - \varphi^{(\bar{n})}(x_j,y^{(\bar{n})}) + O\left(h^{n+1}\right)\right]
\end{aligned}$$

$$+ G(x_i, x_N)\left[\mu_2(x_N, u) - \mu_2^{(\bar{n})}(x_N, u)\right]$$
$$+ G(x_i, x_N)\left[\mu_2^{(\bar{n})}(x_N, u) - \mu_2^{(\bar{n})}(x_N, y^{(\bar{n})})\right],$$

$$z_{\bar{x},i} = -\sum_{j=1}^{N} h_j G_{\bar{x}\bar{\xi}}(x_i, x_j) \left\{ h_j^{n+1} \left[\frac{mh_j \cosh(mh_j)}{\sinh(mh_j)} \psi_1^j(x,u)\Big|_{x=x_{j0}} - \tilde{\psi}_1^j(x,u)\Big|_{x=x_{j0}} \right] \right\}$$
$$+ \frac{h_i^{n+1}}{a(x_i)} \left[\frac{mh_i \cosh(mh_i)}{\sinh(mh_i)} \psi_1^i(x,u)\Big|_{x=x_i} - \tilde{\psi}_1^i(x,u)\Big|_{x=x_i} \right]$$
$$+ \sum_{j=1}^{N-1} \hbar_j G_{\bar{x}}(x_i, x_j) \left[\varphi^{(\bar{n})}(x_j, u) - \varphi^{(\bar{n})}\left(x_j, y^{(\bar{n})}\right) + O\left(h^{n+1}\right) \right]$$
$$+ G_{\bar{x}}(x_i, x_N) \left[\mu_2(x_N, u) - \mu_2^{(\bar{n})}(x_N, u) \right]$$
$$+ G_{\bar{x}}(x_i, x_N) \left[\mu_2^{(\bar{n})}(x_N, u) - \mu_2^{(\bar{n})}\left(x_N, y^{(\bar{n})}\right) \right],$$

where we have set $x_{j0} \stackrel{\text{def}}{=} x_j + 0$. Now, taking into account

$$\sum_{j=1}^{N} h_j G_{\bar{\xi}}(x_i, x_j) = \frac{\sinh(mx_i)\exp(-mx_N)}{m} \le M_3,$$

$$\sum_{j=1}^{N} h_j G_{\bar{x}\bar{\xi}}(x_i, x_j) = \frac{\exp(-mx_N)}{h_i} \int_{x_{i-1}}^{x_i} \cosh(m\eta)\, d\eta - \frac{\sinh(mh_i)}{mh_i} \le M_4$$

we get the estimates

$$|z_i| \le M \max\left\{ h^{n+1}, \left(\frac{1}{x_N}\right)^{n+1} \right\} + q_2 \|z\|_{1,\hat{\omega}_N^+},$$

$$|z_{\bar{x},i}| \le M_1 \max\left\{ h^{n+1}, \left(\frac{1}{x_N}\right)^{n+1} \right\} + q_2 \|z\|_{1,\hat{\omega}_N^+}.$$

For n even, due to (5.60), equality (5.73) can be written as

$$z_i = -\sum_{j=1}^{N} h_j G_{\bar{\xi}}(x_i, x_j) \left\{ h_j^n \frac{mh_j \cosh(mh_j)}{\sinh(mh_j)} \psi_1^j(x,u)\Big|_{x=x_{j0}} \right\}$$
$$+ \sum_{j=1}^{N-1} \hbar_j G(x_i, x_j) \left[\varphi^{(\bar{n})}(x_j, u) - \varphi^{(\bar{n})}\left(x_j, y^{(\bar{n})}\right) + O(h^n) \right]$$

$$+ G(x_i, x_N)\left[\mu_2(x_N, u) - \mu_2^{(\bar{n})}(x_N, u)\right]$$

$$+ G(x_i, x_N)\left[\mu_2^{(\bar{n})}(x_N, u) - \mu_2^{(\bar{n})}\left(x_N, y^{(\bar{n})}\right)\right],$$

$$z_{\bar{x},i} = -\sum_{j=1}^{N} h_j G_{\bar{x}\xi}(x_i, x_j)\left\{h_j^n \frac{mh_j \cosh(mh_j)}{\sinh(mh_j)} \psi_1^j(x,u)\Big|_{x=x_{j0}}\right\}$$

$$+ \frac{h_i^n}{a(x_i)} \frac{mh_i \cosh(mh_i)}{\sinh(mh_i)} \psi_1^i(x,u)\Big|_{x=x_i}$$

$$+ \sum_{j=1}^{N-1} \hbar_j G_{\bar{x}}(x_i, x_j)\left[\varphi^{(\bar{n})}(x_j, u) - \varphi^{(\bar{n})}\left(x_j, y^{(\bar{n})}\right) + O(h^n)\right]$$

$$+ G_{\bar{x}}(x_i, x_N)\left[\mu_2(x_N, u) - \mu_2^{(\bar{n})}(x_N, u)\right]$$

$$+ G_{\bar{x}}(x_i, x_N)\left[\mu_2^{(\bar{n})}(x_N, u) - \mu_2^{(\bar{n})}\left(x_N, y^{(\bar{n})}\right)\right].$$

This implies

$$|z_i| \le M \max\left\{h^n, \left(\frac{1}{x_N}\right)^n\right\} + q_2 \|z\|_{1,\hat{\omega}_N^+},$$

$$|z_{\bar{x},i}| \le M \max\left\{h^n, \left(\frac{1}{x_N}\right)^n\right\} + q_2 \|z\|_{1,\hat{\omega}_N^+}.$$

Thus, we get the estimate

$$\|z\|_{1,\hat{\omega}_N^+} \le \frac{M \max\left\{h^{\bar{n}}, \left(\frac{1}{x_N}\right)^{\bar{n}}\right\}}{1 - q_2},$$

which due to $q_2 < 1$ yields

$$\|z\|_{1,\hat{\omega}_N^+} \le M_1 \max\left\{h^{\bar{n}}, \left(\frac{1}{x_N}\right)^{\bar{n}}\right\}.$$

Since $y_0^{(\bar{n})} = Y_2^0\left(x_0, y^{(\bar{n})}\right)$, $y_j^{(\bar{n})} = Y_1^j\left(x_j, y^{(\bar{n})}\right)$, we have

$$\left|\frac{dz(x_0)}{dx}\right| \le \left|Z_2^{(n)0}\left(x_0, y^{(\bar{n})}\right) - Z_2^0\left(x_0, y^{(\bar{n})}\right)\right| + \left|Z_2^0\left(x_0, y^{(\bar{n})}\right) - Z_2^0(x_0, u)\right|$$

$$+ \frac{m\cosh(mh_1)}{\sinh(mh_1)}\left|Y_2^0\left(x_0, y^{(\bar{n})}\right) - Y_2^{(\bar{n})0}\left(x_0, y^{(\bar{n})}\right)\right|$$

$$\leq M_1 h^{\bar{n}} + \left| \frac{\partial}{\partial u} Z_2^0 (x_0, u) \right|_{u=\tilde{u}} \|z\|_{0,\hat{\omega}_N^+}$$

$$\leq M h^{\bar{n}},$$

$$\left|\frac{dz(x_j)}{dx}\right| \leq \left|Z_1^{(n)j}\left(x_j, y^{(\bar{n})}\right) - Z_1^j\left(x_j, y^{(\bar{n})}\right)\right| + \left|Z_1^j\left(x_j, y^{(\bar{n})}\right) - Z_1^j(x_j, u)\right|$$

$$+ \frac{m\cosh(mh_j)}{\sinh(mh_j)} \left|Y_1^j\left(x_j, y^{(\bar{n})}\right) - Y_1^{(\bar{n})j}\left(x_j, y^{(\bar{n})}\right)\right|$$

$$\leq M_1 h^{\bar{n}} + \left| \frac{\partial}{\partial u} Z_1^j (x_j, u) \right|_{u=\tilde{u}} \|z\|_{0,\hat{\omega}_N^+}$$

$$\leq M h^{\bar{n}}, \quad j = 1, \dots, N,$$

and we get $\|z\|^*_{1,\hat{\omega}_N^+} \leq M \max\left\{h^{\bar{n}}, \left(\frac{1}{x_N}\right)^{\bar{n}}\right\}$, which proves the theorem. ■

The solution of the nonlinear n-TDS (5.55)–(5.58), (5.44) of order of accuracy $\bar{n}$ can be found by fixed point iteration. The following third main result of this chapter gives conditions for convergence of the iteration process and shows the rate of convergence.

Theorem 5.5

Let the assumptions of Theorem 5.4 be satisfied. Then the solution of problem (5.55)-(5.58), (5.44) can be determined by the fixed point iteration

$$\left(a y_{\bar{x}}^{(\bar{n},k)}\right)_{\hat{x},j} - d(x_j)\, y_j^{(\bar{n},k)} = -\varphi^{(\bar{n})}\left(x_j, y^{(\bar{n},k-1)}\right), \quad j = 1, \dots, N-1,$$

$$y_0^{(\bar{n},k)} = \mu_1, \quad -a(x_N)\, y_{\bar{x},N}^{(\bar{n},k)} = \beta_2 y_N^{(\bar{n},k)} - \mu_2^{(\bar{n})}\left(x_N, y^{(\bar{n},k-1)}\right), \tag{5.74}$$

$$y^{(\bar{n},0)}(x_j) = \mu_1 \exp(-m x_j), \quad \frac{dy^{(\bar{n},0)}(x_j)}{dx} = -m\mu_1 \exp(-m x_j),$$

$$j = 0, \dots, N.$$

Moreover, the error estimate

$$\left\|y^{(\bar{n},k)} - u\right\|^*_{1,\hat{\omega}_N^+} \leq M\left(h^{\bar{n}} + q_2^k\right) \tag{5.75}$$

holds, where for $k = 1, 2, \dots$ we have

$$\frac{dy^{(\bar{n},k)}(x_0)}{dx}$$

$$= Z_2^{(n)0}\left(x_0, y^{(\bar{n},k)}\right) + \frac{m\cosh(mh_1)\left(Y_2^{(n)0}\left(x_0, y^{(\bar{n},k)}\right) - y_0^{(\bar{n},k)}\right)}{\sinh(mh_1)},$$

$$\frac{dy^{(\bar{n},k)}(x_j)}{dx} = Z_1^{(n)j}\left(x_j, y^{(\bar{n},k)}\right) + \frac{m\cosh(mh_j)\left(y_j^{(\bar{n},k)} - Y_1^{(\bar{n})j}\left(x_j, y^{(\bar{n},k)}\right)\right)}{\sinh(mh_j)}, \tag{5.76}$$

$j = 1, \ldots, N.$

The constant M does not depend on h, n, k, and $q_2 \stackrel{\text{def}}{=} q + M\,h < 1$.

Proof. Theorem 5.4 yields

$$\begin{aligned} \left\|y^{(\bar{n},k)} - u\right\|^*_{1,\hat{\omega}_N^+} &\le \left\|y^{(\bar{n})} - u\right\|^*_{1,\hat{\omega}_N^+} + \left\|y^{(\bar{n},k)} - y^{(\bar{n})}\right\|^*_{1,\hat{\omega}_N^+} \\ &\le M\,h^{\bar{n}} + \left\|y^{(\bar{n},k)} - y^{(\bar{n})}\right\|^*_{1,\hat{\omega}_N^+}. \end{aligned} \tag{5.77}$$

Moreover, the sequence of iterates

$$y^{(\bar{n},k)}(x) = \Re_h(x, y^{(\bar{n},k-1)}), \quad x \in \hat{\omega}_N^+, \quad k = 1, 2, \ldots,$$

converges (see the proof of Theorem 5.4) with the rate given by

$$\left\|y^{(\bar{n},k)} - y^{(\bar{n})}\right\|_{1,\hat{\omega}_N^+} \le \frac{q_2^k}{1 - q_2}(r + \Delta).$$

Thus, taking into account (5.63) we get

$$\begin{aligned} &\left|\frac{dy^{(\bar{n},k)}(x_0)}{dx} - \frac{dy^{(\bar{n})}(x_0)}{dx}\right| \\ &\le \left|Z_2^{(n)0}\left(x_0, y^{(\bar{n},k)}\right) - Z_2^{(n)0}\left(x_0, y^{(\bar{n})}\right)\right| \\ &\quad + \frac{m\cosh(mh_1)}{\sinh(mh_1)}\left|Y_2^{(\bar{n})0}\left(x_0, y^{(\bar{n},k)}\right) - Y_2^{(\bar{n})0}\left(x_0, y^{(\bar{n})}\right)\right| \\ &\le \left[\frac{m\cosh(mh_1)}{\sinh(mh_1)}\left|\frac{\partial}{\partial u}Y_2^{(\bar{n})0}(x_0, u)\right|_{u=\tilde{y}} + \left|\frac{\partial}{\partial u}Z_2^{(n)0}(x_0, u)\right|_{u=\tilde{y}}\right]\left\|y^{(\bar{n},k)} - y^{(\bar{n})}\right\|_{0,\hat{\omega}_N^+} \\ &\le M\left\|y^{(\bar{n},k)} - y^{(\bar{n})}\right\|_{1,\hat{\omega}_N^+}, \end{aligned}$$

$$\begin{aligned} &\left|\frac{dy^{(\bar{n},k)}(x_j)}{dx} - \frac{dy^{(\bar{n})}(x_j)}{dx}\right| \\ &\le \left|Z_1^{(n)j}\left(x_j, y^{(\bar{n},k)}\right) - Z_1^{(n)j}\left(x_j, y^{(\bar{n})}\right)\right| \end{aligned}$$

$$+\frac{m\cosh(mh_j)}{\sinh(mh_j)}\left|Y_1^{(\bar{n})j}\left(x_j, y^{(\bar{n},k)}\right) - Y_1^{(\bar{n})j}\left(x_j, y^{(\bar{n})}\right)\right|$$

$$\leq \left[\frac{m\cosh(mh_j)}{\sinh(mh_j)}\left|\frac{\partial}{\partial u}Y_1^{(\bar{n})j}(x_j,u)\right|_{u=\tilde{y}} + \left|\frac{\partial}{\partial u}Z_1^{(n)j}(x_j,u)\right|_{u=\tilde{y}}\right]\left\|y^{(\bar{n},k)} - y^{(\bar{n})}\right\|_{0,\hat{\omega}_N^+}$$

$$\leq M\left\|y^{(\bar{n},k)} - y^{(\bar{n})}\right\|_{1,\hat{\omega}_N^+}, \quad j = 1,\ldots,N.$$

This implies

$$\left\|y^{(\bar{n},k)} - y^{(\bar{n})}\right\|_{1,\hat{\omega}_N^+}^* \leq Mq_2^k. \tag{5.78}$$

The inequalities (5.77) and (5.78) yield the estimate (5.75). Thus, the proof is complete. ∎

In order to construct an algorithm which automatically generates a grid providing a given tolerance ε of the approximate solution, one can use various strategies. We discuss briefly only the following two possibilities based on the theory developed above.

The first possibility is the classical technique which was proposed by Runge. The estimate (5.75) via standard considerations implies the following *a posteriori* h-$h/2$-strategy to arrive with a given tolerance ε (for a fixed $\bar{n}$). The tolerance ε is achieved if

$$\left\|y_N^{(\bar{n})} - y_{4N}^{(\bar{n})}\right\|_{1,\omega_N^+}^* \leq (2^{\bar{n}} - 1)\varepsilon,$$

where $y_N^{(\bar{n})}$ is the solution of the n-TDS on the uniform grid

$$\bar{\omega}_N \stackrel{\text{def}}{=} \{x_j,\ j = 0,1,\ldots,N,\ h = 1/\sqrt{N}\}$$

and $y_{4N}^{(\bar{n})}$ is the solution on the grid $\bar{\omega}_{4N}$. The main drawback of this strategy is that the grid for the difference scheme (5.55) can be uniform or quasiuniform only.

The second approach to the automatic grid generation is based on the following simple idea. Due to Theorem 5.4 the difference scheme (5.55)–(5.58) possesses order of accuracy $\bar{n}$, which is an even integer number. In order to obtain TDS of the orders $\bar{n}$ and $\bar{n}+2$ by our method one should solve the IVPs (5.51) by one-step methods (5.44) of the corresponding orders. Here we use the embedded Runge-Kutta-Nystrom methods [14, 31] of orders $\bar{n}$ and $\bar{n}+2$ ($\mathtt{RKN}(\bar{n}+2)(\bar{n})$). An *a posteriori* error estimate for the difference scheme (5.55) is then given by $\|y^{(\bar{n})} - y^{(\bar{n}+2)}\|_{1,\hat{\omega}_N^+}^*$, and we have

$$\left\|y^{(\bar{n})} - y^{(\bar{n}+2)}\right\|_{1,\hat{\omega}_N^+}^* = \left\|y^{(\bar{n})} - u\right\|_{1,\hat{\omega}_N^+}^* + O\left(h^{\bar{n}+2}\right).$$

Thus, we can compute the approximate solution of problem (5.1) by the difference scheme (5.55) with a given tolerance ε provided that $\|y^{(\bar{n})} - y^{(\bar{n}+2)}\|_{1,\hat{\omega}_N^+}^* < \varepsilon$.

The following algorithm generates a non-uniform grid $\hat{\hat{\omega}}_N$ and computes an approximate solution of the problems (5.42). The relations (5.45), (5.46) and Theorem 5.4 guarantee that the error of the approximate solution of the BVP (5.1) is within a given tolerance ε, provided that for each $j = 2 - \alpha, \ldots, N + 1 - \alpha$ and $\alpha = 1, 2$ the solutions of the IVPs (5.42) are given with tolerance $h_\gamma \, \varepsilon$. Using the well-known idea of embedded Runge-Kutta methods we can then construct a non-uniform grid $\hat{\hat{\omega}}_N$ such that the IVPs (5.42) are solved with tolerance $h_\gamma \, \varepsilon$ and that the conditions (5.20), (5.21) are fulfilled. Having in mind the IVPs (5.42), the error of the Runge-Kutta-Nystrom method of order $\bar{n}$ is

$$Y^j_\alpha(x_j, u) - Y^{(\bar{n})j}_\alpha(x_j, u) = \psi^j_\alpha(x_\beta, u) \left[(-1)^{\alpha+1} h_\gamma\right]^{\bar{n}+1} + O(h_\gamma^{\bar{n}+2}),$$

$$Z^j_\alpha(x_j, u) - Z^{(\bar{n})j}_\alpha(x_j, u) = \tilde{\psi}^j_\alpha(x_\beta, u) \left[(-1)^{\alpha+1} h_\gamma\right]^{\bar{n}+1} + O(h_\gamma^{\bar{n}+2}),$$

i.e., for the embedded Runge-Kutta-Nystrom methods of orders $\bar{n}$ and $\bar{n} + 2$ we have the following *a posteriori* estimates (neglecting the terms of order $O(h_\gamma^{\bar{n}+2})$)

$$Y^{(\bar{n}+2)j}_\alpha(x_j, u) - Y^{(\bar{n})j}_\alpha(x_j, u) \approx \psi^j_\alpha(x_\beta, u) \left[(-1)^{\alpha+1} h_\gamma\right]^{\bar{n}+1},$$

$$Z^{(\bar{n}+2)j}_\alpha(x_j, u) - Z^{(\bar{n})j}_\alpha(x_j, u) \approx \tilde{\psi}^j_\alpha(x_\beta, u) \left[(-1)^{\alpha+1} h_\gamma\right]^{\bar{n}+1}.$$

In Step 4 of the algorithm it is checked whether the solutions of the "left" and the "right" IVPs could be determined on the interval $[x_{j-1}, x_j]$ within the prescribed tolerance. If this is not the case the previous step-size is divided by two.

A doubling of the step-size h_γ changes the main terms of the error as follows:

$$2^{\bar{n}+1} \psi^j_\alpha(x_\beta, u) \left[(-1)^{\alpha+1} h_\gamma\right]^{\bar{n}+1}, \qquad 2^{\bar{n}+1} \tilde{\psi}^j_\alpha(x_\beta, u) \left[(-1)^{\alpha+1} h_\gamma\right]^{\bar{n}+1}.$$

This fact is used in Step 6 for a possible increase of the next step-size provided that on the previous subinterval these terms satisfy the prescribed tolerance. In Steps 8–10 values for N, h_N and x_N are provided which satisfy the inequalities (5.21). More precisely, if the condition in Step 7 is not fulfilled, then N (see Step 8), h_N and x_N (see Step 5) are chosen such that the left part of (5.21) is satisfied.

Now, the following two cases are possible:

1) We have $x_N \le 1/h_{min}$. Then both inequalities in (5.21) are satisfied.

2) We have $x_N > 1/h_{min}$. Then the inequalities in (5.21) are not satisfied and the values h_N and x_N must be decreased. This is realized in Step 9.

Algorithm AG($\hat{\omega}_N$, ε, $\bar{n}$, h_1, RKN($\bar{n}+2$)($\bar{n}$), m, f, u, u')

Input: An error tolerance ε, the order of accuracy $\bar{n}$ of the TDS (an even natural number), an initial step-size h_1, an embedded Runge-Kutta-Nystrom IVP-solver $\mathtt{RKN}(\bar{n}+2)(\bar{n})$, the problem data m and f as well as the initial values u, u' for the IVPs (5.42).

Output: A non-uniform grid $\hat{\omega}_N$ on which the numerical solution of (5.42),

$$Y_\alpha^{(l)j}(x_j, u), \quad Z_\alpha^{(l)j}(x_j, u), \quad l = \bar{n}, \bar{n}+2,$$

$$x_j \in \hat{\omega}_N, \quad j = 2-\alpha, \ldots, N+1-\alpha, \quad \alpha = 1, 2,$$

is determined within the given tolerance.

1) Set $j := 0, \quad x_0 := 0, \quad n := \bar{n}.$

2) Set $j := j + 1.$

3) Solve the "left" and the "right" IVPs (5.42) on the interval $[x_{j-1}, x_j]$ using $\mathtt{RKN}(\bar{n}+2)(\bar{n})$ and compute

$$e_2 \stackrel{\text{def}}{=} \max\left\{ \frac{\left|Y_2^{(\bar{n})j-1}(x_{j-1}, y^{(\bar{n})}) - Y_2^{(\bar{n}+2)j-1}(x_{j-1}, y^{(\bar{n})})\right|}{\max\left\{\left|Y_2^{(\bar{n}+2)j-1}(x_{j-1}, y^{(n)})\right|, 1\right\}}, \right.$$

$$\left. \frac{\left|Z_2^{(\bar{n})j-1}(x_{j-1}, y^{(\bar{n})}) - Z_2^{(\bar{n}+2)j-1}(x_{j-1}, y^{(\bar{n})})\right|}{\max\left\{\left|Z_2^{(\bar{n}+2)j-1}(x_{j-1}, y^{(\bar{n})})\right|, 1\right\}} \right\},$$

$$e_1 \stackrel{\text{def}}{=} \max\left\{ \frac{\left|Y_1^{(\bar{n})j}(x_j, y^{(\bar{n})}) - Y_1^{(\bar{n}+2)j}(x_j, y^{(\bar{n})})\right|}{\max\left\{\left|Y_1^{(\bar{n}+2)j}(x_j, y^{(\bar{n})})\right|, 1\right\}}, \right.$$

$$\left. \frac{\left|Z_1^{(\bar{n})j}(x_j, y^{(\bar{n})}) - Z_1^{(\bar{n}+2)j}(x_j, y^{(\bar{n})})\right|}{\max\left\{\left|Z_1^{(\bar{n}+2)j}(x_j, y^{(\bar{n})})\right|, 1\right\}} \right\}.$$

4) **if** $\max\{e_1, e_2\} > h_j\varepsilon$ **then begin** $h_j := h_j/2$; **go to** Step 3 **end.**

5) Set $x_j := x_{j-1} + h_j.$

6) **if** $2^{\bar{n}+1}\max\{e_1, e_2\} \le h_j\varepsilon$ **then begin** $h_{j+1} := 2h_j$ **end else** $h_{j+1} := h_j.$

7) **if** $x_j < \dfrac{1}{\max_{1\le i\le j} h_i}$ **then go to** Step 2.

8) Set $N := j; \quad h_{min} := \min_{1\le i\le N} h_i.$

9) **if** $x_N > \dfrac{1}{h_{min}}$ **then begin**

$h_N := \dfrac{1}{h_{min}} - x_{N-1}; \quad x_N := \dfrac{1}{h_{min}};$

Solve the IVPs (5.42) on the interval $[x_{N-1}, x_N]$ using $\mathtt{RKN}(\bar{n}+2)(\bar{n})$ **end.**

10) **Stop.**

Theorem 5.4 is the basis of the following algorithm which computes for the BVP (5.1) the corresponding solution $y_j^{(\bar{n}+2)}$, $j = 1, \ldots, N$, of the TDS (5.55) of order of accuracy $\bar{n}+2$ on a given grid $\hat{\hat{\omega}}_N$.

Algorithm A($y^{(\bar{n}+2)}$, $\hat{\hat{\omega}}_N$, ε, $\bar{n}$, h_1, A, $\mathtt{RKN}(\bar{n}+2)(\bar{n})$, μ_1, m, f)

Input: The grid $\hat{\hat{\omega}}_N$, an error tolerance ε, the order of accuracy $\bar{n}$ of the TDS (an even natural number), an initial step-size h_1, the vector $A \overset{\text{def}}{=} (A_1, \ldots, A_{\bar{n}-1})$ of coefficients which are defined in (5.49), an embedded Runge-Kutta-Nystrom method $\mathtt{RKN}(\bar{n}+2)(\bar{n})$, as well as the problem data μ_1, m , f of problem (5.1).
Output: The solution $y^{(\bar{n}+2)}(x_j)$, $x_j \in \hat{\hat{\omega}}_N$, of the BVP (5.1) with an error smaller than the given tolerance ε.

1) Set $\varepsilon_{it} := 0.25\varepsilon$.

2) Determine the starting values

$$u^{(0)}(x) = \mu_1 e^{-mx}, \qquad \frac{du^{(0)}}{dx} = -m\mu_1 e^{-mx}.$$

3) Compute the grid $\hat{\hat{\omega}}_N$ using the algorithm

$\mathrm{AG}(\hat{\hat{\omega}}_N, \varepsilon, \bar{n}, h_1, \mathtt{RKN}(\bar{n}+2)(\bar{n}), m, f, u^{(0)}, \dfrac{du^{(0)}}{dx})$.

Set $k := 0; \quad l := \bar{n}$.

4) Set $k := k+1; \quad n := l$.

5) Compute $\varphi^{(l)}(x_j, y^{(l,k-1)})$ in accordance with (5.56) for all $j = 1, \ldots, N-1$ and $\mu_2^{(l)}(x_N, y^{(l,k-1)})$ by (5.57) provided that at least one of the coefficients $A_i \neq 0$ or by (5.58) otherwise.

6) Find $y_j^{(l,k)}$, $j = 1, \ldots, N$, by solving the system of linear algebraic equations (5.74) with a tridiagonal system matrix.

7) Find

$$Y_\alpha^{(l)j}(x_j, y^{(l,k)}), \quad Z_\alpha^{(l)j}(x_j, y^{(l,k)}), \quad x_j \in \hat{\omega}_N,$$

$$j = 2-\alpha, \ldots, N+1-\alpha, \quad \alpha = 1,2,$$

by an embedded Runge-Kutta-Nystrom method RKN$(l+2)(l)$ and compute $\frac{dy^{(l,k)}(x_j)}{dx}$, $j = 0, \ldots, N$, by (5.76).

8) **if** $\left\| \frac{y^{(l,k)} - y^{(l,k-1)}}{max\left(\left|y^{(l,k)}\right|, 1\right)} \right\|^*_{1,\hat{\omega}_N^+} > \varepsilon_{it}$ **then go to** Step 4.

9) **if** $l = \bar{n}$ **then begin** $k := 0$; $l := \bar{n}+2$; **go to** Step 4 **end**.

10) **if** $\left\| \frac{y^{(\bar{n}+2)} - y^{(\bar{n})}}{max\left(\left|y^{(\bar{n}+2)}\right|, 1\right)} \right\|^*_{1,\hat{\omega}_N^+} > \varepsilon$ **then begin**

Find new starting values by interpolating the values

$$y_i^{(\bar{n}+2)}, \quad \frac{dy^{(\bar{n}+2)}(x_i)}{dx}, \quad i = 0, \ldots, N,$$

using the formulas

$$u^{(0)}(x) = \frac{1}{x_i - x_{i-1}} \left[y_i^{(\bar{n}+2)}(x - x_{i-1}) + y_{i-1}^{(\bar{n}+2)}(x_i - x) \right],$$

$$\frac{du^{(0)}}{dx} = \frac{1}{x_i - x_{i-1}} \left[\frac{dy^{(\bar{n}+2)}(x_i)}{dx}(x - x_{i-1}) + \frac{dy^{(\bar{n}+2)}(x_{i-1})}{dx}(x_i - x) \right],$$

$$x_{i-1} \le x \le x_i, \quad i = 1, \ldots, N,$$

$$u^{(0)}(x) = y_N^{(\bar{n}+2)} \exp\left(-m(x - x_N)\right),$$

$$\frac{du^{(0)}}{dx} = \frac{dy^{(\bar{n}+2)}(x_N)}{dx} \exp\left(m(x - x_N)\right), \quad x \in [x_N, \infty);$$

go to Step 3 **end**.

11) Stop.

5.4 Numerical examples

In this section we give two numerical examples confirming the theory above. For each example we have done three experiments.

Example 5.3. The BVP

$$\begin{aligned}
&\frac{d^2u}{dx^2} - 4u = 2u^3 + 6u^2, \quad x \in (0,\infty), \\
&u(0) = -1, \quad \lim_{x\to\infty} u(x) = 0,
\end{aligned} \tag{5.79}$$

possesses the exact solution $u(x) = \tanh(x) - 1$ which decays exponentially as $x \to \infty$.

Let us illustrate Theorem 5.2 by the BVP (5.79). Here we have

$$f(x,u) = -2u^3 - 6u^2, \quad m = 2, \quad L(x) = 6[1 - (\exp(-2x) - 1)^2],$$

$$p(x) = -1 + \tanh(x) \leq -\exp(-2x) = r(x),$$

$$\int_0^\infty G(x,\xi)\, L(\xi) d\xi = \frac{1}{2}\,(\exp(-4x) - \exp(-2x)) + 3x\exp(-2x) < 1.$$

Therefore, we can conclude that this problem has a unique solution which can be determined by a fixed point iteration.

The function $f(x,u) = -2u^3 - 6u^2$ is analytic in a neighbourhood of the point $(\infty, 0)$ and $f'_u(x,0) = 0$. Moreover, the ODE is autonomous, which implies $A_i = 0$, $i \in \mathbb{N}$, and we can use a difference scheme of type (5.55), (5.56), (5.58).

We apply the difference scheme of order of accuracy 4 on the uniform grid

$$\bar{\omega}_N \stackrel{\text{def}}{=} \{x_j, \quad j = 0,1,\dots,N, \quad h = 1/\sqrt{N}\}.$$

The numerical solution of the IVPs (5.42) was obtained by a fourth-order Runge-Kutta-Nystrom method (see [31], Table 14.2). The solution $y_j^{(\bar{n})}$, $j = 1,\dots,N$, of the difference scheme (5.55) has been determined by the fixed point iteration (5.74). The solution $y_j^{(\bar{n},k)}$, $j = 1,\dots,N$, of the system of linear algebraic equations (5.74) with a three-diagonal matrix was computed by the Thomas algorithm (double sweep algorithm, see e.g. [12]). To estimate experimentally the rate of convergence we have determined the *computed error* er and the *computed order of accuracy* p by the formulas

$$\mathtt{er} \stackrel{\text{def}}{=} \|z\|^*_{1,\hat{\omega}_N^+} = \left\|y^{(4)} - u\right\|^*_{1,\hat{\omega}_N^+}, \quad \mathtt{p} \stackrel{\text{def}}{=} \log_2 \frac{\|z\|^*_{1,\hat{\omega}_N^+}}{\|z\|^*_{1,\hat{\omega}_{4N}^+}}. \tag{5.80}$$

The corresponding numerical results are presented in Table 5.1. They show excellent agreement with our theoretical considerations.

In Table 5.2 we give the difference `Error` between the approximate and the exact solution and the number of ODE calls `NFUN` for a given tolerance `EPS` using the difference scheme (5.55) – (5.57) and the h-$h/2$-strategy.

N	er	p
32	$0.2135\,E-4$	
128	$0.1270\,E-5$	4.1
512	$0.7848\,E-7$	4.0
2048	$0.4727\,E-8$	4.1
8192	$0.2522\,E-9$	4.2
32768	$0.1440\,E-10$	4.1

Table 5.1: Numerical results for Example 5.3

EPS	N	NFUN	Error
$1.0\,E-3$	32	6144	$0.286\,E-3$
$1.0\,E-5$	32	13056	$0.285\,E-5$
$1.0\,E-7$	128	71040	$0.164\,E-7$
$1.0\,E-9$	2048	1276800	$0.234\,E-9$

Table 5.2: Numerical results for problem (5.3): Runge strategy

Finally, Table 5.3 presents the results obtained with Algorithm $A(\cdot)$ using the embedded Runge-Kutta-Nystrom method `RKN6(4)` (see [14], Table 5). The grid has been generated automatically with Algorithm $AG(\cdot)$. For example, the generated grid for `EPS`$=10^{-5}$ is

$$\hat{\omega}_N = \{0, 0.125, 0.25, 0.375, 0.5, 0.75, 1, 1.25, 1.5, 1.75, 2, 2.25, 2.5, 2.75, 3.25, 3.75, 4.25, 5.25, 6.25, 8\}.$$

EPS	N	NFUN	Error
$1.0\,E-3$	4	1320	$0.552\,E-3$
$1.0\,E-5$	19	5280	$0.704\,E-6$
$1.0\,E-7$	75	46164	$0.352\,E-8$
$1.0\,E-9$	417	225948	$0.398\,E-10$

Table 5.3: Numerical results for problem (5.3): Algorithm $A(\cdot)$ □

Example 5.4. The problem

$$\begin{aligned}
&\frac{d^2u}{dx^2} - 4u = -\frac{4}{(1+x)^2} + \frac{7}{(1+x)^4} - u^2, \quad x \in (0,\infty), \\
&u(0) = 1, \quad \lim_{x\to\infty} u(x) = 0,
\end{aligned} \tag{5.81}$$

possesses the exact solution $u(x) = \dfrac{1}{(1+x)^2}$.

Here we have

$$f(x,u) = \frac{4}{(1+x)^2} - \frac{7}{(1+x)^4} + u^2, \quad m = 2,$$

$$f(x,0) = \frac{4}{(1+x)^2} - \frac{7}{(1+x)^4}, \quad f'_u(x,0) = 0.$$

We represent the solution of this problem in the form

$$u(x) = \int_0^\infty G(x,\xi)\, u^2(\xi) d\xi + u^{(0)}(x),$$

where

$$u^{(0)}(x) = \frac{1}{(1+x)^2} - \int_0^\infty \frac{G(x,\xi)}{(1+\xi)^4} d\xi.$$

Let us now introduce the operator

$$\Re(x, u(\cdot)) \stackrel{\text{def}}{=} \int_0^\infty G(x,\xi)\, u^2(\xi) d\xi + u^{(0)}(x),$$

and show that this operator transforms the set

$$\Omega([0,\infty),1) \stackrel{\text{def}}{=} \left\{ u(x) \in \mathbb{C}[0,\infty) : \left\| u - u^{(0)} \right\|_{0,[0,\infty)} \le 1 \right\}$$

into itself.

Since $0 \le u^{(0)}(x) \le 1$, it holds that $\left|v(x) + u^{(0)}(x)\right| \le 3$ for all $x \in \Omega([0,\infty),1)$. Thus, we have

$$\begin{aligned}
&\left|\Re(x,u(\cdot)) - u^{(0)}(x)\right| \\
&= \int_0^\infty G(x,\xi) \left\{ v^2(\xi) - \left[u^{(0)}(\xi)\right]^2 \right\} d\xi + \int_0^\infty G(x,\xi) \left[u^{(0)}(\xi)\right]^2 d\xi
\end{aligned}$$

$$\le 4 \int_0^\infty G(x,\xi)\, d\xi = 1 - \exp(-2x) \le 1.$$

We see that the operator $\Re(x, v(\cdot))$ transforms $\Omega([0,\infty), 1)$ into itself and is contractive, since

$$|\Re(x,u(\cdot)) - \Re(x,v(\cdot))|$$

$$\le \int_0^\infty G(x,\xi) \left|u^2(\xi) - v^2(\xi)\right| d\xi \le 3 \int_0^\infty G(x,\xi)\, d\xi \left\|u - u^{(0)}\right\|_{0,[0,\infty)}$$

$$\le q \left\|u - u^{(0)}\right\|_{0,[0,\infty)},$$

where $q \stackrel{\text{def}}{=} \frac{3}{4}(1 - \exp(-2x)) < 1$. We have solved this problem by the difference scheme (5.55) – (5.57) with $\mu_1 = 1$ and

$$\mu_2^{(4)}(x_N, u) = \frac{mA_1}{x_N} + \frac{mA_2 - A_1}{x_N^2} + \frac{mA_3 - A_2}{x_N^3}$$

$$- Z_1^{(4)N}(x_N, u) + \frac{m\left(\cosh(mh_N)\, Y_1^{(4)N}(x_N, u) - u_{N-1}\right)}{\sinh(mh_N)}.$$

The functions $Z_\alpha^{(4)j}(x_j, u)$, $Y_\alpha^{(4)j}(x_j, u)$, $j = 2 - \alpha, \ldots, N + 1 - \alpha, \alpha = 1, 2$, are the numerical solutions of (5.42) which were solved by a Runge-Kutta-Nystrom method of order of accuracy 4 (see [31],Table 14.2).

For this example we have

$$0 = \tilde{F}(t, \{A\})$$

$$= \sum_{i=1}^\infty [(i-2)(i-1)A_{i-2} - 4A_i]\, t^i + \frac{4t^2}{(1+t)^2} - \frac{7t^4}{(1+t)^4} + \left[\sum_{i=1}^\infty A_i t^i + r\left(\frac{1}{t}\right)\right]^2$$

$$+ r''\left(\frac{1}{t}\right) - 4r\left(\frac{1}{t}\right).$$

Solving systems (5.48) exactly, we get $A_1 = 0$, $A_2 = 1$, $A_3 = -2$. In Tables 5.4 – 5.6 we present the results for the BVP (5.81) which have been obtained with the same three experiments described in Example 5.3.

Here, the automatically generated grid for EPS $= 10^{-3}$ is

$$\hat{\omega}_N = \{0, 0.05, 0.15, 0.35, 0.55, 0.95, 1.35, 2.15, 2.95, 3.75, 4.55, 5.35, 6.15, 6.95, 7.75, 8.55, 9.35, 10.15, 10.95, 11.75, 12.55, 13.35, 14.15, 14.95, 16.55, 18.15, 19.75, 20\}.$$

N	er	p
8	$0.5757\,E-1$	
32	$0.4486\,E-2$	3.7
128	$0.3184\,E-3$	3.8
512	$0.2130\,E-4$	3.9
2048	$0.1379\,E-5$	3.9
8192	$0.8779\,E-7$	4.0
32768	$0.5537\,E-8$	4.0

Table 5.4: Numerical results for Example 5.4

EPS	N	NFUN	Error
$1.0\,E-3$	512	56544	$0.485\,E-4$
$1.0\,E-5$	512	84480	$0.142\,E-5$
$1.0\,E-7$	2048	410256	$0.920\,E-7$

Table 5.5: Numerical results for problem (5.4): Runge strategy

Remark 5.1. For an often used test problem (see e.g. [37], [4])

$$\begin{aligned} &\frac{d^2u}{dx^2} - m^2u = 0, \quad x \in (0,\infty), \\ &u(0) = \mu_1, \qquad \lim_{x\to\infty} u(x) = 0 \end{aligned} \tag{5.82}$$

our numerical scheme (5.55) provides the exact solution, i.e.,

$$y^{(\bar{n})}(x_j) = \mu_1 \exp(-mx_j), \; x_j \in \hat{\bar{\omega}}_N.$$

Indeed, it is easy to see from (5.56) – (5.58) that in the linear case (where $f(x,u) \equiv 0$) we have $\varphi^{(\bar{n})}(x_j,u) = 0$, $j = 1,\ldots,N-1$, $\mu_2^{(\bar{n})}(x_N,u) = 0$, i.e. the scheme (5.55) has the form

$$\begin{aligned} &\left(ay_{\bar{x}}^{(\bar{n})}\right)_{\hat{x},j} - d(x_j)\,y_j^{(\bar{n})} = 0, \quad j = 1,\ldots,N-1, \\ &y_0^{(\bar{n})} = \mu_1, \quad -a(x_N)\,y_{\bar{x},N}^{(\bar{n})} = \beta_2 y_N^{(\bar{n})}, \end{aligned} \tag{5.83}$$

with coefficients which can be computed by the explicit formulas (5.29). Due to $\varphi(x_j,u) = 0$, $j = 1,\ldots,N-1$, $\mu_2(x_N,u) = 0$ (see (5.30)), the EDS (5.27) coincides with (5.83) for the BVP (5.82). □

EPS	N	NFUN	Error
$1.0\,E-3$	27	2604	$0.380\,E-5$
$1.0\,E-5$	117	14052	$0.940\,E-7$
$1.0\,E-7$	573	82524	$0.985\,E-9$

Table 5.6: Numerical results for problem (5.4): Algorithm $A(\cdot)$

Remark 5.2. Analogous to the discussions above one can construct an EDS and an n-TDS for BVPs on the whole axis $(-\infty, \infty)$. □

Remark 5.3. The solution of the system of nonlinear equations representing the n-TDS can also be determined by other iterative methods, for example, by Newton's method. □

Chapter 6

Exercises and solutions

> You never solve a problem by putting it on ice.
>
> *Winston Spencer Churchill (1974–1965)*

In this last chapter we present a variety of mathematical exercises and the corresponding sample solutions by which the reader can test and deepen the knowledge acquired in the previous chapters of this book.

6.1 Exercises

Exercise 6.1. Check that the vector-function

$$\boldsymbol{u}(x) = \begin{pmatrix} u_1(x) \\ u_2(x) \end{pmatrix} = \begin{pmatrix} \ln(ax+b) \\ \dfrac{a}{ax+b} \end{pmatrix} \tag{6.1}$$

is the solution of the following BVP for a system of two first-order ODEs

$$\begin{aligned} \boldsymbol{u}'(x) &= \boldsymbol{f}(\boldsymbol{u}), \\ B_0\,\boldsymbol{u}(0) + B_1\,\boldsymbol{u}(1) &= \boldsymbol{\beta}, \end{aligned} \tag{6.2}$$

with

$$\begin{aligned} &\boldsymbol{f}(\boldsymbol{u}) = \begin{pmatrix} f_1(u_1,u_2) \\ f_2(u_1,u_2) \end{pmatrix} = \begin{pmatrix} u_2 \\ -u_2^2 \end{pmatrix}, \\ &B_0 = \begin{pmatrix} 1 & 0 \\ 0 & 0 \end{pmatrix}, \quad B_1 = \begin{pmatrix} 0 & 0 \\ 1 & 0 \end{pmatrix}, \quad \boldsymbol{\beta} = \begin{pmatrix} \ln b \\ \ln(a+b) \end{pmatrix}. \end{aligned} \tag{6.3}$$

Prove that on the uniform grid $\overline{\omega}_h$ the difference scheme

$$\begin{aligned} \boldsymbol{u}_{\overline{x},i} &= \boldsymbol{f}_h(\boldsymbol{u}_{i-1}), \quad i = 0, 1, 2, \ldots, N, \\ B_0\, \boldsymbol{u}(0) &+ B_1\, \boldsymbol{u}(1) = \boldsymbol{\beta}, \end{aligned} \tag{6.4}$$

is exact, where

$$\boldsymbol{u}_i \stackrel{\text{def}}{=} \boldsymbol{u}(x_i), \quad \boldsymbol{u}_{\overline{x},i} \stackrel{\text{def}}{=} h^{-1}(\boldsymbol{u}_i - \boldsymbol{u}_{i-1}),$$

$$\boldsymbol{f}_h(\boldsymbol{u}) = \begin{pmatrix} f_{1,h}(u_1, u_2) \\ f_{2,h}(u_1, u_2) \end{pmatrix} \stackrel{\text{def}}{=} \begin{pmatrix} h^{-1} \ln(1 + h u_{2,i-1}) \\ -\dfrac{u_2^2(x_{i-1})}{1 + h u_{2,i-1}} \end{pmatrix}. \tag{6.5}$$

Exercise 6.2. Construct the EDS for the BVP (6.2) on an arbitrary grid $\hat{\overline{\omega}}_h$ following the theory of Chapter 2.

Exercise 6.3. Construct a 4-TDS for the BVP (6.2) on an arbitrary grid $\hat{\overline{\omega}}_h$ following the theory of Chapter 2 and using an explicit 4-stage Runge-Kutta method of order 4.

Exercise 6.4. Check that the ODE

$$u' - a x^n (u^2 + 1) = 0, \quad n \neq -1 \tag{6.6}$$

has the general solution

$$u(x) = \tan\left(\frac{a}{n+1} x^{n+1} + C\right), \tag{6.7}$$

where C is an arbitrary real constant.

Find a first-order difference equation (which represents the EDS) whose solution coincides on an arbitrary grid ω_h, $h \stackrel{\text{def}}{=} \max\limits_{i=1,2,\ldots,N} h_i$, $h_i \stackrel{\text{def}}{=} x_i - x_{i-1}$, with the exact solution of the BVP defined by the ODE (6.6) and the periodic boundary condition

$$u(0) = -u(1). \tag{6.8}$$

Exercise 6.5. Develop an EDS for the BVP (6.6) using the algorithm of Section 2.4.

Exercise 6.6. Using the EDS from the previous exercise construct TDS of orders of accuracy 2 and 3.

Exercise 6.7. Check that the following ODE of the mass interaction

$$u' = (Au - a)(Bu - b), \tag{6.9}$$

with $\Delta \stackrel{\text{def}}{=} aB - bA \neq 0$, possesses the general solution

$$u(x) = \frac{C_1 b e^{\Delta x} - C_2 a}{C_1 B e^{\Delta x} - C_2 A}, \tag{6.10}$$

where $C_1^2 + C_2^2 > 0$.

Find the EDS on an arbitrary grid ω_h, $h \stackrel{\text{def}}{=} \max_{i=1,2,\ldots,N} h_i$, $h_i \stackrel{\text{def}}{=} x_i - x_{i-1}$, for the BVP defined by the ODE (6.9) and the periodic boundary condition (6.8).

Exercise 6.8. Construct an EDS for the BVP (6.9), (6.8) following the theory of Chapter 2.

Exercise 6.9. Construct a 3-TDS for the BVP (6.9), (6.8) following the theory of Chapter 2.

Exercise 6.10. Show that on a uniform grid with step-size h and with respect to the ODE

$$u'' = e^u, \; x \in (0,1), \tag{6.11}$$

the difference equations

$$y_{\bar{x}x} = e^y \tag{6.12}$$

and

$$y_{\bar{x}x} - \frac{h^2}{12}\left[e^y\right]_{\bar{x}x} = e^y \tag{6.13}$$

possess truncation errors of order $\mathcal{O}(h^2)$ provided that $u \in \mathbb{C}^4[0,1]$, and $\mathcal{O}(h^4)$ provided that $u \in \mathbb{C}^6[0,1]$, respectively.

Exercise 6.11. Develop an algorithm of order 4 with the input data L, c_1, c_2 and $f(x)$ for the BVP

$$u'' = e^u - e^{-u} + f(x), \quad x \in (0,L), \quad u(0) = c_1, \quad u(L) = c_2, \quad L > 0. \tag{6.14}$$

Note, such problems often arise in the modelling of semiconductor devices.

Exercise 6.12. Check that the function

$$u(x) = -\ln\left(\cos^2 \frac{x}{\sqrt{2}}\right) \tag{6.15}$$

is the solution of the BVP

$$u''(x) = e^u, \quad x \in (0,1), \quad u(0) = 0, \quad u(1) = -\ln\left(\cos^2 \frac{1}{\sqrt{2}}\right). \tag{6.16}$$

1) Show that the difference scheme

$$\begin{aligned} y_{\bar{x}x} &- \frac{h^2}{12}e^y\left[(y_{\overset{\circ}{x}})^2\left(1 - \frac{h^2}{3}e^y\right) + e^y\right]_{\bar{x}x} \\ &- \frac{2h^4}{6!}\left\{e^y\left(y_{\overset{\circ}{x}}\right)^4 + 11e^{2y}\left(y_{\overset{\circ}{x}}\right)^2 + 4e^{3y}y_{\overset{\circ}{x}}\right\} = e^y \end{aligned} \tag{6.17}$$

has a truncation error $\mathcal{O}(h^6)$.

2) Solve this BVP with the TDS of order 2 and compare the result with the exact solution.

3) Solve this BVP with the TDS of order 4 and compare the result with the exact solution.

4) Find experimentally the *a posteriori* error by the Runge principle.

Exercise 6.13. Check that the function

$$u(x) = c \sinh(ax + b), \tag{6.18}$$

where c is an arbitrary and a, b are positive constants, satisfies the BVP

$$\begin{aligned} &u''(x) = 2 \sinh u + f(x), \quad x \in (0, L), \\ &u(0) = c \sinh(b), \quad u(L) = c \sinh(aL + b), \quad L > 0. \end{aligned} \tag{6.19}$$

Here, $f(x) \overset{\text{def}}{=} c(2 - a^2) \sinh(ax + b)$. Develop a three-point EDS on 1) a uniform and 2) a non-equidistant grid.

Exercise 6.14. Check that the function

$$u(x) = \frac{1}{1+x} \tag{6.20}$$

is the exact solution of the BVP

$$u''(x) = 2\, u^3(x), \quad x \in (0, 1), \quad u(0) = 1, \quad u(1) = 0.5. \tag{6.21}$$

On the basis of the facts

$$u_{\bar{x}x}(x) - u''(x) = 2 \sum_{k=2}^{\infty} \frac{u^{(2k)}(x)}{(2k)!} h^{2k-2} \tag{6.22}$$

on the equidistant grid ω_h and $u^{(2k)}(x) = (-1)^k (2k)! u^{2k+1}(x)$ find the EDS and a TDS of order $2n$ for an arbitrary n.

Exercise 6.15. Check that the BVP

$$\begin{aligned} &u''(x) + \omega^2 u + u^2(x) = \frac{A^2}{2} \left(1 - \cos(2\omega x + 2\varphi)\right), \\ &u(0) = A \sin\varphi, \quad u(1) = A \sin(\omega + \varphi), \end{aligned} \tag{6.23}$$

where A, ω and φ are arbitrary constants, has the exact solution

$$u(x) = A \sin(\omega x + \varphi). \tag{6.24}$$

Find on the uniform grid ω_h a second-order difference equation (EDS) and a TDS of order $2n$ for an arbitrary n.

Exercise 6.16. Check that the BVP

$$u''(x) - b\,\frac{u\ln u}{x} = 0, \quad u(1) = e^b, \quad u(2) = e^{2b}, \tag{6.25}$$

where b is an arbitrary constant, has the exact solution

$$u(x) = e^{bx}. \tag{6.26}$$

Find the 3-point EDS for this BVP on uniform grid with step-size $h = 1/N$ and a TDS of order $2n$ for an arbitrary n.

Exercise 6.17. Check that the BVP

$$u''(x) = -\frac{2a}{c}\,u\,u', \quad u(0) = \frac{c}{b}, \quad u(1) = \frac{c}{a+b}, \tag{6.27}$$

where a, b and c are arbitrary real constants, has the exact solution

$$u(x) = \frac{c}{ax+b}. \tag{6.28}$$

Find the 3-point EDS for this BVP on the uniform grid $\overline{\omega}_h$ and a TDS of order $2n$ for an arbitrary n.

Exercise 6.18. Check that the BVP

$$u''(x) = \frac{a^2 m(m+1)}{\sqrt[m]{c^2}}\,u^{\frac{m+2}{m}}, \quad u(0) = \frac{c}{b^m}, \quad u(1) = \frac{c}{(a+b)^m}, \tag{6.29}$$

where a, b, c are arbitrary positive real constants and m is a natural number, has the exact solution

$$u(x) = \frac{c}{(ax+b)^m}. \tag{6.30}$$

Find the 3-point EDS for this BVP on the uniform grid $\overline{\omega}_h$ and a TDS of order $2n$ for an arbitrary n.

Exercise 6.19. Check that the function

$$u(x) = \frac{1}{ax+b}, \tag{6.31}$$

where a and b are arbitrary positive constants, is the solution of the BVP

$$\begin{aligned} &u''(x) = 2a^2u^3(x), \quad x \in (0,1),\\ &u(0) = \frac{1}{b}, \quad u(1) = \frac{1}{a+b}. \end{aligned} \tag{6.32}$$

1) Find the 3-point EDS on an arbitrary non-uniform grid $\hat{\omega}$, i.e. determine its coefficients explicitly.

2) Derive the 3-TDS of order n for an arbitrary $n \in \mathbb{N}$ with explicitly given coefficients and solve the BVP by this 3-TDS with $n = 4$.

3) Solve the BVP (6.32) by the algorithm presented in Section 3.3 using the implicitly defined TDS of order 4 and compare the results with 2).

Exercise 6.20. Check that the function

$$u(x) = \ln(ax+b), \tag{6.33}$$

where a and b are arbitrary positive constants, is the solution of the BVP

$$\begin{aligned} u''(x) &= -\left(u'\right)^2(x), \quad x \in (0,1), \\ u(0) &= \ln b, \quad u(1) = \ln(a+b). \end{aligned} \tag{6.34}$$

1) Find the three-point EDS on a uniform grid $\overline{\omega}$, i.e. determine its coefficients explicitly.

2) Derive the TDS of the order $2n$ for an arbitrary $n \in \mathbb{N}$ with explicitly given coefficients and solve the BVP with $n = 4$.

3) Solve the BVP by the algorithm presented in Section 3.3 using the implicitly defined TDS of order 4 and compare the results with 2).

Exercise 6.21. Prove that the difference scheme

$$\begin{aligned} &u_{\overline{x}x}(x) \\ &= -u_{\overset{\circ}{x}}^2(x) + \frac{1}{h^2}\ln\left(\frac{(ax+b)^2 - a^2h^2}{(ax+b)^2}\right) + \frac{1}{4h^2}\ln^2\left(\frac{ax+b+ah}{ax+b-ah}\right), \quad x \in \omega_h, \\ &u(0) = \ln b, \quad u(1) = \ln(a+b) \end{aligned} \tag{6.35}$$

is exact for the BVP

$$u''(x) = -\left(u'\right)^2(x), \quad x \in (0,1), \quad u(0) = \ln b, \quad u(1) = \ln(a+b). \tag{6.36}$$

Exercise 6.22. Prove that for the right-hand side of the EDS from the previous exercise it holds that

$$\begin{aligned} &\frac{1}{h^2}\ln\left(\frac{(ax+b)^2 - a^2h^2}{(ax+b)^2}\right) \\ &+ \frac{1}{4h^2}\ln^2\left(\frac{ax+b+ah}{ax+b-ah}\right) = \frac{a^4}{6(ax+b)^4}h^2 + \frac{8a^6}{45(ax+b)^6}h^4 + \cdots. \end{aligned} \tag{6.37}$$

Exercise 6.23. Let us consider the BVP

$$\frac{d^2u}{dx^2} = -\frac{du}{dx}f(x,u), \quad x \in (0,1), \quad u(0) = \gamma_0, \quad u(1) = \gamma_1, \tag{6.38}$$

and suppose that its solution possesses the property

$$u^{(k)}(x) = \varphi_k(x)u(x), \quad k = 1, 2, \ldots. \tag{6.39}$$

Show that the difference scheme

$$u_{\bar{x},x}(x)$$

$$= -u_{\overset{\circ}{x}} f(x,u) + 2\sum_{k=2}^{\infty} \frac{\varphi_{2k}(x)}{(2k)!} h^{2k-2} + 2f(x,u) \sum_{k=2}^{\infty} \frac{\varphi_{2k}(x)}{(2k)!} h^{2k-2} u^k(x), \qquad (6.40)$$

$$u(0) = \gamma_0, \quad u(1) = \gamma_1,$$

is exact (provided that the series converges). What is the TDS of order $\mathcal{O}(h^{2n+1})$ for an arbitrary $n \in \mathbb{N}$?

Exercise 6.24. Check that the function

$$u(x) = cx^n \qquad (6.41)$$

is the solution of the BVP

$$u''(x) = c^2(n-1)\frac{u'(x)}{\sqrt[n]{u^2(x)}}, \quad c > 0,$$

$$u(a) = ca^n, \quad u(b) = cb^n, \quad b > a > 0, \qquad (6.42)$$

and find an EDS and TDS for this BVP.

Exercise 6.25. Develop an EDS for the BVP

$$\frac{d^2u}{dx^2} = \frac{1}{u^3}, \quad 0 < x < 1, \quad u(0) = u(1) = 1, \qquad (6.43)$$

whose exact solution is $u(x) = (2x^2 - 2x + 1)^{-3/2}$.

Exercise 6.26. Construct a 6-TDS for BVP (6.43).

Exercise 6.27. Construct an EDS for the problem

$$u''(x) = \sinh u, \quad x \in (0, L),$$

$$u(0) = \ln\cosh b, \quad u(L) = \ln\cosh(aL + b), \qquad (6.44)$$

where a, b and L are positive real constants (mathematical problems of this type arise in the modelling of semiconductor devices).

Exercise 6.28. Check that in the domain $\Omega \stackrel{\text{def}}{=} \{(x,y) : 0 \le x, y \le 1\}$ the function

$$u(x,y) = \frac{a}{bx + cy + d}, \quad a, b, c, d \in \mathbb{R}, \qquad (6.45)$$

is the solution of the following BVP for the nonlinear partial differential equation

$$\frac{\partial^2 u(x,y)}{\partial x^2} + \frac{\partial^2 u(x,y)}{\partial y^2} = \frac{2(b^2 + c^2)}{a^2} u^3(x,y), \quad (x,y) \in \Omega,$$

$$u(0,y) = \frac{a}{cy + d}, \quad u(1,y) = \frac{a}{b + cy + d}, \qquad (6.46)$$

$$u(x,0) = \frac{a}{bx + d}, \quad u(x,1) = \frac{a}{bx + c + d}.$$

Propose an EDS for this BVP on the rectangular grid

$$\omega_{h_1,h_2} \stackrel{\text{def}}{=} \{(x_i, y_j) : x_i = ih_1,\ i = 0, 1, \ldots, N_1,\ h_1 = 1/N_1; \\ y_j = jh_2,\ j = 0, 1, \ldots, N_2,\ h_2 = 1/N_2\}.$$

Is this EDS uniquely determined?

Exercise 6.29. Check that in the domain Ω the function

$$u(x, y) = a \sin(bx + cy + d), \quad a, b, c, d \in \mathbb{R}, \tag{6.47}$$

is the solution of the following BVP for the nonlinear partial differential equation

$$\begin{aligned} &\frac{\partial^2 u(x,y)}{\partial x^2} + \frac{\partial^2 u(x,y)}{\partial y^2} + 2xyu^2(x,y) + a(b^2 + c^2)u \\ &= 2a^2 xy \sin^2(bx + cy + d), \quad (x, y) \in \Omega, \\ &u(0, y) = c \sin(by + d), \quad u(1, y) = c \sin(a + by + d), \\ &u(x, 0) = c \sin(ax + d), \quad u(x, 1) = c \sin(ax + b + d). \end{aligned} \tag{6.48}$$

Propose an EDS for this BVP on the rectangular grid ω_{h_1,h_2}.

Exercise 6.30. An ODE of the form $u'(x) = f(u)$ is called autonomous. Suppose that an algorithm to calculate $F_k(u) = f^{(k)}(u)$ for all $k = 1, 2, \ldots$, is given. Construct an EDS for the IVP

$$u'(x) = f(u), \quad u(0) = u_0,$$

on an arbitrary non-uniform grid with N nodes. What is the algorithm for a TDS of order $\mathcal{O}(h^n)$, where $h = \max_i h_i$, $h_i = x_i - x_{i-1}$?

Exercise 6.31. Suppose that an algorithm to calculate $F_k(u) = f^{(k)}(u)$ for all $k = 1, 2, \ldots$, is given. Construct an EDS for the BVP

$$u''(x) = f(u), \quad u(0) = u_0, \quad u(1) = u_1,$$

on the equidistant grid with the step-size $h = 1/N$. What is the algorithm for a TDS of order $\mathcal{O}(h^n)$?

Exercise 6.32. The BVP

$$\begin{aligned} &u''(x) + \lambda\, u(x) = 0, \quad x \in (0, 1), \\ &u(0) = 0, \quad u(1) = 0 \end{aligned} \tag{6.49}$$

has the exact solution

$$u_n(x) = \sin(n\pi x), \quad \lambda_n = (n\pi)^2, \quad n = 1, 2, \ldots . \tag{6.50}$$

Show that on the grid

$$\overline{\omega}_h \stackrel{\text{def}}{=} \omega_h \cup \{x_0 = 0,\ x_{N+1} = 0\}, \quad \omega_h \stackrel{\text{def}}{=} \{x_i = \frac{i}{N+1}, \quad i = 1, \dots, N\} \tag{6.51}$$

the eigenvalues of the difference scheme (in non-indexed form)

$$\begin{aligned} &u_{\overline{x}x}(x) + \lambda \left[\frac{\sin(\sqrt{\lambda}h/2)}{(\sqrt{\lambda}h/2)} \right]^2 u(x) = 0, \quad x \in \omega_h, \\ &u(0) = 0, \quad u(1) = 0 \end{aligned} \tag{6.52}$$

or (in indexed form)

$$\begin{aligned} &\frac{1}{h^2}[u(x_{i+1}) - 2u(x_i) + u(x_{i-1})] + \lambda \left[\frac{\sin(\sqrt{\lambda}h/2)}{(\sqrt{\lambda}h/2)} \right]^2 u(x_i) = 0, \\ &u(x_0) = 0, \quad u(x_{N+1}) = 0, \quad i = 1, \dots, N, \end{aligned} \tag{6.53}$$

coincide with the first N exact eigenvalues and the eigenfunctions (eigenvectors) are the projections of the first N exact eigenfunctions onto the grid .

Exercise 6.33. Show that

- the eigenvalues λ_k^h of the difference eigenvalue problem

$$\begin{aligned} &y_{\overline{x}x}(x) + \lambda^h y(x) = 0, \quad x \in \omega, \\ &y(0) = 0, \quad y(1) = 0 \end{aligned} \tag{6.54}$$

  approximate the exact eigenvalues λ_k of (6.49) (k fixed and independent of N) with second order of accuracy $\mathcal{O}(h^2)$, and

- the eigenfunctions $y_k(x)$ are the projections of the first N exact eigenfunctions $u_k(x)$ onto the grid.

Exercise 6.34. For the Sturm-Liouville BVP

$$\begin{aligned} &u''(x) + [\lambda - q(x)]u(x) = 0, \quad x \in (0,1), \\ &u(0) = 0, \quad u(1) = 0, \end{aligned} \tag{6.55}$$

with

$$q(x) \stackrel{\text{def}}{=} \begin{cases} 0, & 0 \le x \le \frac{1}{2}, \\ 1, & \frac{1}{2} < x \le 1, \end{cases}$$

construct the EDS on the equidistant grid

$$\overline{\omega}_h \stackrel{\text{def}}{=} \left\{ x_i = ih, \quad i = 0, 1, \dots, 2N+2, \quad h = \frac{1}{2(N+1)} \right\}.$$

Exercise 6.35. (See [20, 54])
Let the eigenvalue problem

$$\begin{aligned} &Lu(x) + \lambda u(x) = 0, \quad x \in (a,b), \\ &|u(a)| \neq \infty, \quad |u(b)| \neq \infty, \end{aligned} \tag{6.56}$$

be given, where

$$Lu(x) \stackrel{\text{def}}{=} A(x)u''(x) + B(x)u'(x), \quad A(x) \stackrel{\text{def}}{=} a_2(x-a)(x-b), \quad B(x) \stackrel{\text{def}}{=} b_1 x + b_0.$$

Determine an equidistant grid ω_h with $2N+1$ nodes and specify conditions under which the eigenvalues of the following difference scheme

$$\begin{aligned} &L_h y(x) + \lambda_h y(x) = 0, \quad x \in \omega_h, \\ &u(x_{-N}) \neq \infty, \quad u(x_N) \neq \infty, \end{aligned} \tag{6.57}$$

where

$$L_h y(x) \stackrel{\text{def}}{=} A(x) y_{\bar{x}x} + B(x) y_{\overset{\circ}{x}},$$

coincide with the first $2N+1$ eigenvalues of the exact problem (6.56). Which is the approximation order of the corresponding eigenfunctions with respect to h?

Hint: If we substitute

$$x = \frac{1}{2}[(b-a)t + a + b] \tag{6.58}$$

into (6.56) and define $y(t) \stackrel{\text{def}}{=} u(\frac{1}{2}[(b-a)t + a + b])$ then problem (6.56) reads

$$\begin{aligned} &(1-t^2)y''(t) - \frac{1}{a_2}\left[\frac{1}{2}b_1(b-a)t + \frac{1}{2}b_1(b+a) + b_0\right]y'(t) - \frac{\lambda}{a_2}y(t) = 0, \\ &|y(-1)| \neq \infty, \quad |y(1)| \neq \infty, \quad t \in (-1,1). \end{aligned} \tag{6.59}$$

The solution of (6.59) is a polynomial of degree n iff $\lambda = -a_2 n(n + \frac{b_1(b-a)}{2a_2} - 1)$. This solution is given by $u(x) = P_n(x) = P_n^{(\alpha,\beta)}(\frac{2x-a-b}{b-a})$, where $P_n^{(\alpha,\beta)}(z)$ are the Jacobi polynomials with $\beta = -\frac{b_0}{2a_2} - 1$ and $\alpha = -1 + \frac{b_1(b-a)}{2a_2} + \frac{b_0}{2a_2}$ (see e.g. [6]).

Exercise 6.36. (See [20, 54])
After the change of variables

$$U = r^{-1/2}(\sin\theta)^m V, \quad x = \ln r, \quad y = \frac{1+\cos\theta}{2}$$

the Dirichlet problem for the Poisson equation in spherical coordinates ($m = 0$ is the case of axial symmetry)

$$LU = L_r U + L_\theta U = \frac{\partial}{\partial r}\left(r^2 \frac{\partial U}{\partial r}\right) + \frac{1}{\sin\theta}\frac{\partial}{\partial\theta}\left(\sin\theta \frac{\partial U}{\partial\theta}\right) - \frac{m^2 U}{\sin\theta} = f(r, \cos\theta),$$

$$0 < r_1 < r < r_2, \quad 0 < \theta < \pi, \quad m = 0, 1, 2, \ldots,$$

$$U(r_1, \theta) = \varphi_1(\theta), \quad U(r_2, \theta) = \varphi_2(\theta)$$

is transformed into the problem

$$LV = (L_x + L_y)V = F(x, y), \quad (x, y) \in G,$$

$$V(x_i, y) = \Phi_i(y), \quad i = 1, 2,$$

where

$$L_x V = \frac{\partial^2 V}{\partial x^2} + \left(m + \frac{1}{2}\right)^2 V, \quad L_y V = y(1-y)\frac{\partial^2 V}{\partial y^2} - (m+1)(2y-1)\frac{\partial V}{\partial y},$$

$$F(x, y) = e^{x/2}\Big(4y(1-y)\Big)^{-m/2} f(e^x, 2y - 1).$$

Approximate the operator L_y by a difference operator with an exact spectrum similar to Exercise 6.35.

Exercise 6.37. Given the Sturm-Liouville problem

$$\begin{aligned} &Lu(x) + \lambda u(x) = [(1 - x^2)u'(x)]' + \lambda u(x) = 0, \quad x \in (-1, 1), \\ &|u(-1)| \neq \infty, \quad |u(1)| \neq \infty, \end{aligned} \tag{6.60}$$

and let $P_n(x)$ be the Legendre polynomials. To construct the 3-point difference scheme which has the exact spectrum of (6.60)

$$\lambda_n = n(n+1), \quad u_n(x; \lambda_n) = P_n(x), \quad x \in (-1, 1), \quad n = 0, 1, \ldots, \tag{6.61}$$

use the stencil functions v_1^i and v_2^i which on an arbitrary grid

$$\omega_h = \{-1 < x_{-N} < x_{-N+1} < \cdots x_N < 1\} \tag{6.62}$$

satisfies the equations

$$\begin{aligned} &Lv_\alpha^i(x) + \lambda v_\alpha^i(x) = 0, \quad x \in (x_{i-1}, x_{i+1}), \quad \alpha = 1, 2, \\ &v_1^i(x_{i-1}) = 0, \quad \frac{dv_1^i(x_{i-1})}{dx} = 1, \\ &v_2^i(x_{i+1}) = 0, \quad \frac{dv_2^i(x_{i+1})}{dx} = -1. \end{aligned} \tag{6.63}$$

Exercise 6.38. The Sturm-Liouville problem

$$\begin{aligned} &u''(x) - 2xu(x) + \lambda u(x) = 0, \quad x \in (-\infty, \infty), \\ &\int_{-\infty}^{\infty} e^{-x^2} u^2(x) dx < \infty \end{aligned} \tag{6.64}$$

possesses the exact solution

$$\lambda_n = 2n, \quad u_n(x) = u_n(x; \lambda_n) = H_n(x), \quad x \in (-\infty, \infty), \quad n = 0, 1, \ldots,$$

where $H_n(x)$ are the Hermite polynomials [6]. Show that on the finite grid

$$\omega_h = \{x_i = ih : i = -N, \ldots, N, \quad h = 1/\sqrt{N}\}$$

the difference scheme

$$\begin{aligned} &y_{\overline{x},x}(x) - 2xy_{\overset{\circ}{x}}(x) + \lambda^h y(x) = 0, \quad x \in \omega_h, \\ &y(x_{-N-1}) \neq \infty, \quad y(x_{N+1}) \neq \infty \end{aligned} \tag{6.65}$$

has the exact eigenvalues of the given problem (6.64), i.e. the eigenvalues

$$\lambda_n^h = 2n, \quad n = 0, 1, \ldots, 2N + 1.$$

6.2 Solutions

Solution 6.1. Check that the exact solution satisfies the EDS.

Solution 6.2. The EDS on the grid $\hat{\bar{\omega}}_h$ can be written in the form

$$\boldsymbol{u}_j = \boldsymbol{Y}^j(x_j, \boldsymbol{u}_{j-1}), \quad j = 1, 2, \ldots, N, \quad B_0 \boldsymbol{u}_0 + B_1 \boldsymbol{u}_N = \boldsymbol{\beta},$$

where

$$\boldsymbol{Y}^j(x, \boldsymbol{u}_{j-1}) = \begin{pmatrix} Y_1^j(x, \boldsymbol{u}_{j-1}) \\ Y_2^j(x, \boldsymbol{u}_{j-1} \end{pmatrix}$$

is the solution of the IVP

$$\begin{aligned} &\frac{dY_1^j(x, \boldsymbol{u}_{j-1})}{dx} = Y_2^j(x, \boldsymbol{u}_{j-1}), \\ &\frac{dY_2^j(x, \boldsymbol{u}_{j-1})}{dx} = -\left[Y_2^j(x, \boldsymbol{u}_{j-1})\right]^2, \quad x \in (x_{j-1}, x_j], \\ &Y_1^j(x_{j-1}, \boldsymbol{u}_{j-1}) = u_{1,j-1}, \\ &Y_2^j(x_{j-1}, \boldsymbol{u}_{j-1}) = u_{2,j-1}, \quad j = 1, 2, \ldots, N. \end{aligned} \tag{6.66}$$

The general solution of the ODE is

$$Y_1^j(x, \boldsymbol{u}_{j-1}) = \ln|x + C_1| + C_2, \quad Y_2^j(x, \boldsymbol{u}_{j-1}) = \frac{1}{x + C_1}.$$

The relations

$$Y^j(x_{j-1}, \boldsymbol{u}_{j-1}) = \ln|x_{j-1} + C_1| + C_2 = u_{1,j-1},$$

$$Y_2^j(x, \boldsymbol{u}_{j-1}) = \frac{1}{x_{j-1} + C_1} = u_{2,j-1},$$

yield

$$C_1 = \frac{1}{u_{2,j-1}} - x_{j-1}, \quad C_2 = u_{1,j-1} - \ln\frac{1}{|u_{2,j-1}|}.$$

Thus, the solution of the IVP (6.66) is the function

$$\boldsymbol{Y}^j(x, \boldsymbol{u}_{j-1}) = \begin{pmatrix} \ln|1 + (x - x_{j-1})\, u_{2,j-1}| + u_{1,j-1} \\ \dfrac{u_{2,j-1}}{1 + (x - x_{j-1}) u_{2,j-1}} \end{pmatrix},$$

and we obtain the two-point EDS

$$\boldsymbol{u}_j = \begin{pmatrix} \ln|1 + h_j u_{2,j-1}| + u_{1,j-1} \\ \dfrac{u_{2,j-1}}{1 + h_j u_{2,j-1}} \end{pmatrix}.$$

Solution 6.3. In accordance with our theory we obtain the following 4-EDS:

$$\boldsymbol{y}_j^{(4)} = \boldsymbol{Y}^{(4)j}\left(x_j, \boldsymbol{y}_{j-1}^{(4)}\right), \quad j = 1, 2, \ldots, N, \quad B_0 \boldsymbol{y}_0^{(4)} + B_1 \boldsymbol{y}_N^{(4)} = \boldsymbol{\beta}, \tag{6.67}$$

where by the classical Runge-Kutta method of order 4 we have

$$\boldsymbol{y}_j^{(4)} \approx \boldsymbol{u}_j, \quad \boldsymbol{Y}^{(4)j}\left(x_j, \boldsymbol{y}_{j-1}^{(4)}\right) = \boldsymbol{y}_{j-1}^{(4)} + \frac{h_j}{6}\left(\boldsymbol{k}_1 + 2\boldsymbol{k}_2 + 2\boldsymbol{k}_3 + \boldsymbol{k}_4\right),$$

$$\boldsymbol{k}_1 = \boldsymbol{f}\left(x_{j-1}, \boldsymbol{y}_{j-1}^{(4)}\right), \quad \boldsymbol{k}_2 = \boldsymbol{f}\left(x_{j-1} + \frac{h_j}{2}, \boldsymbol{y}_{j-1}^{(4)} + \frac{h_j \boldsymbol{k}_1}{2}\right),$$

$$\boldsymbol{k}_3 = \boldsymbol{f}\left(x_{j-1} + \frac{h_j}{2}, \boldsymbol{y}_{j-1}^{(4)} + \frac{h_j \boldsymbol{k}_2}{2}\right), \quad \boldsymbol{k}_4 = \boldsymbol{f}\left(x_{j-1} + h_j, \boldsymbol{y}_{j-1}^{(4)} + h_j \boldsymbol{k}_3\right),$$

$$\boldsymbol{f}(x, \boldsymbol{u}) \equiv \begin{pmatrix} u_2 \\ -u_2^2 \end{pmatrix}.$$

Solution 6.4. We have from (6.7)

$$u(x_{i-1}) = \tan\left(\frac{a}{n+1}\, x_{i-1}^{n+1} + C\right). \tag{6.68}$$

Furthermore, by evident transformations we obtain successively

$$
\begin{aligned}
&C = \arctan\left(u(x_{i-1})\right) - \frac{a}{n+1}\, x_{i-1}^{n+1}, \\
&u(x_i) = \tan\left(\frac{a}{n+1}\, x_i^{n+1} + \arctan\left(u(x_{i-1})\right) - \frac{a}{n+1}\, x_{i-1}^{n+1}\right) \\
&= \frac{\tan\left(\dfrac{a}{n+1}\,(x_i - x_{i-1})\right) + u(x_{i-1})}{1 - (x_i - x_{i-1})\tan\left(\dfrac{a}{n+1}\,(x_i - x_{i-1})\right)}, \\
&\frac{u(x_i) - u(x_{i-1})}{h_i} = \frac{1}{h_i}\left[\frac{\tan\left(\dfrac{a}{n+1}\,(x_i - x_{i-1})\right) + u(x_{i-1})}{1 - u(x_{i-1})\tan\left(\dfrac{a}{n+1}\,(x_i - x_{i-1})\right)} - u(x_{i-1})\right] \\
&= \frac{\tan\left(\dfrac{a}{n+1}\,(x_i - x_{i-1})\right) + u^2(x_{i-1})\tan\left(\dfrac{a}{n+1}\,(x_i - x_{i-1})\right)}{h_i\left[1 - u(x_{i-1})\tan\left(\dfrac{a}{n+1}\,(x_i - x_{i-1})\right)\right]},
\end{aligned}
\tag{6.69}
$$

and therefore the two-point EDS is

$$
\begin{aligned}
&u_{\overline{x}_i} = \frac{\tan\left(\dfrac{a}{n+1}\,(x_i - x_{i-1})\right)}{h_i}\;\frac{1 + u^2(x_{i-1})}{1 - u(x_{i-1})\tan\left(\dfrac{a}{n+1}\,(x_i - x_{i-1})\right)}, \\
&u(x_0) = -u(x_N).
\end{aligned}
\tag{6.70}
$$

Solution 6.5. The EDS on the grid $\hat{\overline{\omega}}_h$ can be written in the form

$$
u_j = Y^j(x_j, u_{j-1}), \quad j = 1, 2, \ldots, N, \quad u_0 = -u_N,
$$

where $Y^j(x, u_{j-1})$ is the solution of the IVP

$$
\begin{aligned}
&\frac{dY^j(x, u_{j-1})}{dx} = ax^n\left(Y^j(x, u_{j-1}) + 1\right), \quad x \in (x_{j-1}, x_j], \\
&Y^j(x_{j-1}, u_{j-1}) = u_{j-1}, \quad j = 1, 2, \ldots, N.
\end{aligned}
\tag{6.71}
$$

The general solution of the ODE in (6.71) is

$$
Y^j(x, u_{j-1}) = \tan\left(\frac{a}{n+1}\, x^{n+1} + C\right).
$$

Thus we have

$$
Y^j(x_{j-1}, u_{j-1}) = \tan\left(\frac{a}{n+1}\, x_{j-1}^{n+1} + C\right) = u_{j-1},
$$

$$C = \arctan u_{j-1} - \frac{a}{n+1}\, x_{j-1}^{n+1},$$

and the solution of the IVP (6.71) is

$$Y^j(x, u_{j-1}) = \tan\left(\frac{a}{n+1}\,\left(x^{n+1} - x_{j-1}^{n+1}\right) + \arctan u_{j-1}\right).$$

Now the two-point EDS can be represented in the form

$$u_j = \tan\left(\frac{a}{n+1}\,\left(x^{n+1} - x_{j-1}^{n+1}\right) + \arctan u_{j-1}\right), \quad j = 1, 2, \dots N,$$

$$u_0 = -u_N.$$

Solution 6.6. In accordance with our theory we obtain the 2-TDS

$$y_j^{(2)} = Y^{(2)j}\left(x_j, y_{j-1}^{(2)}\right), \quad j = 1, 2, \dots, N, \quad y_0^{(2)} = -y_N^{(2)}, \tag{6.72}$$

where by the two-stage Runge-Kutta method of order 2 (the so-called modified polygon method or Heun's method) we have

$$y_j^{(2)} \approx u_j, \quad Y^{(2)j}\left(x_j, y_{j-1}^{(2)}\right) = y_{j-1}^{(2)} + h_j k_2,$$

$$k_1 = f\left(x_{j-1}, y_{j-1}^{(2)}\right), \quad k_2 = f\left(x_{j-1} + \frac{h_j}{2}, y_{j-1}^{(2)} + \frac{h_j k_1}{2}\right),$$

$$f(x, u) \equiv a x^n (u^2 + 1).$$

Here one can use any Runge-Kutta method of order 2. The EDS (6.72) is a nonlinear system of algebraic equations.

Analogously one can develop the 3-TDS

$$y_j^{(3)} = Y^{(3)j}\left(x_j, y_{j-1}^{(3)}\right), \quad j = 1, 2, \dots, N, \quad y_0^{(3)} = -y_N^{(3)},$$

where e.g. the following three-stage Runge-Kutta formulas of order 3 (see e.g. [35], p. 29) is used

$$Y^{(3)j}\left(x_j, y_{j-1}^{(3)}\right) = y_{j-1}^{(3)} + \frac{h_j}{6}\,(k_1 + 4k_2 + k_3),$$

$$k_1 = f\left(x_{j-1}, y_{j-1}^{(3)}\right), \quad k_2 = f\left(x_{j-1} + \frac{h_j}{2}, y_{j-1}^{(3)} + \frac{h_j k_1}{2}\right),$$

$$k_3 = f\left(x_{j-1} + h_j, y_{j-1}^{(3)} - h_j k_1 + 2 h_j k_2\right).$$

Solution 6.7. We have from (6.10)

$$u(x_{i-1}) = \frac{b\, e^{\Delta x_{i-1}} - Ca}{B\, e^{\Delta x_{i-1}} - CA}, \quad C \stackrel{\text{def}}{=} \frac{C_2}{C_1}, \quad C_1 \neq 0. \tag{6.73}$$

Furthermore, by evident transformations we obtain successively

$$C[a - Au_{i-1}] = -e^{\Delta x_{i-1}}[Bu_{i-1} - b], \quad u_{i-1} = u(x_{i-1}),$$

$$C = e^{\Delta x_{i-1}} \frac{b - Bu_{i-1}}{a - Au_{i-1}},$$

$$u_i = \frac{b\, e^{\Delta x_i} - a\, e^{\Delta x_{i-1}} \dfrac{b - Bu_{i-1}}{a - Au_{i-1}}}{B\, e^{\Delta x_i} - A\, e^{\Delta x_{i-1}} \dfrac{b - Bu_{i-1}}{a - Au_{i-1}}}, \tag{6.74}$$

$$\frac{u_i - u_{i-1}}{h_i} = \frac{e^{\Delta x_i} - e^{\Delta x_{i-1}}}{h_i} \times \frac{(b - Bu_{i-1})(a - Au_{i-1})}{B\, a\, e^{\Delta x_i} - A\, b\, e^{\Delta x_{i-1}} + A\, Bu_{i-1}(e^{\Delta x_i} - e^{\Delta x_{i-1}})}.$$

Now the EDS reads

$$\begin{aligned} u_{\bar{x}_i} &= \frac{e^{\Delta x_i} - e^{\Delta x_{i-1}}}{h_i} \\ &\quad \times \frac{(b - Bu_{i-1})(a - Au_{i-1})}{B\, a\, e^{\Delta x_i} - A\, b\, e^{\Delta x_{i-1}} + A\, B\, u_{i-1}(e^{\Delta x_i} - e^{\Delta x_{i-1}})}, \\ u_0 &= u_N. \end{aligned} \tag{6.75}$$

Solution 6.8. The EDS on the grid $\hat{\bar{\omega}}_h$ can be written in the form

$$u_j = Y^j(x_j, u_{j-1}), \quad j = 1, 2, \ldots, N, \quad u_0 = u_N,$$

where $Y^j(x, u_{j-1})$ is the solution of the IVP

$$\begin{aligned} \frac{dY^j(x, u_{j-1})}{dx} &= (AY^j(x, u_{j-1}) - a)(BY^j(x, u_{j-1}) - b), \quad x \in (x_{j-1}, x_j], \\ Y^j(x_{j-1}, u_{j-1}) &= u_{j-1}, \quad j = 1, 2, \ldots, N. \end{aligned} \tag{6.76}$$

The general solution of the ODE in (6.76) is

$$Y^j(x, u_{j-1}) = \frac{b \exp(\Delta x) - Ca}{B \exp(\Delta x) - CA}.$$

Thus we have

$$Y^j(x_{j-1}, u_{j-1}) = \frac{b \exp(\Delta x_{j-1}) - Ca}{B \exp(\Delta x_{j-1}) - CA} = u_{j-1},$$

$$C = \frac{(b - u_{j-1}B) \exp(\Delta x_{j-1})}{a - u_{j-1}A}.$$

The solution of the IVP (6.76) is the function

$$Y^j(x,u_{j-1}) = \frac{b(a-u_{j-1}A)\exp(\Delta x) - a(b-u_{j-1}B)\exp(\Delta x_{j-1})}{B(a-u_{j-1}A)\exp(\Delta x) - A(b-u_{j-1}B)\exp(\Delta x_{j-1})},$$

and we obtain the two-point EDS in the form

$$u_j = \frac{b(a-u_{j-1}A)\exp(\Delta x_j) - a(b-u_{j-1}B)\exp(\Delta x_{j-1})}{B(a-u_{j-1}A)\exp(\Delta x_j) - A(b-u_{j-1}B)\exp(\Delta x_{j-1})}, \qquad u_0 = u_N.$$

Solution 6.9. In accordance with our theory we obtain the 3-EDS

$$y_j^{(3)} = Y^{(3)j}\left(x_j, y_{j-1}^{(3)}\right), \quad j = 1,2,\ldots,N, \quad y_0^{(3)} = y_N^{(3)},$$

where the Runge-Kutta method of order 3 which we presented in Solution 6.6 is again used.

Solution 6.10. For $u \in \mathbb{C}^4[0,1]$ it holds that

$$u_{\overline{x}x}(x) - e^{u(x)} - \left[u''(x) - e^{u(x)}\right] = 2\sum_{k=1}^{\infty} \frac{h^{2k}u^{(2k+2)}(x)}{(2k+2)!} = \frac{1}{12}u^{(4)}(\tilde{x}), \tag{6.77}$$

with some $\tilde{x} \in (x-h, x+h)$.

For $u \in \mathbb{C}^6[0,1]$ we have

$$\begin{aligned}
&u_{\overline{x}x}(x) - \frac{h^2}{12}\left[e^{u(x)}\right]_{\overline{x}x} - e^{u(x)} - \left[u''(x) - e^{u(x)}\right] \\
&= u_{\overline{x}x}(x) - u''(x) - \frac{h^2}{12}\left[u''(x)\right]_{\overline{x}x} - e^{u(x)} \\
&= \frac{h^2}{12}u^{(4)}(x) + \frac{2}{6!}h^4 u^{(6)}(\tilde{x}) - \frac{h^2}{12}\left[u^{(4)}(x) + \frac{h^2}{12}u^{(6)}(\tilde{\tilde{x}})\right] \\
&= \mathcal{O}(h^6),
\end{aligned} \tag{6.78}$$

with some $\tilde{\tilde{x}} \in (x-h, x+h)$.

Solution 6.11. Analogously to Solution 6.10.

Solution 6.12. Analogously to Solution 6.10.

Solution 6.13. Use the fact that $u^{(k)}(x) = a^k u(x)$.

Solution 6.14. We have

$$\begin{aligned}
&u_{\overline{x}x}(x) - 2u^3(x) - \left[u''(x) - 2u^3(x)\right] \\
&= 2\sum_{k=2}^{\infty} \frac{u^{(2k)}(x)}{(2k)!}h^{2k-2} = 2\sum_{k=2}^{\infty} u^{2k+1}(x)h^{2k-2}.
\end{aligned} \tag{6.79}$$

Therefore the difference scheme

$$u_{\overline{x}x}(x) - 2u^3(x) = 2\sum_{k=2}^{\infty} u^{2k+1}(x)h^{2k-2}, \quad x \in \omega_h,$$
$$u(0) = 1, \quad u(1) = 0.5, \tag{6.80}$$

or, which is the same,

$$u_{\overline{x}x}(x) - 2u^3(x) = 2h^2 \frac{u^5(x)}{1-(hu(x))^2}, \quad x \in \omega_h,$$
$$u(0) = 1, \quad u(1) = 0.5, \tag{6.81}$$

is exact and the difference scheme

$$y_{\overline{x}x}(x) - 2y^3(x) - 2\sum_{k=2}^{n} y^{2k+1}(x)h^{2k-2} = 0, \quad x \in \omega_h,$$
$$y(0) = 1, \quad y(1) = 0.5 \tag{6.82}$$

possesses a truncation error of order $\mathcal{O}(h^{2n})$.

Solution 6.15. We have

$$u_{\overline{x}x}(x) + \omega^2 u + u^2(x) - \frac{A^2}{2}\left(1 - \cos(2\omega x + \varphi)\right)$$
$$- \left[u''(x) + \omega^2 u + u^2(x) - \frac{A^2}{2}\left(1-\cos(2\omega x+\varphi)\right)\right] \tag{6.83}$$
$$= 2\sum_{k=2}^{\infty} \frac{u^{(2k)(x)}}{(2k)!} h^{2k-2} = 2\sum_{k=2}^{\infty} (-1)^k \omega^{2k} u(x) h^{2k-2}.$$

Therefore the difference scheme

$$u_{\overline{x}x}(x) + \omega^2 u + u^2(x) - \left[2\sum_{k=2}^{\infty}(-1)^k \omega^{2k} h^{2k-2}\right] u(x)$$
$$= \frac{A^2}{2}\left(1-\cos(2\omega x+\varphi)\right), \quad x \in \omega_h, \tag{6.84}$$
$$u(0) = A\sin(\varphi), \quad u(1) = A\sin(\omega+\varphi),$$

or, which is the same,

$$u_{\overline{x}x}(x) + \omega^2 u + u^2(x) - 2h^2 \frac{\omega^4}{1+(\omega h)^2} u(x)$$
$$= \frac{A^2}{2}\left(1-\cos(2\omega x+\varphi)\right), \quad x \in \omega_h, \tag{6.85}$$
$$u(0) = A\sin(\varphi), \quad u(1) = A\sin(\omega+\varphi),$$

is exact and the difference scheme

$$y_{\overline{x}x}(x) + \omega^2 y + y^2(x) - \left[2\sum_{k=2}^{n}(-1)^k\omega^{2k}h^{2k-2}\right]y(x)$$
$$= \frac{A^2}{2}\left(1 - \cos\left(2\omega x + \varphi\right)\right), \quad x \in \omega_h, \tag{6.86}$$
$$y(0) = A\sin(\varphi), \quad y(1) = A\sin(\omega + \varphi),$$

has a truncation error of order $\mathcal{O}(h^{2n})$.

Solution 6.16. We have

$$u^{(k)}(x) = b^k e^{bx}, \quad k = 1, 2, \dots . \tag{6.87}$$

Thus,

$$u_{\overline{x}x}(x) - b\frac{u\ln u}{x} - \left[u''(x) - b\frac{u\ln u}{x}\right] = 2\sum_{k=2}^{\infty}\frac{u^{(2k)}(x)}{(2k)!}h^{2k-2}$$
$$= 2\sum_{k=2}^{\infty} b^{2k}u(x)h^{2k-2} = b^4h^2\frac{1}{1-(bh)^2}u(x). \tag{6.88}$$

The three-point difference scheme

$$u_{\overline{x}x}(x) - b\frac{u\ln u}{x} - 2\sum_{k=2}^{\infty} b^{2k}u(x)h^{2k-2} = 0, \quad x \in \omega_h,$$
$$u(1) = e^b, \quad u(2) = e^{2b}, \tag{6.89}$$

or, which is the same (provided that $|bh| < 1$),

$$u_{\overline{x}x}(x) - b\frac{u\ln u}{x} - \frac{2b^4h^2}{1-(bh)^2}u(x) = 0, \quad x \in \omega_h,$$
$$u(1) = e^b, \quad u(2) = e^{2b}, \tag{6.90}$$

is exact.

The difference scheme

$$y_{\overline{x}x}(x) - b\frac{y\ln y}{x} - 2y(x)\sum_{k=2}^{n} b^{2k}h^{2k-2} = 0, \quad x \in \omega_h,$$
$$y(1) = e^b, \quad y(1) = e^{2b}, \tag{6.91}$$

or

$$y_{\overline{x}x}(x) - b\frac{y\ln y}{x} - 2b^4h^2\frac{1-(bh)^{2n-3}}{1-(bh)^2}y(x) = 0, \quad x \in \omega_h,$$
$$y(1) = e^b, \quad y(1) = e^{2b}, \tag{6.92}$$

possesses a truncation error of order $\mathcal{O}(h^{2n})$.

Solution 6.17. It can be easily seen that

$$u^{(k)}(x) = (-1)^k k! c\, a^k \frac{1}{(ax+b)^{k+1}} = (-1)^k k! \left(\frac{a}{c}\right)^k u^{k+1}(x), \quad k = 1, 2, \dots .$$

Furthermore, we have (provided that h is sufficiently small)

$$\begin{aligned}
& u_{\bar{x}x}(x) + \frac{2a}{c} u\, u_{\overset{\circ}{x}} - \left[u''(x) + \frac{2a}{c} u\, u'\right] \\
&= [u_{\bar{x}x}(x) - u''(x)] + \frac{2a}{c} u \left[u''(x) - u_{\overset{\circ}{x}}\right] \\
&= 2\sum_{k=2}^{\infty} \frac{u^{(2k)}(x)}{(2k)!} h^{2k-2} + \frac{2au}{c} \sum_{k=1}^{\infty} \frac{u^{(2k+1)}(x)}{(2k+1)!} h^{2k} \\
&= 2\sum_{k=2}^{\infty} \left(\frac{a}{c}\right)^{2k} h^{2k-2} u^{2k+1}(x) - \frac{2au}{c} \sum_{k=1}^{\infty} \left(\frac{a}{c}\right)^{2k+1} h^{2k} u^{2k+2}(x) \\
&= \frac{2a^4 h^2}{c^4} \frac{u^5(x)}{1 - \left(\dfrac{ahu(x)}{c}\right)^2} - \frac{2a^4 h^2}{c^4} \frac{u^5(x)}{1 - \left(\dfrac{ahu(x)}{c}\right)^2} = 0.
\end{aligned} \tag{6.93}$$

Thus the three-point difference scheme

$$\begin{aligned}
& u_{\bar{x}x}(x) + \frac{2a}{c} u(x)\, u_{\overset{\circ}{x}}(x) = 0, \quad x \in \omega_h, \\
& u(0) = \frac{c}{b}, \quad u(1) = \frac{c}{a+b},
\end{aligned} \tag{6.94}$$

is exact.

It is interesting to note that this difference scheme is exact on the exact solution but in general it has only a truncation error of second order if u belongs to the class $\mathbb{C}^4[0,1]$.

On the other hand we have

$$\begin{aligned}
& u_{\bar{x}x}(x) + \frac{2a}{c} u\, u_x - \left[u''(x) + \frac{2a}{c} u\, u'\right] \\
&= [u_{\bar{x}x}(x) - u''(x)] + \frac{2a}{c} u \left[u_x - u'(x)\right] \\
&= 2\sum_{k=2}^{\infty} \frac{u^{(2k)}(x)}{(2k)!} h^{2k-2} \\
&+ \frac{2au}{c} \sum_{k=2}^{\infty} \frac{u^{(k)}(x)}{k!} h^{k-1}
\end{aligned} \tag{6.95}$$

$$= 2\sum_{k=2}^{\infty}\left(\frac{a}{c}\right)^{2k} h^{2k-2}u^{2k+1}(x) + \frac{2au}{c}\sum_{k=2}^{\infty}(-1)^k\left(\frac{a}{c}\right)^k h^{k-1}u^{k+1}(x)$$

$$= \frac{2a^4h^2}{c^4}\frac{u^5(x)}{1-\left(\dfrac{a\,hu(x)}{c}\right)^2} + \frac{2a^3h}{c^3}\frac{u^3(x)}{1+\dfrac{a\,hu(x)}{c}}.$$

Therefore the three-point difference scheme

$$u_{\bar{x}x}(x) + \frac{2a}{c}u(x)\,u_x(x) - \left[\frac{2a^4h^2}{c^4}\frac{u^5(x)}{1-\left(\dfrac{a\,hu(x)}{c}\right)^2} + \frac{2a^3h}{c^3}\frac{u^3(x)}{1+\dfrac{a\,hu(x)}{c}}\right] = 0,$$

$$x \in \omega_h, \quad u(0) = \frac{c}{b}, \quad u(1) = \frac{c}{a+b} \tag{6.96}$$

is exact, too.

The difference scheme

$$y_{\bar{x}x}(x) + \frac{2a}{c}y(x)\,y_x(x) - 2\sum_{k=2}^{n}\left(\frac{a}{c}\right)^{2k} h^{2k-2}y^{2k+1}(x)$$

$$+ \frac{2au}{c}\sum_{k=2}^{n}(-1)^k\left(\frac{a}{c}\right)^k h^{k-1}u^{k+1}(x) = 0, \quad x \in \omega_h, \tag{6.97}$$

$$y(0) = \frac{c}{b}, \quad y(1) = \frac{c}{a+b},$$

or

$$y_{\bar{x}x}(x) + \frac{a}{c}y\,y_x - 2a^4h^2\frac{1-(ah)^{2n-2}}{1-(ah)^2}y(x)$$

$$- \frac{2a^3hu^3(x)}{c^3}\frac{1-\left(-\dfrac{a\,hu(x)}{c}\right)^{n+1}}{1-(a\,h)^2} = 0, \quad x \in \omega_h, \tag{6.98}$$

$$y(0) = \frac{c}{b}, \quad y(1) = \frac{c}{a+b},$$

possesses a truncation error of order $\mathcal{O}(h^{n+1})$.

Solution 6.18. It can easily be seen that

$$u^{(k)}(x) = (-1)^k ca^k m(m+1)\cdots(m+k-1)\frac{1}{(ax+b)^{k+m}}$$

$$= (-1)^k ca^k m(m+1)\cdots(m+k-1)\frac{1}{(ax+b)^k}u(x)$$

$$= (-1)^k c a^k (m)_k \frac{1}{(ax+b)^k} u(x),$$

where $(m)_k \stackrel{\text{def}}{=} m(m+1)\cdots(m+k-1)$ is Pochhammer's symbol. Further we have

$$u_{\bar{x}x}(x) + \frac{a^2 m(m+1)}{\sqrt[m]{c^2}} u^{\frac{m+2}{m}}(x) - \left[u''(x) + \frac{a^2 m(m+1)}{\sqrt[m]{c^2}} u^{\frac{m+2}{m}}(x) \right]$$

$$= [u_{\bar{x}x}(x) - u''(x)] = 2\sum_{k=2}^{\infty} \frac{u^{(2k)}(x)}{(2k)!} h^{2k-2} \tag{6.99}$$

$$= 2c \sum_{k=2}^{\infty} a^{2k} \frac{(m)_{2k}}{(2k)!} h^{2k-2} \frac{1}{(ax+b)^{2k}} u(x).$$

Thus the three-point difference scheme

$$u_{\bar{x}x}(x) = \frac{a^2 m(m+1)}{\sqrt[m]{c^2}} u^{\frac{m+2}{m}}(x) + 2c \sum_{k=2}^{\infty} a^{2k} \frac{(m)_{2k}}{(2k)!} h^{2k-2} \frac{1}{(ax+b)^{2k}} u(x),$$

$$x \in \omega_h, \quad u(0) = \frac{c}{b^m}, \quad u(1) = \frac{c}{(a+b)^m}, \tag{6.100}$$

is exact and the difference scheme

$$y_{\bar{x}x}(x) = \frac{a^2 m(m+1)}{\sqrt[m]{c^2}} y^{\frac{m+2}{m}}(x) + 2c \left[\sum_{k=2}^{n} a^{2k} h^{2k-2} \frac{(m)_{2k}}{(2k)!} \frac{1}{(ax+b)^{2k}} \right] y(x),$$

$$x \in \omega_h, \quad y(0) = \frac{c}{b^m}, \quad y(1) = \frac{c}{(a+b)^m}, \tag{6.101}$$

has a truncation error of order $\mathcal{O}(h^{2n})$.

Solution 6.19. It can easily be seen that the exact solution satisfies

$$u^{(k)}(x) = \frac{(-1)^k k! a^k}{(ax+b)^{k+1}} = (-1)^k k! a^k u^{k+1}(x). \tag{6.102}$$

Using Taylor's expansion and the formula for the sum of the infinite geometric progression we obtain

$$u_{\overline{x}\hat{x},j} - 2a^2u_j^3 - \left[u_j^{(2)} - 2a^2u_j^3\right] = \sum_{k=3}^{\infty}\frac{h_{j+1}^{k-1}u_j^{(k)}}{k!} - \sum_{k=3}^{\infty}\frac{(-1)^k h_j^{k-1}u_j^{(k)}}{k!}$$

$$\begin{aligned} &- \sum_{k=3}^{\infty}(-1)^k h_{j+1}^{k-1}u_j^{k+1} - \sum_{k=3}^{\infty}h_j^{k-1}u_j^{k+1} \\ &= -h_{j+1}^2a^3u_j^4(1 - h_{j+1}au_j + (h_{j+1}au_j)^2 + \cdots) \\ &\quad - h_j^2a^3u_j^4(1 + h_jau_j + (h_{j+1}au_j)^2 + \cdots) \\ &= -\frac{h_{j+1}^2a^3u_j^4}{1 + h_{j+1}au_j} - \frac{h_j^2a^3u_j^4}{1 - h_jau_j}, \end{aligned} \tag{6.103}$$

where $u_j \stackrel{\text{def}}{=} u(x_j)$, $x_j \in \hat{\omega}$, and $h_j \stackrel{\text{def}}{=} x_j - x_{j-1}$.

Thus the three-point difference scheme

$$\begin{aligned} &u_{\overline{x}\hat{x},j} = 2a^2u_j^3 - \frac{h_{j+1}^2a^3u_j^4}{1 + h_{j+1}au_j} - \frac{h_j^2a^3u_j^4}{1 - h_jau_j}, \quad j = 1, 2, \ldots, N-1, \\ &u_0 = \frac{1}{b}, \quad u_N = \frac{1}{a+b}, \end{aligned} \tag{6.104}$$

is exact and the TDS

$$\begin{aligned} &y_{\overline{x}\hat{x},j} = 2a^2y_j^3 + \sum_{k=3}^{n}(-1)^k h_{j+1}^{k-1}y_j^{k+1} - \sum_{k=3}^{n}h_j^{k-1}y_j^{k+1}, \quad j = 1, 2, \ldots, N-1, \\ &y_0 = \frac{1}{b}, \quad y_N = \frac{1}{a+b}, \end{aligned} \tag{6.105}$$

or

$$\begin{aligned} y_{\overline{x}\hat{x},j} = 2a^2y_j^3 &- \frac{h_{j+1}^2a^3u_j^4(1 + (h_{j+1}au_j)^{n+1})}{1 + h_{j+1}au_j} \\ &- \frac{h_j^2a^3u_j^4(1 - (h_jau_j)^{n+1})}{1 - h_jau_j}, \quad j = 1, 2, \ldots, N-1, \\ y_0 = \frac{1}{b}, &\quad y_N = \frac{1}{a+b}, \end{aligned} \tag{6.106}$$

possesses a truncation error of order $\mathcal{O}(h^n)$, with $h \stackrel{\text{def}}{=} \max_{j=1,\ldots,N} h_j$.

Solution 6.20. Since

$$u^{(k)} = \frac{(-1)^{k-1}(k-1)!a^{k-1}}{(ax+b)^{k-1}}u'(x) \tag{6.107}$$

we have

$$u_{\bar{x}x}(x) + u_{\overset{\circ}{x}}^2(x) - \left[u''(x) + (u'(x))^2\right]$$

$$= \left[u_{\bar{x}x}(x) - u''(x)\right] + \left[u_{\overset{\circ}{x}}(x) - u'(x)\right]\left[u_{\overset{\circ}{x}}(x) + u'(x)\right]$$

$$= 2\sum_{k=2}^{\infty}\frac{u^{(2k)}(x)}{(2k)!}h^{2k-2} + \left[\sum_{k=1}^{\infty}\frac{u^{(2k+1)}(x)}{(2k+1)!}h^{2k}\right]\left[\sum_{k=1}^{\infty}\frac{u^{(2k+1)}(x)}{(2k+1)!}h^{2k} + 2u'(x)\right]$$

$$= -2\sum_{k=2}^{\infty}\frac{a^{2k}}{(2k)(ax+b)^{2k}}h^{2k-2}u'(x)$$

$$+ (u'(x))^2\left[\sum_{k=1}^{\infty}\frac{a^{2k}}{(2k+1)(ax+b)^{2k}}h^{2k}\right]\left[\sum_{k=1}^{\infty}\frac{a^{2k}}{(2k+1)(ax+b)^{2k}}h^{2k} + 2\right]$$

$$= -\sum_{k=2}^{\infty}\frac{a^{2k}}{k(ax+b)^{2k}}h^{2k-2}$$

$$+ \left[\sum_{k=1}^{\infty}\frac{a^{2k+1}}{(2k+1)(ax+b)^{2k+1}}h^{2k}\right]\left[\sum_{k=1}^{\infty}\frac{a^{2k+1}}{(2k+1)(ax+b)^{2k+1}}h^{2k} + \frac{2a}{ax+b}\right].$$

Thus the difference scheme

$$u_{\bar{x}x}(x) = -u_{\overset{\circ}{x}}^2(x) - \sum_{k=2}^{\infty}\frac{a^{2k}}{k(ax+b)^{2k}}h^{2k-2}$$

$$+ \left[\sum_{k=1}^{\infty}\frac{a^{2k+1}}{(2k+1)(ax+b)^{2k+1}}h^{2k}\right]\left[\sum_{k=1}^{\infty}\frac{a^{2k+1}}{(2k+1)(ax+b)^{2k+1}}h^{2k} + \frac{2a}{ax+b}\right],$$

$$u(0) = \ln b, \quad u(1) = \ln(a+b), \quad x \in \omega_h, \tag{6.108}$$

is exact and the difference scheme

$$y_{\bar{x}x}(x) = -y_{\overset{\circ}{x}}^2(x) - \sum_{k=2}^{n}\frac{a^{2k}}{k(ax+b)^{2k}}h^{2k-2}$$

$$+ \left[\sum_{k=1}^{n}\frac{a^{2k+1}}{(2k+1)(ax+b)^{2k+1}}h^{2k}\right]\left[\sum_{k=1}^{n}\frac{a^{2k+1}}{(2k+1)(ax+b)^{2k+1}}h^{2k} + \frac{2a}{ax+b}\right],$$

$$y(0) = \ln b, \quad y(1) = \ln((a+b)), \quad x \in \omega_h, \tag{6.109}$$

possesses a truncation error of order $\mathcal{O}(h^{2n})$.

In order to be able to write these schemes more compactly let us consider the series

$$F'(q) \stackrel{\text{def}}{=} \sum_{k=1}^{\infty} q^{2k} \tag{6.110}$$

and

$$F(q) \stackrel{\text{def}}{=} \sum_{k=1}^{\infty} \frac{q^{2k+1}}{(2k+1)}, \tag{6.111}$$

with $q \stackrel{\text{def}}{=} \dfrac{ah}{ax+b}$. It can easily be seen that $F'(q) = \dfrac{q^2}{1-q^2}$. Therefore

$$\begin{aligned} F(q) &= \int_0^1 \frac{q^2}{1-q^2}\,dq = -\int_0^1 dq + \frac{1}{2}\int_0^1 \frac{dq}{1-q} + \frac{1}{2}\int_0^1 \frac{dq}{1+q} \\ &= -q + \frac{1}{2}\ln\left(\frac{1+q}{1-q}\right) = -\frac{ah}{ax+b} + \frac{1}{2}\ln\left(\frac{ax+b+ah}{ax+b-ah}\right). \end{aligned} \tag{6.112}$$

Analogously we obtain

$$\begin{aligned} F_1(q) &= \sum_{k=2}^{\infty} \frac{q^{2k}}{2k} = \int_0^q \left(\sum_{k=2}^{\infty} q^{2k-1}\right) dq = \int_0^q \frac{q^3 dq}{1-q^2} \\ &= -\frac{q^2}{2} - \frac{1}{2}\ln\left(1-q^2\right) = -\frac{a^2h^2}{2(ax+b)^2} - \frac{1}{2}\ln\left(\frac{(ax+b)^2 - a^2h^2}{(ax+b)^2}\right). \end{aligned} \tag{6.113}$$

Now, using the formulas (6.111) and (6.113) we can reformulate the EDS (6.108) in the compact form

$$\begin{aligned} &u_{\bar{x}x}(x) = -u_{\overset{\circ}{x}}^2(x) + \frac{1}{2}\ln\left(\frac{(ax+b)^2 - a^2h^2}{(ax+b)^2}\right) + \frac{1}{2}\ln\left(\frac{ax+b+ah}{ax+b-ah}\right), \\ &u(0) = \ln b, \quad u(1) = \ln(a+b), \quad x \in \omega_h. \end{aligned} \tag{6.114}$$

Solution 6.21. Use the Taylor expansion.

Solution 6.22. Using Solution 6.9 check that the difference scheme

$$\begin{aligned} &u_{\bar{x}x}(x) + \frac{a}{c}u(x)u_{\overset{\circ}{x}}(x) = 0, \quad x \in \omega, \\ &u_x(0) + \frac{a}{b}u(0) - \frac{ca^2h}{b^2(ah+b)} = 0, \quad u(1) = \frac{c}{a+b}, \end{aligned} \tag{6.115}$$

is exact.

The left boundary condition can be written in the form

$$\begin{aligned} &u_x(0) + \frac{a}{b}u(0) - \frac{ca^2h}{b^2(ah+b)} \\ &= u_x(0) + \frac{a}{b}u(0) - \frac{ca^2h}{b^3}\sum_{k=0}^{\infty}\left(-\frac{ah}{b}\right)^k = 0. \end{aligned} \tag{6.116}$$

The TDS

$$\begin{aligned} &y_{\bar{x}x}(x) + \frac{a}{c} y(x) y_{\overset{\circ}{x}}(x) = 0, \quad x \in \omega, \\ &y_x(0) + \frac{a}{b} y(0) - \frac{ca^2 h}{b^3} \sum_{k=0}^{n} \left(-\frac{ah}{b}\right)^k = 0, \quad y(1) = \frac{c}{a+b} \end{aligned} \tag{6.117}$$

or

$$\begin{aligned} &y_{\bar{x}x}(x) + \frac{a}{c} y(x) y_{\overset{\circ}{x}}(x) = 0, \quad x \in \omega, \\ &y_x(0) + \frac{a}{b} y(0) - \frac{ca^2 h}{b^3} \frac{1 - \left(-\frac{ah}{b}\right)^{n+1}}{1 + \frac{ah}{b}} = 0, \quad y(1) = \frac{c}{a+b} \end{aligned} \tag{6.118}$$

has a truncation error of order $\mathcal{O}(h^{n+2})$.

Solution 6.23. We have

$$\begin{aligned} &u_{\bar{x},x}(x) - u'' + f(x,u)\left[u_{\overset{\circ}{x}}(x) - u'(x)\right] \\ &= 2\sum_{k=2}^{\infty} \frac{u^{(2k)}(x)}{(2k)!} h^{2k-2} + 2f(x,u) \sum_{k=1}^{\infty} \frac{u^{(2k+1)}(x)}{(2k+1)!} h^{2k} \\ &= 2\sum_{k=2}^{\infty} \frac{\varphi_{2k}(x)}{(2k)!} h^{2k-2} u(x) + 2f(x,u) \sum_{k=1}^{\infty} \frac{\varphi_{(2k+1)}(x)}{(2k+1)!} h^{2k} u(x). \end{aligned} \tag{6.119}$$

Thus the three-point difference scheme

$$\begin{aligned} &u_{\bar{x},x}(x) = -u_{\overset{\circ}{x}} f(x,u) \\ &+ \left[2\sum_{k=2}^{\infty} \frac{\varphi_{2k}(x)}{(2k)!} h^{2k-2} + 2f(x,u) \sum_{k=1}^{\infty} \frac{\varphi_{(2k+1)}(x)}{(2k+1)!} h^{2k}\right] u(x), \quad x \in \omega, \\ &u(0) = \gamma_0, \quad u(1) = \gamma_1, \end{aligned} \tag{6.120}$$

is exact.

The truncation error of the following TDS is of order $\mathcal{O}(h^{2n+1})$ for an arbitrary n:

$$\begin{aligned} &y_{\bar{x},x}(x) = -y_{\overset{\circ}{x}} f(x, y(x)) \\ &+ \left[2\sum_{k=2}^{n} \frac{\varphi_{2k}(x)}{(2k)!} h^{2k-2} + 2f(x,u) \sum_{k=1}^{n} \frac{\varphi_{(2k+1)}(x)}{(2k+1)!} h^{2k}\right] u(x), \quad x \in \omega, \\ &y(0) = \gamma_0, \quad y(1) = \gamma_1. \end{aligned} \tag{6.121}$$

Solution 6.24. Use the fact that $u^{(k)}(x) = \varphi_k(x)u(x)$ with

$$\varphi_k(x) = \begin{cases} n(n-1)\cdots(n-k+1)x^{-k}, & \text{if } k \le n, \\ 0, & \text{if } k > n \end{cases}$$

and the results of the previous exercise.

Solution 6.25. Let the grid $\hat{\bar{\omega}}_h$ be given. There exists the EDS

$$u_{\bar{x}\hat{x},j} = -\hbar_j^{-1} \sum_{\alpha=1}^{2} (-1)^\alpha \left[Z_\alpha^j(x_j, u) + (-1)^\alpha \frac{Y_\alpha^j(x_j, u) - u_\beta}{V_\alpha^j(x_j)} \right],$$

$$u_0 = \mu_1, \quad u_N = \mu_2, \quad j = 1, 2, \dots, N-1,$$

where $Y_\alpha^j(x, u)$, $Z_\alpha^j(x, u)$, $\alpha = 1, 2$, are the solutions of the IVPs

$$\frac{dY_\alpha^j(x,u)}{dx} = Z_\alpha^j(x,u), \quad \frac{dZ_\alpha^j(x,u)}{dx} = \frac{1}{\left[Y_\alpha^j(x,u)\right]^3}, \quad x \in e_\alpha^j,$$

$$Y_\alpha^j(x_\beta, u) = u_\beta, \quad Z_\alpha^j(x_\beta, u) = k(x)\left.\frac{du}{dx}\right|_{x=x_\beta}, \tag{6.122}$$

$$j = 2-\alpha,\ 3-\alpha, \dots, N+1-\alpha, \quad \alpha = 1, 2.$$

These IVPs possess the exact solutions

$$Y_\alpha^j(x,u) = \sqrt{\frac{(C_1 x + C_2)^2 + 1}{C_1}}, \quad Z_\alpha^j(x,u) = (C_1 x + C_2)\sqrt{\frac{C_1}{(C_1 x + C_2)^2 + 1}},$$

$$C_1 = \left[\frac{du}{dx}\right]^2_{x=x_\beta} + \frac{1}{u_\beta^2}, \quad C_2 = u_\beta \left[\frac{du}{dx}\right]_{x=x_\beta} - x_\beta C_1.$$

Therefore the EDS is of the form

$$u_{\bar{x}\hat{x},j} = -\hbar_j^{-1} \sum_{\alpha=1}^{2} (-1)^\alpha \left[(C_1 x_j + C_2)\sqrt{\frac{C_1}{(C_1 x_j + C_2)^2 + 1}} \right.$$

$$\left. + \frac{(-1)^\alpha}{h_\gamma} \left(\sqrt{\frac{(C_1 x_j + C_2)^2 + 1}{C_1}} - u_\beta \right) \right],$$

$$u_0 = \mu_1, \quad u_N = \mu_2, \quad j = 1, 2, \dots, N-1,$$

Solution 6.26. In accordance with our theory we obtain the 6-EDS

$$y^{(6)}_{\bar{x}x,j} = -\varphi^{(6)}(x_j, y^{(6)}), \quad j = 1, 2, \dots, N-1, \quad y_0^{(6)} = \mu_1, \quad y_N^{(6)} = \mu_2,$$

$$\varphi^{(6)}(x_j, u) = h^{-1} \sum_{\alpha=1}^{2} (-1)^\alpha \left[Z_\alpha^{(6)j}(x_j, u) + (-1)^\alpha \frac{Y_\alpha^{(6)j}(x_j, u) - u_\beta}{h} \right], \tag{6.123}$$

where $Y_\alpha^{(6)j}(x_j,u)$, $Z_\alpha^{(6)j}(x_j,u)$ is the numerical solution of the IVP (6.122), obtained by the Runge-Kutta-Nystrom method of order 6 (see [14, 31])

$$Y_\alpha^{(6)j}(x_j,u) = u_\beta + (-1)^{\alpha+1}hu'_\beta + h^2\bigg(\frac{151}{2142}k_1 + \frac{5}{116}k_2 + \frac{385}{1368}k_3$$
$$+ \frac{55}{168}k_4 - \frac{6250}{28101}k_5\bigg),$$
$$Z_\alpha^{(6)j}(x_j,u) = u'_\beta + (-1)^{\alpha+1}h\bigg(\frac{151}{2142}k_1 + \frac{25}{522}k_2 + \frac{275}{684}k_3 + \frac{275}{252}k_4$$
$$- \frac{78125}{112404}k_5 + \frac{1}{12}k_6\bigg),$$
$$k_1 = -f(x_\beta, u_\beta),$$
$$k_2 = -f\big(x_\beta + 0.1\,(-1)^{\alpha+1}h, u_\beta + 0.1\,(-1)^{\alpha+1}hu'_\beta + 0.005h^2k_1\big),$$
$$k_3 = -f\big(x_\beta + 0.3\,(-1)^{\alpha+1}h, u_\beta + 0.3\,(-1)^{\alpha+1}hu'_\beta - \frac{1}{2200}h^2k_1 + \frac{1}{22}h^2k_2\big),$$
$$k_4 = -f\big(x_\beta + 0.7\,(-1)^{\alpha+1}h, u_\beta + 0.7\,(-1)^{\alpha+1}hu'_\beta + \frac{637}{6600}h^2k_1$$
$$- \frac{7}{110}h^2k_2 + \frac{7}{33}h^2k_3\big),$$
$$k_5 = -f\big(x_\beta + \frac{17}{25}(-1)^{\alpha+1}h, u_\beta + \frac{17}{25}(-1)^{\alpha+1}hu'_\beta + \frac{225437}{1968750}h^2k_1 - \frac{30073}{281250}h^2k_2$$
$$+ \frac{65569}{281250}h^2k_3 - \frac{9367}{984375}h^2k_4\big),$$
$$k_6 = -f\big(x_\beta + (-1)^{\alpha+1}h, u_\beta + (-1)^{\alpha+1}hu'_\beta + \frac{151}{2142}h^2k_1 + \frac{5}{116}h^2k_2 + \frac{385}{1368}h^2k_3$$
$$+ \frac{55}{168}h^2k_4 - \frac{6250}{28101}h^2k_5\big),$$
$$f(x,u) = -\frac{1}{u_3}, \quad \mu_1 = 0, \quad \mu_2 = -\ln\cos^2\left(\frac{1}{\sqrt{2}}\right).$$

Solution 6.27. We have

$$\begin{aligned} u_{x^k} &= \frac{(-1)^k k! a b^k}{(bx+cy+d)^{k+1}} = \frac{(-1)^k k! b^k}{a^k} u^{k+1}, \\ u_{y^k} &= \frac{(-1)^k k! a c^k}{(bx+cy+d)^{k+1}} = \frac{(-1)^k k! c^k}{a^k} u^{k+1}, \quad k = 1, 2, \dots, \end{aligned} \tag{6.124}$$

and for $(x,y) \in \omega_{h_1,h_2}$ it holds that

$$\begin{aligned}
u_{\overline{x},x}(x,y) &= \frac{u(x+h_1,y)-2u(x,y)+u(x-h_1,y)}{h_1^2} \\
&= \frac{\partial^2 u(x,y)}{\partial x^2} + 2\sum_{k=2}^{\infty}\frac{1}{(2k)!}\frac{\partial^{2k}u(x,y)}{\partial x^{2k}}h_1^{2k-2} \\
&= \frac{\partial^2 u(x,y)}{\partial x^2} + 2\sum_{k=2}^{\infty} h_1^{2k-2}\left(\frac{b}{a}\right)^{2k} u^{2k+1} \\
&= \frac{\partial^2 u(x,y)}{\partial x^2} + 2h_1^2\left(\frac{b}{a}\right)^4 u^5 \frac{1}{1-h_1^2(b/a)^2u^2}, \\
u_{\overline{y},y}(x,y) &= \frac{u(x,y+h_2)-2u(x,y)+u(x,y-h_2)}{h_2^2} \\
&= \frac{\partial^2 u(x,y)}{\partial y^2} + 2\sum_{k=2}^{\infty}\frac{1}{(2k)!}\frac{\partial^{2k}u(x,y)}{\partial y^{2k}}h_2^{2k-2} \\
&= \frac{\partial^2 u(x,y)}{\partial y^2} + 2\sum_{k=2}^{\infty} h_2^{2k-2}\left(\frac{c}{a}\right)^{2k} u^{2k+1} \\
&= \frac{\partial^2 u(x,y)}{\partial y^2} + 2h_2^2\left(\frac{c}{a}\right)^4 u^5 \frac{1}{1-h_2^2(c/a)^2u^2}.
\end{aligned} \tag{6.125}$$

Thus the difference scheme

$$\begin{aligned}
&u_{\overline{x},x}(x,y) + u_{\overline{y},y}(x,y) \\
&= 2\left[h_1^2\left(\frac{b}{a}\right)^4 \frac{1}{1-h_1^2(b/a)^2u^2} + h_2^2\left(\frac{c}{a}\right)^4 \frac{1}{1-h_2^2(c/a)^2u^2}\right]u^5(x,y) \\
&\quad + 2\frac{b^2+c^2}{a^2}u^3(x,y), \quad (x,y)\in\omega_{h_1,h_2}, \\
&u(0,y) = \frac{a}{cy+d}, \quad u(1,y) = \frac{a}{b+cy+d}, \\
&u(x,0) = \frac{a}{bx+d}, \quad u(x,1) = \frac{a}{bx+c+d}
\end{aligned}$$

is exact. The TDS

$$\begin{aligned}
u_{\overline{x},x}(x,y) + u_{\overline{y},y}(x,y) &= 2\sum_{k=2}^{N_x} h_1^{2k-2}\left(\frac{b}{a}\right)^{2k} u^{2k+1} + 2\sum_{k=2}^{N_y} h_2^{2k-2}\left(\frac{c}{a}\right)^{2k} u^{2k+1} \\
&+ 2\frac{b^2+c^2}{a^2}u^3(x,y), \quad (x,y)\in\omega_{h_1,h_2},
\end{aligned}$$

$$u(0,y) = \frac{a}{cy+d}, \quad u(1,y) = \frac{a}{b+cy+d},$$

$$u(x,0) = \frac{a}{bx+d}, \quad u(x,1) = \frac{a}{bx+c+d},$$

or

$$u_{\overline{x},x}(x,y) + u_{\overline{y},y}(x,y) = 2h_1^2(b/a)^4 u^5(x,y)\frac{1-(hbu/a)^{2N_x-3}}{1-(h_1bu/a)^2}$$

$$+ 2h_2^2(c/a)^4 u^5(x,y)\frac{1-(hcu/a)^{2N_y-3}}{1-(h_2cu/a)^2}$$

$$+ 2\frac{b^2+c^2}{a^2}u^3(x,y), \quad (x,y) \in \omega_{h_1,h_2},$$

$$u(0,y) = \frac{a}{cy+d}, \quad u(1,y) = \frac{a}{b+cy+d},$$

$$u(x,0) = \frac{a}{bx+d}, \quad u(x,1) = \frac{a}{bx+c+d},$$

has a truncation error of order $\mathcal{O}(h_1^{2N_x-1} + h_2^{2N_y-1})$.

Solution 6.28. This problem can be solved analogously to the solution of the previous exercise.

Solution 6.29. Use the Taylor expansion for the finite difference and the fact that

$$u^{(k)}(x) = F_k(u(x)), \quad k = 1, 2, \ldots,$$

where e.g.,

$$F_0(u) = f(u), \quad F_1(u) = f'(u)\,f(u), \quad \text{etc.}$$

Solution 6.30. Use the finite difference $u_{\overline{x}x}$ to approximate the second derivative and the fact that the Taylor expansion of $u_{\overline{x}x} - u''$ contains only derivatives of even order which can be expressed through the derivatives of the right-hand side $f(u)$.

Solution 6.31. This problem can be solved analogously to the solution of the previous exercise.

Solution 6.32. We transform problem (6.53) into

$$\begin{aligned} &u(x_{i+1}) - 2\cos(h\sqrt{\lambda})u(x_i) + u(x_{i-1}) = 0, \\ &u(0) = 0, \quad u(1) = 0, \quad i = 1, \ldots, N. \end{aligned} \tag{6.126}$$

This is the well-known recurrence formula for the orthogonal Chebyshev polynomials of first and second kind,

$$T_i(z) = \cos(i \arccos(z)), \quad U_i(z) = \frac{\sin((i+1)\arccos z)}{\sqrt{1-z^2}},$$

where $z \stackrel{\text{def}}{=} \cos(h\sqrt{\lambda})$. The left boundary condition in (6.126) yields

$$u(x_i) = a\,U_i(h\sqrt{\lambda}) = a\frac{\sin(ih\sqrt{\lambda})}{\sin(h\sqrt{\lambda})},$$

where a is an arbitrary constant. We choose this constant subject to the condition

$$\frac{a}{\sin(h\sqrt{\lambda})} = 1$$

and obtain

$$u(x_i) = \sin(ih\sqrt{\lambda}), \quad i = 0, 1, \ldots, N+1.$$

The right boundary condition is satisfied provided that λ satisfies

$$u(1) = u(x_{N+1}) = \sin(\sqrt{\lambda}) = 0,$$

i.e.,

$$\lambda = \lambda_n = (n\pi)^2, \quad u_n(x_i) = \sin(n\pi x_i) = \sin(in\pi/(N+1)),$$
$$i = 0, \ldots, N+1, \quad n = 1, \ldots, N.$$

Solution 6.33. The difference equation (6.54) in indexed form reads

$$y_{i+1} - (2 - h^2\lambda^h)y_i + y_{i-1}y_i = 0, \quad i = 0, 1, \ldots, N+1, \tag{6.127}$$

and the solution satisfying the left boundary condition is (compare with the previous exercise)

$$y(x_i) = \sin\big(i\arccos(1 - \frac{\lambda^h h^2}{2})\big), \quad i = 0, 1, \ldots, N+1.$$

The right boundary condition is fulfilled provided that λ^h satisfies

$$y(1) = y(x_{N+1}) = \sin\big((N+1)\arccos(1 - \frac{\lambda^h h^2}{2})\big) = 0.$$

This implies

$$\arccos(1 - \frac{\lambda^h h^2}{2}) = \frac{k\pi}{N+1}, \quad k = 1, \ldots, N$$

or

$$1 - \frac{\lambda^h h^2}{2} = \cos\left(\frac{k\pi}{N+1}\right).$$

Thus, we have

$$\lambda_k^h = 2\left(1 - \cos\left(\frac{k\pi}{N+1}\right)\right)(N+1)^2 = \left[\frac{\sin\left(\frac{k\pi}{2(N+1)}\right)}{\frac{k\pi}{2(N+1)}}\right]^2 (k\pi)^2.$$

For a fixed k independent of N we have from the Taylor expansion

$$\lambda_k^h = \frac{\sin\left(\frac{k\pi}{2(N+1)}\right)}{\frac{k\pi}{2(N+1)}}(k\pi)^2 = (k\pi)^2\left[1 - \frac{1}{3!}\left(\frac{k\pi}{N+1}\right)^2 + \cdots\right]$$
$$= (k\pi)^2 + \mathcal{O}(h^2).$$

Note that the first eigenfunction $y_1(x) = \sin(\pi x)$, $x \in \overline{\omega}$ coincides with the projection of the first exact eigenfunction $u_1(x) = \sin(\pi x)$ onto the grid.

Solution 6.34. The EDS for (6.55) is

$$\begin{aligned}
&u_{\overline{x}x}(x_i) + \lambda_n\left[\frac{\sin\frac{i\sqrt{\lambda_n}}{2(N+1)}}{\frac{i\sqrt{\lambda_n}}{2(N+1)}}\right]u(x_i) = 0, \quad i = 1, 2, \ldots, N,\\
&u_{\overline{x}x}(x_i) + \lambda_n\left[\frac{\sin\frac{i\sqrt{\lambda_n - 1}}{2(N+1)}}{\frac{i\sqrt{\lambda_n - 1}}{2(N+1)}}\right]u(x_i) = 0, \quad i = N+2, \ldots, 2N+1,\\
&u(1/2) = A\sin(\sqrt{\lambda_n}h) + u(1/2 - h)\cos(\sqrt{\lambda_n}h),
\end{aligned} \tag{6.128}$$

where

$$\begin{aligned}
A &\stackrel{\text{def}}{=} -\frac{\sqrt{(\lambda_n - 1)\lambda_n}}{\Delta_h}\Big\{u(1/2 - h)\Big[\sqrt{\lambda_n - 1}\cos(h\sqrt{\lambda_n})\cos(h\sqrt{\lambda_n - 1})\\
&\quad - \sqrt{\lambda_n - 1}\sin(h\sqrt{\lambda_n})\sin(\sqrt{\lambda_n - 1}h)\Big] - u(1/2 + h)\sqrt{\lambda_n - 1}\Big\},\\
\Delta_h &\stackrel{\text{def}}{=} \sqrt{\lambda_n}\Big[(\lambda_n - 1)\sin(h\sqrt{\lambda_n})\cos(h\sqrt{\lambda_n - 1})\\
&\quad + \sqrt{\lambda_n(\lambda_n - 1)}\cos(h\sqrt{\lambda_n})\sin(h\sqrt{\lambda_n - 1})\Big]
\end{aligned} \tag{6.129}$$

and λ_n is the n-th root of the equation

$$f(\lambda) \stackrel{\text{def}}{=} \sqrt{\lambda - 1}\sin\frac{\sqrt{\lambda}}{2}\cos\frac{\sqrt{\lambda - 1}}{2} + \sqrt{\lambda}\cos\frac{\sqrt{\lambda}}{2}\sin\frac{\sqrt{\lambda - 1}}{2} = 0, \tag{6.130}$$

where the roots are ordered as follows:

$$0 < \lambda_1 < \lambda_2 < \cdots \lambda_n < \cdots . \tag{6.131}$$

The exact solution of (6.55) is

$$u_n(x) = \begin{cases} \sin\left(\sqrt{\lambda_n}x\right), & 0 \le x \le \dfrac{1}{2}, \\ \dfrac{\sin\left(\sqrt{\lambda_n - 1}(1-x)\right)}{\sin\left(\sqrt{\lambda_n - 1}/2\right)} \sin\left(\sqrt{\lambda_n}/2\right), & \dfrac{1}{2} < x \le 1, \end{cases} \tag{6.132}$$

where λ_n is the n-th root of equation (6.130). This can be easily proved by the substitution of (6.132) into equation (6.55) for $x \in (0, 1/2)$ and $x \in (1/2, 1)$ (compare with Exercise 6.32) and into the continuity conditions

$$\begin{aligned} [u_n(x)]_{x=1/2} &= u_n(1/2+0) - u_n(1/2-0) = 0, \\ [u_n'(x)]_{x=1/2} &= u_n'(1/2+0) - u_n'(1/2-0) = 0. \end{aligned} \tag{6.133}$$

That fact that (6.132) satisfies the equation (6.128) can be checked by substitution of (6.132) into (6.128), elementary transformations and taking into account that λ_n is a root of equation (6.130). Note that the first eigenvalues are

$$\begin{aligned} \lambda_1 &= 10.36327300443007\ldots, \\ \lambda_2 &= 39.98316575543942\ldots, \\ \lambda_3 &= 89.32573613877946\ldots. \end{aligned} \tag{6.134}$$

Solution 6.35 We consider problem (6.59) on a grid

$$\omega_h = \{x_i = ih : i = -N, \ldots, N\}, \tag{6.135}$$

where h must be determined. The truncation error of the difference operator for functions $v \in C^\infty(a, b)$ is

$$\Psi(x) = L_h v(x) - Lv(x) = 2\sum_{k=1}^{\infty} h^{2k} L^{(k)} v,$$

where

$$L^{(k)} v = \frac{A(x)}{(2k+2)!} v^{(2k+2)} + \frac{B(x)(k+1)}{(2k+2)!} v^{(2k+1)}.$$

We look for the solution of the eigenvalue problem (6.57) as the projection of the function

$$v(x, h) = \sum_{k=0}^{\infty} h^{2k} v_k(x)$$

on the grid (6.135). Substituting this expression into (6.57) and comparing the coefficients in front of the powers of h, we obtain the following system of equations for $v_k(x)$:

$$Lv_k + \lambda^h v_k = -2\sum_{l=1}^{k} L^{(l)} v_{k-l}, \quad k = 0, 1, \ldots. \tag{6.136}$$

Note, the right-hand side is equal to zero for $k = 0$. The solution of (6.136) with $k = 0$ is a polynomial $v_0(x) = P_n(x)$ of degree n. For $v_1(x)$ we have

$$\begin{aligned} &Lv_1(x) + \lambda^h v_1(x) = -2L^{(1)}v_0(x),\\ &|v_1(-1)| \neq \infty, \quad |v_1(1)| \neq \infty, \end{aligned} \tag{6.137}$$

where the parameter λ^h is the same as for $v_0(x) = P_n(x)$, namely

$$\lambda^h = \lambda = -a_2 n(n + \frac{b_1(b-a)}{2a_2} - 1).$$

Since $v_0(x)$ is a polynomial of degree n, the right-hand side of (6.137) is a polynomial of degree $n-3$. This inhomogeneous problem is solvable iff the right-hand side is orthogonal to the solution $P_n(x)$ of the homogeneous equation. Here, the orthogonality is defined with a weight function, as is usual for the Jacobi polynomials. This condition is fulfilled since $P_n(x)$ is orthogonal to all polynomials of degree less than or equal to $n-1$. Therefore this problem has a polynomial $v_1(x) = P_{n-3}(x)$ of degree $n-3$ as a particular solution which is orthogonal to $P_n(x)$. If we continue this process, then there exists a j_0 such that $v_j(x) = 0$ for all $j > j_0$. More precisely, the equation (in fact it is a system of $2N-1$ equations for the $2N+1$ unknowns $y_N, y_{-N+1}, \dots, y_N$)

$$L_h y(x) + \lambda^h\, y(x) = 0, \quad x \in \omega_h \tag{6.138}$$

possesses a solution of the form

$$y(x) = v(x,h) = v_0(x) + h^2 v_1(x) + \cdots + h^{2j_0} v_{j_0}(x), \quad x \in \omega_h.$$

To bring the number of equations of the system of linear algebraic equations (6.138) into agreement with the number of unknowns, we demand that the coefficients in front of y_{-N} and y_N vanish, i.e.,

$$\frac{1-(N-1)^2h^2}{h^2} + \frac{1}{2a_2 h}\left[\frac{1}{2}b_1(b-a)(N-1)h + \frac{1}{2}b_1(b+a) + b_0\right] = 0,$$

$$\frac{1-(N-1)^2h^2}{h^2} - \frac{1}{2a_2 h}\left[\frac{1}{2}b_1(b-a)[-(N-1)h] + \frac{1}{2}b_1(b+a) + b_0\right] = 0.$$

One can easily see that these equations coincide provided that $\frac{1}{2}b_1(b+a) + b_0 = 0$ and we have

$$h = \sqrt{\frac{4a_2}{4a_2(N-1)^2 + b_1(b-a)(N-1)}}. \tag{6.139}$$

Thus, we have constructed the finite difference scheme (6.57) on the equidistant grid (6.135) with mesh size (6.139) and we have shown that its eigenvalues coincide with the first $2N-1$ exact eigenvalues

$$\lambda_n^h = \lambda = -a_2 n(n + \frac{b_1(b-a)}{2a_2} - 1), \quad n = 0, 1, \dots, 2N-2.$$

The difference eigenfunctions approximate the exact ones with accuracy $\mathcal{O}(h^2)$.

Solution 6.36. After the change of variables $y = (z+1)/2$ the operator L_y will be transformed into the Legendre differential operator which analogously as above can be approximated by a difference operator with the exact spectrum on an equidistant grid with the mesh size $h = 1/\sqrt{N(N+1)}$.

Solution 6.37. Using the Legendre functions $P_\nu(x)$ and $Q_\nu(x)$ with the parameter $\nu = -\dfrac{1}{2} + \sqrt{\dfrac{1}{4} + \lambda}$ we can write

$$v_1^i(x) = v_1^i(x;\lambda) = (1 - x_{i-1}^2)[-P_\nu(x)Q_\nu(x_{i-1}) + Q_\nu(x)P_\nu(x_{i-1})],$$

$$v_2^i(x) = v_2^i(x;\lambda) = (1 - x_{i+1}^2)[-P_\nu(x)Q_\nu(x_{i+1}) + Q_\nu(x)P_\nu(x_{i+1})].$$

The exact 3-point relation for the solution of the spectral problem (without the boundary conditions) is

$$u(x_i) = \frac{v_1^i(x_i;\lambda)}{v_1^i(x_{i+1};\lambda)} u(x_{i+1}) + \frac{v_2^i(x_i;\lambda)}{v_2^i(x_{i-1};\lambda)} u(x_{i-1}), \qquad (6.140)$$

$$i = -N+1, \dots, N-1.$$

Next, in order to obtain the exact boundary conditions and to add two more equations to the system (6.140), let us determine the solution of the differential equation (6.60) on the interval $[x_{N-1}, 1)$. Taking into account the second boundary condition in (6.60) we get

$$u(x) = \frac{P_\nu(x)}{P_\nu(x_{N-1})} u(x_{N-1}),$$

from which we obtain the exact boundary condition

$$u(x_N) = \frac{P_\nu(x_N)}{P_\nu(x_{N-1})} u(x_{N-1}). \qquad (6.141)$$

Analogously we obtain the other exact boundary condition

$$u(x_{-N}) = \frac{P_\nu(x_{-N})}{P_\nu(x_{-N+1})} u(x_{-N+1}). \qquad (6.142)$$

The exact spectral 3-point difference eigenvalue problem (6.140), (6.141) and (6.142) possesses the exact spectrum

$$\lambda_n = n(n+1), \quad u_n(x;\lambda_n) = P_n(x), \quad x \in \omega_h, \quad n = 0, 1, \dots, 2N+1,$$

where $P_n(x)$ are the Legendre polynomials.

Solution 6.38. Difference equation (6.65) can be rewritten in the equivalent indexed form

$$(N-i)y_{i+1} - 2Ny_i + (N+i)y_{i-1} + \lambda^h y_i = 0, \quad i = -N, \dots, N,$$

$$y_{-N-1} \neq \infty, \quad y_{N+1} \neq \infty.$$

Comparing this equation with the difference equation for Kravchuk's polynomials with $p = 1/2$, $q = 1/2$ (see e.g. [6]) and with the explicit representation

$$k_n(x) = \frac{(-1)^n x!(2N-x)!}{n!2^n}\Delta^n\left(\frac{1}{(x-n)!(2N-x)!}\right), \quad x = 0, \dots, 2N,$$

$$n = 0, 1, \dots, \quad \Delta f(x) = f(x+1) - f(x) \quad \text{for all } f(x),$$

one can see that the exact difference eigenfunctions are

$$y_i = y_n(x_i) = k_n(i+N), \quad i = -N, \dots, N.$$

On the relationship between Kravchuk's polynomials and Hermite polynomials see e.g. [6]. The solution is completely analogous to the solution of Exercise 6.35.

Index

Bibliography

[1] R. P. Agarwal and D. O'Regan. *Infinite Interval Problems for Differential, Difference and Integral Equations.* Kluwer Academic Publishers, 2001.

[2] U. M. Ascher, R. M. M. Mattheij, and R. D. Russell. *Numerical Solution of Boundary Value Problems for Ordinary Differential Equations.* Prentice-Hall Series in Computational Mathematics. Prentice-Hall, Englewood Cliffs, 1988.

[3] A. Ashyralyev and P.E. Sobolevskii. *New Difference Schemes for Partial Differential Equations.* Birkhauser Verlag, Basel-Boston-Berlin, 2006.

[4] W. Auzinger, O. Koch, J. Petrickovic, and E. Weinmüller. Numerical solution of boundary value problems with an essential singularity. Technical Report 3/03, TU Wien, Institute for Applied Mathematics, 2003.

[5] W. Auzinger, O. Koch, D. Praetorius, and E. Weinmüller. New aposteriori error estimates for singular boundary value problems. *Numerical Algorithms*, 40:79–100, 2005.

[6] G. Bateman and A. Erday. *Higher Transcendental Functions [Russian translation]*, volume Vol. 2. Nauka, Moscow, 1974.

[7] J. C. Butcher. On Runge-Kutta processes of high order. *Journal Austral. Math. Soc.*, 4:179–194, 1964.

[8] J. C. Butcher. *Numerical Methods for Ordinary Differential Equations.* John Wiley & Sons, Ltd., Chichester, West Sussex, 2003.

[9] C. Carathéodory. *Vorlesungen über reelle Funktionen.* Dover (reprint), 1948.

[10] E. A. Coddington and N. Levinson. *Theory of Ordinary Differential Equations.* McGraw-Hill, New York, 1955.

[11] L. Collatz. *The Numerical Treatment of Differential Equations.* Springer Verlag, Berlin, 1960.

[12] S. D. Conte and C. deBoor. *Elementary Numerical Analysis.* McGraw-Hill, New York, 1972.

[13] E. J. Doedel. The constructions to ordinary differential equations. *SIAM J. Numer. Analysis*, 15:450–465, 1978.

[14] J. R. Dormand, E. A. El-Mikkawy, and P. J. Prince. Families of Runge-Kutta-Nystrom formulae. *IMA J. of Numer. Analysis*, 7:235–250, 1987.

[15] Yu. Eidelman, V. Milman, and A. Tsolomitis. *Functional Analysis - An Introduction.* American Mathematical Society, Providence, Rhode Island, 2004.

[16] R. Fazio. A novel approach to the numerical solution of boundary value problems on infinite intervals. *SIAM J. Numer. Anal.*, 33:1473–1483, 1996.

[17] L. Fox. *Numerical Solution of Two-Point Boundary Value Problems in Ordinary Differential Equations.* Clarendon Press, Oxford, 1957.

[18] S. Fučik and A. Kufner. *Nonlinear Differential Equations.* Elsevier, Amsterdam, Oxford, New York, 1980.

[19] E. C. Jr. Gartland. Strong stability of compact discrete boundary value problems via exact discretizations. *SIAM J. Numer. Analysis*, 25:111–123, 1988.

[20] I. P. Gavrilyuk. *Grid schemes with exact and explicit spectra (in Russian).* PhD thesis, Taras Shevchenko National University of Kiev, 1977.

[21] I. P. Gavrilyuk. Algorithm for solving a class of one-dimensional variational inequalities. *J. Sov. Math.*, 66:2250–2255, 1993.

[22] I. P. Gavrilyuk. A class of one-dimensional variational inequalities in difference schemes of arbitrary given degree of accuracy. *Z. Anal. Anwend.*, 12:751–758, 1993.

[23] I. P. Gavrilyuk. Exact difference schemes and difference schemes of arbitrary given degree of accuracy for generalized one-dimensional third boundary value problems. *Z. Anal. Anwend.*, 12:549–566, 1993.

[24] I. P. Gavrilyuk, M. Hermann, M. Kutniv, and I. L. Makarov. Two-point difference schemes of an arbitrary given order of accuracy for nonlinear BVPs. *Applied Mathematics Letters*, 23(5):585–590, 2010.

[25] I. P. Gavrilyuk, M. Hermann, M. V. Kutniv, and V.L. Makarov. New methods for nonlinear BVPs on the half-axis using Runge-Kutta IVP-solvers. Technical Report 05-18, Friedrich Schiller University Jena, Department of Mathematics and Computer Science, 2005.

[26] I. P. Gavrilyuk, M. Hermann, M. V. Kutniv, and V.L. Makarov. Difference schemes for nonlinear BVPs on the semiaxis. *Comput. Methods Appl. Math.*, 7:25–47, 2007.

[27] I. P. Gavrilyuk, M. Hermann, M.V. Kutniv, and V.L. Makarov. Difference schemes for nonlinear BVPs using Runge-Kutta IVP-solvers. *Advances in Difference Equations*, 2006:1–29, 2006. Article ID 12167.

[28] L. B. Gnativ and M. V. Kutniv. Modified three-point difference schemes of high-order accuracy for systems of the second-order ordinary differential equations with monotone operator(in Ukrainian). *Mathematical Methods and Physicomechanical Fields*, 47:32–42, 2004.

[29] L. B. Gnativ, M. V. Kutniv, and V. L. Makarov. Generalized three-point difference schemes of high-order accuracy for systems of second order nonlinear ordinary differential equations. *Differential Equations*, 45(7):998–1019, 2009.

[30] G. H. Golub and C. F. Van Loan. *Matrix Computations.* The John Hopkins University Press, Baltimore and London, 1996.

[31] E. Hairer, S. P. Nørsett, and G. Wanner. *Solving Ordinary Differential Equations I, Nonstiff Problems.* Springer Verlag, Berlin, Heidelberg, New York, 1993.

[32] E. Hairer and G. Wanner. *Solving Ordinary Differential Equations II, Stiff and Differential-Algebraic Problems.* Springer Verlag, Berlin, Heidelberg, New York, 1996.

[33] Ph. Hartman. *Ordinary Differential Equations.* Birkhäuser Verlag, Boston, Basel, Stuttgart, 1982.

[34] M. Hermann and D. Kaiser. Shooting methods for two-point BVPs with partially separated endconditions. *ZAMM*, 75:651–668, 1995.

[35] Martin Hermann. *Numerik gewöhnlicher Differentialgleichungen, Anfangs- und Randwertprobleme.* Oldenbourg Verlag, München, Wien, 2004.

[36] F. R. de Hoog and R. Weiss. The numerical solution of boundary value problems with an essential singularity. *SIAM J. Numer. Anal.*, 16:637–669, 1979.

[37] F. R. de Hoog and R. Weiss. An approximation theory for boundary value problems on infinite intervals. *Computing*, 24:227–239, 1980.

[38] S. R. K. Iyengar and A. C. R. Pillai. Difference schemes of polynomial and exponential orders. *Applied Mathematical Modelling*, 13:58–62, 1989.

[39] H. B. Keller. *Numerical Methods for Two-Point Boundary-Value Problems.* Blaisdell Publishing Company, Waltham, Massachusetts, Toronto, London, 1968.

[40] H. B. Keller. *Numerical Solution of Two-Point Boundary Value Problems.* SIAM, Philadelphia, 1976.

[41] H. B. Keller and Jr. White, A. B. Difference methods for boundary value problems in ordinary differential equations. *SIAM J. Numer. Anal.*, (12):791–802, 1975.

[42] M. V. Kutniv. Accurate three-point difference schemes for second-order monotone ordinary differential equations and their implementation. *Comput. Math. Math. Phys.*, 40:368–382, 2000.

[43] M. V. Kutniv. Three-point difference schemes of high accuracy order for systems of nonlinear ordinary differential equations of second order. *Comput. Math. Math. Phys.*, 41:860–873, 2001.

[44] M. V. Kutniv. High-order accurate three-point difference schemes for systems of second-order ordinary differential equations with a monotone operator. *Comput. Math. Math. Phys.*, 42(5):724–738, 2002.

[45] M. V. Kutniv. Three-point difference schemes of high accuracy order for second order nonlinear ordinary differential equations with the boundary conditions of the third type (in Ukrainian). *Visnyk of Lviv University. Series Applied Mathematics and Computer Science*, (4):61–66, 2002.

[46] M. V. Kutniv. Modified three-point difference schemes of high-accuracy order for second order nonlinear ordinary differential equations. *Comput. Methods Appl. Math.*, 3:287–312, 2003.

[47] M. V. Kutniv. Modified three-point difference schemes of high accuracy order for the second-order monotone ordinary differential equations (in Ukrainian). *Mathematical Methods and Physicomechanical Fields*, 46:120–129, 2003.

[48] M. V. Kutniv. Numerical solution of three-point difference schemes (in Ukrainian). *Visnyk of Lviv University. Series Applied Mathematics and Computer Science*, (6):68–73, 2003.

[49] M. V. Kutniv, V. L. Makarov, and A. A. Samarskii. Accurate three-point difference schemes for second-order nonlinear ordinary differential equations and their implementation. *Comput. Math. Math. Phys.*, 39:45–60, 1999.

[50] M. Laspinska-Chrzczonowicz and P. Matus. Exact difference schemes for parabolic equations. *Int. J. Numer. Anal. Mod.*, 5(2):303–319, 2008.

[51] M. Lentini and H. B. Keller. Boundary value problems on semi-infinite intervals and their numerical solutions. *SIAM J. Numer. Anal.*, 17:577–604, 1980.

[52] R.E. Lynch and J.R. Rice. A high-order difference method for differential equations. *Math. of Comp.*, 34:333–372, 1980.

[53] Chawla M. M. A sixth order tri-diagonal finite difference method for general non-linear two point boundary value problems. *J. Inst. Math. Appl.*, 24:35–42, 1979.

[54] V. L. Makarov. *Orthogonal polynomials and difference schemes with exact and explicit spectrum (in Russian)*. PhD thesis, Taras Shevchenko National University of Kiev, 1974.

[55] V. L. Makarov, I. P. Gavrilyuk, M. V. Kutniv, and M. Hermann. A two-point difference scheme of arbitrary given accuracy order for BVPs for systems of first order nonlinear ODEs. *Comput. Methods Appl. Math.*, 4:464–493, 2004.

[56] V. L. Makarov and S. G. Gocheva. Finite difference schemes of arbitrary accuracy for second-order differential equations on a half-line. *Differ. Equ.*, 17:367–377, 1981.

[57] V. L. Makarov and V. V. Guminskii. A three-point difference scheme of a high order of accuracy for a system of second-order ordinary differential equations (the nonselfadjoint case). *Differ. Equ.*, 30:457–465, 1994.

[58] V. L. Makarov, I. L. Makarov, and V. G. Prikazchikov. Exact difference schemes and schemes of any order of accuracy for systems of second-order differential equations (in Russian). *Differ. Uravn.*, 15:1194–1205, 1979.

[59] V. L. Makarov and A. A. Samarskii. Exact three-point difference schemes for second-order nonlinear ordinary differential equations and their implementation. *Soviet Math. Dokl.*, 41:495–500, 1991.

[60] V. L. Makarov and A. A. Samarskii. Realization of exact three point difference schemes for second-order ordinary differential equations with piecewise smooth coefficients. *Soviet Math. Dokl.*, 41:463–467, 1991.

[61] P. A. Markowich. A theory for the approximation of solution of boundary value problems on infinite intervals. *SIAM J. Math. Anal.*, 13:484–513, 1982.

[62] P. A. Markowich. Analysis of boundary value problems on infinite intervals. *SIAM J. Math. Anal.*, 14:11–37, 1983.

[63] P. A. Markowich and C. A. Ringhofer. Collocation methods for boundary value problems on "long" intervals. *Math. Comp.*, 40:123–150, 1983.

[64] P. Matus, U. Irkhin, and M. Lapinska-Chrzczonowicz. Exact difference schemes for time-dependent problems. *Computational Methods in Applied Mathematics*, 5:422–448, 2005.

[65] P. Matus and A. Kolodynska. Exact difference schemes for hyperbolic equations. *Comput. Meth. Appl. Math.*, 7(4):341–364, 2007.

[66] P. P. Matus, U. Irkhin, M. Lapinska-Chrzczonowicz, and Lemeshevsky S. V. About exact difference schemes for hyperbolic and parabolic equations (in russian). *Differ. Uravn.*, 43(7):978–986, 2006.

[67] N. S. Nedialkov and J. D. Pryce. Solving differential-algebraic equations by Taylor series (i): Computing Taylor coefficients. *BIT Numerical Mathematics*, 45:561–591, 2005.

[68] J. M. Ortega and W. C. Rheinboldt. *Iterative Solution of Nonlinear Equations in Several Variables.* Academic Press, New York, London, 1970.

[69] A. Paradzinska and P. Matus. High accuracy difference schemes for nonlinear transfer equation $\frac{\partial u}{\partial t} + u\frac{\partial u}{\partial x} = f(u)$. *Math. Model. Anal.*, 12(4):469–482, 2007.

[70] M. Ronto and A. M. Samoilenko. *Numerical-Analytic Methods in the Theory of Boundary-Value Problems.* Word Scientific, Singapore, 2000.

[71] A. A. Samarskii. *Introduction to the Theory of Finite Difference Schemes.* Nauka, Moscow, 1972.

[72] A. A. Samarskii. *The Theory of Difference Schemes.* Marcel Dekker Inc., New York and Basel, 2001.

[73] A. A. Samarskii and V. L. Makarov. Realization of exact three-point difference schemes for second-order ordinary differential equations with piecewise-smooth coefficients. *Differ. Equ.*, 26:922–930, 1991.

[74] A. A. Samarskii and B.S. Nikolaev. *Solution Methods for Grid Equations(in Russian).* Nauka, Moscow, 1978.

[75] A. A. Samarskii and E. S. Nikolaev. *Numerical Methods for Grid Equations, Vol.1.* Birkhäuser Verlag, Basel, 1989.

[76] M. Schechter. *Principles of Functional Analysis. Second edition.* American Mathematical Society, Providence, Rhode Island, 2002.

[77] C. Schmeiser. Approximate solution of boundary value problems on infinite intervals by collocation methods. *Math. Comp.*, 46:479–490, 1986.

[78] M. R. Scott and H. A. Watts. A systemalized collection of codes for solving two-point boundary value problems. In A. K. Aziz, editor, *Numerical Methods for Differential Systems*, pages 197 – 227, New York and London, 1976. Academic Press.

[79] J. Stoer and R. Bulirsch. *Introduction to Numerical Analysis.* Springer Verlag, New York, Berlin, Heidelberg, 2002.

[80] A. N. Tihonov and A. A. Samarskii. Homogeneous difference schemes (in Russian). *Zh. Vychisl. Mat. i Mat. Fiz.*, 1:5–63, 1961.

[81] A. N. Tihonov and A. A. Samarskii. Homogeneous difference schemes of a high order of accuracy on non-uniform nets (in Russian). *Zh. Vychisl. Mat. i Mat. Fiz.*, 1:425–440, 1961.

[82] V.A. Trenogin. *Functional Analysis(in Russian).* Nauka, Moscow, 1980.

[83] B. A. Troesch. A simple approach to a sensitive two-point boundary value problem. *J. Comput. Phys.*, 21:279–290, 1976.

[84] W. Wallisch and M. Hermann. *Schießverfahren zur Lösung von Rand- und Eigenwertaufgaben.* Teubner-Texte zur Mathematik, Bd. 75. Teubner Verlag, Leipzig, 1985.

[85] A. I. Zadorin. Numerical solution of a boundary value problem for a system of equations with a small parameter. *Comput. Math. Math. Phys.*, 38(8):1201–1211, 1998.

[86] E. Zeidler. *Nonlinear Functional Analysis and its Applications*, volume I. Springer-Verlag, New York et al., 1986.